Ulrich Walter
Astronautics

Related Titles

Mark, H. (ed.)

Encyclopedia of Space Science and Technology
2 Volume Set

1258 pages
Hardcover
ISBN-13: 978-0-471-32408-9
ISBN-10: 0-471-32408-6

Szebehely, V. G., Mark, H.

Adventures in Celestial Mechanics

320 pages with 86 figures
1998
Hardcover
ISBN-13: 978-0-471-13317-9
ISBN-10: 0-471-13317-5

David Darling

The Complete Book of Spaceflight: From Apollo 1 to Zero Gravity

544 pages
2002
Hardcover
ISBN-13: 978-0-471-05649-2
ISBN-10: 0-471-05649-9

Jacob, D. / Sachs, G. / Wagner, S. (eds.):

Basic Research and Technologies for Two-Stage-to-Orbit Vehicles
Collaborative Research Centres (Series: Sonderforschungsbereiche (DFG))

666 pages
2005
Hardcover
ISBN-13: 978-3-527-27735-3
ISBN-10: 3-527-27735-8

Ulrich Walter

Astronautics

WILEY-VCH Verlag GmbH & Co. KGaA

The Author

Prof. Dr. Ulrich Walter
Technical University Munich
Institute of Astronautics
Munich, Germany
Email: U.Walter@lrt.mw.tum.de

Cover

MightySat, a 320 kg U.S. Air Force/Phillips Laboratory satellite was ejected from a canister in the shuttle's cargo bay on Dec. 15, 1998
(Used with permission of NASA)

All books published by **Wiley-VCH** are carefully produced. Nevertheless, authors, editors, and publisher do not warrant the information contained in these books, including this book, to be free of errors. Readers are advised to keep in mind that statements, data, illustrations, procedural details or other items may inadvertently be inaccurate.

Library of Congress Card No.:
applied for

British Library Cataloguing-in-Publication Data
A catalogue record for this book is available from the British Library.

Bibliographic information published by the Deutsche Nationalbibliothek
The Deutsche Nationalbibliothek lists this publication in the Deutsche Nationalbibliografie; detailed bibliographic data are available on the Internet at http://dnb.d-nb.de.

© 2008 WILEY-VCH Verlag GmbH & Co. KGaA, Weinheim

Every effort has been made to trace copyright holders and secure permission prior to publication. If notified, the publisher will rectify any error or omission at the earliest opportunity.

All rights reserved (including those of translation into other languages). No part of this book may be reproduced in any form – by photoprinting, microfilm, or any other means – nor transmitted or translated into a machine language without written permission from the publishers. Registered names, trademarks, etc. used in this book, even when not specifically marked as such, are not to be considered unprotected by law.

Typesetting Uwe Krieg, Berlin
Printing betz-druck GmbH, Darmstadt
Binding Litges & Dopf GmbH, Heppenheim

Printed in the Federal Republic of Germany
Printed on acid-free paper

ISBN: 978-3-527-40685-2

Contents

Preface *XIII*

List of Symbols *XVII*

1	**Rocket Fundamentals** *1*	
1.1	The Rocket Principle *2*	
1.2	Rocket Thrust *4*	
1.2.1	Pressure Becomes Thrust *4*	
1.2.2	Momentum Thrust and Pressure Thrust *9*	
1.2.3	Continuity Equation *10*	
1.3	Rocket Performance *11*	
1.4	Rocket Equation of Motion *13*	
	Problems *15*	
2	**Rocket Flight** *17*	
2.1	General Considerations *17*	
2.2	Rocket in Free Space *19*	
2.3	Impulsive Maneuvers *20*	
2.4	Rocket in a Gravitational Field *20*	
2.4.1	Brief Thrust *21*	
2.4.2	Gravitational Loss *21*	
2.5	Delta-v Budget *23*	
2.6	Fuel Demand – Star Trek Plugged *24*	
2.7	Rocket Efficiency *26*	
2.8	Payload Considerations *28*	
2.9	Relativistic Rocket *30*	
2.9.1	Space Flight Dynamics *31*	
2.9.2	Relativistic Rocket Equation *34*	
2.9.3	Exhaust Considerations *36*	
2.9.4	External Efficiency *38*

Astronautics. Ulrich Walter
Copyright © 2008 WILEY-VCH Verlag GmbH & Co. KGaA, Weinheim
ISBN: 978-3-527-40685-2

2.9.5	Space-Time Transformations	*39*
	Problems	*41*

3 Rocket Staging *43*
3.1	Serial Staging – General Considerations	*44*
3.1.1	Definitions	*44*
3.1.2	Rocket Equation	*47*
3.2	Stage Optimization	*47*
3.2.1	Road to Stage Optimization	*47*
3.2.2	General Optimization	*48*
3.3	Analytical Solutions	*50*
3.3.1	Uniform Staging	*51*
3.3.2	Uniform Exhaust Velocities	*53*
3.3.3	Uneven Staging	*54*
3.4	Stage Number Determination	*56*
3.5	Parallel Staging	*60*
3.6	Other Types of Staging	*62*
	Problems	*63*

4 Thermal Propulsion *65*
4.1	Engine Thermodynamics	*66*
4.1.1	Physics of Propellant Gases	*66*
4.1.2	Flow Velocity	*70*
4.1.3	Mass Flow Density	*71*
4.1.4	Flow at the Throat	*72*
4.1.5	Flow in the Nozzle	*73*
4.1.6	Ideal Nozzle Adaptation	*77*
4.1.7	Engine Performance	*80*
4.1.8	Engine Thrust	*82*
4.2	Thermal Engine Design	*85*
4.2.1	Combustion Chamber Design	*85*
4.2.2	Nozzle Design	*85*
4.2.3	Thrust Coefficient C_f	*86*
4.2.4	Thrust Performance	*87*
4.2.5	Nozzle Efficiency C_n	*88*
4.2.6	Nozzle Shape	*90*
	Problems	*93*

5 Electric Propulsion *95*
5.1	Overview	*95*
5.2	Ion Engine	*96*
5.2.1	Ion Acceleration	*97*

5.2.2	Thrust of an Ion Engine 99
5.2.3	Internal Efficiency 101
5.3	Electric Propulsion Optimization 102
	Problems 106

6 Ascent Flight 107
6.1	Earth's Atmosphere 107
6.1.1	Density Master Equation 107
6.1.2	Homosphere (Barometric Formula) 110
6.1.3	Heterosphere 111
6.1.4	Piecewise-Exponential Atmospheric Model 113
6.2	Equations of Motion 114
6.3	Ascent Phases 119
6.4	Optimum-Ascent Problem 121
6.4.1	Formulation of Problem 121
6.4.2	Gravity-Turn Maneuver 124
6.4.3	Pitch Maneuver 125
6.4.4	Constant-Pitch-Rate Maneuver 126
6.4.5	Optimum-Ascent Trajectory 128

7 Orbits 131
7.1	Equation of Motion 131
7.1.1	Gravitational Potential 131
7.1.2	Gravitational Field 134
7.1.3	Conservation Laws 135
7.1.4	Newton's Laws and Equation of Motion 136
7.1.5	Real Two-Body Problem 138
7.2	Conservation Laws in a Gravitational Field 141
7.2.1	Angular Momentum Conservation 141
7.2.2	Motion in a Plane 142
7.2.3	Kepler's Second Law 143
7.2.4	Energy Conservation 144
7.2.5	Rotational Potential 145
7.3	Motion in a Gravitational Field 146
7.3.1	Orbit Equation 146
7.3.2	Orbit Velocity, Flight Path Angle 149
7.3.3	Orbital Energy and Angular Momentum 151
7.3.4	Orbital Elements (Keplerian Elements) 153
7.3.5	Position on the Orbit 155
7.4	Keplerian Orbits 156
7.4.1	Circular Orbit 156
7.4.2	Parabolic Orbit 157
7.4.3	Elliptical Orbit 159

7.4.4	Hyperbolic Orbit	*166*
7.4.5	Rectilinear Orbit	*168*
7.5	Life in Other Universes?	*172*
7.5.1	Equation of Motion in n Dimensions	*172*
7.5.2	The $n=3$ Universe	*175*
7.5.3	The $n=4$ Universe	*175*
7.5.4	Universes with $n \geq 5$	*177*
7.5.5	Universes with $n \leq 2$	*178*
	Problems	*180*

8	**Orbit Transitions**	*185*
8.1	Two-Impulse Transfer (Hohmann Transfer)	*185*
8.1.1	General Considerations	*186*
8.1.2	Transfer between Circular Orbits	*187*
8.1.3	Transfer between Nearly Circular Orbits	*192*
8.1.4	Sensitivity Analysis	*193*
8.2	Continuous Thrust Transfer	*195*
8.3	Three-Impulse Transfer	*197*
8.4	One-Impulse Maneuvers	*200*
8.4.1	General Considerations	*200*
8.4.2	Orbit Correction Maneuvers	*202*
	Problems	*207*

9	**Interplanetary Flight**	*209*
9.1	Patched-Conics Method	*210*
9.1.1	Sphere of Influence	*210*
9.1.2	Patched Conics	*212*
9.2	Departure Orbits	*214*
9.3	Transit Orbits	*217*
9.3.1	Hohmann Transfers	*217*
9.3.2	Non-Hohmann Transfers	*220*
9.4	Arrival Orbit	*226*
9.5	Flyby Maneuvers	*228*
9.5.1	Basic Considerations	*228*
9.5.2	Flyby Framework	*230*
9.5.3	Flyby Analysis in the Planetocentric System	*232*
9.5.4	Flyby Analysis in the Heliocentric System	*238*
9.5.5	Change of Orbital Elements	*242*
9.6	Earth–Moon Orbits	*244*
9.6.1	Rapprochement Orbits	*244*
9.6.2	Free-Return Trajectories	*246*
9.7	Weak Stability Boundary Transfers	*247*
	Problems	*251*

10 Reentry *253*

10.1 Introduction *253*
10.1.1 Thermal Problem Setting *253*
10.1.2 Entry Interface *255*
10.1.3 Deorbit (Phase A) *255*
10.2 Equations of Motion *260*
10.2.1 Normalized Equations of Motion *261*
10.2.2 Reduced Equations of Motion *263*
10.3 Preliminary Considerations *268*
10.3.1 Reentry – Phase B *269*
10.3.2 Ballistic Reentry Without Perturbations *270*
10.3.3 Maximum Heating for Ballistic Reentries *273*
10.3.4 Reentry with Lift *274*
10.4 Second-Order Solutions *276*
10.4.1 Flight Path Angle *276*
10.4.2 Critical Deceleration *278*
10.5 Low-Lift Reentry (First-Order Solutions) *279*
10.5.1 Velocity *279*
10.5.2 Entry Trajectory *280*
10.5.3 Critical Deceleration *282*
10.6 Reflection and Skip Reentry *285*
10.6.1 Reflection *285*
10.6.2 Skip Reentry *288*
10.6.3 Phygoid Modes *291*
10.7 Lifting Reentry *295*
10.7.1 Equations of Motion and L/D Control Law *296*
10.7.2 Critical Deceleration Parameters *298*
10.7.3 Maximum Heat Load *299*
10.8 Space Shuttle Reentry *300*
10.8.1 From Deorbit Burn to Entry Interface *300*
10.8.2 From Entry Interface down to 80 km *302*
10.8.3 Blackout Phase *303*
10.8.4 Aerodynamic Flight Phase *304*
Problems *305*

11 Three-Body Problem *307*

11.1 Overview *307*
11.2 Synchronous Orbits *309*
11.2.1 Isomass Configurations *310*
11.2.2 Euler Configuration *311*
11.2.3 Lagrange Configuration *316*
11.3 Restricted Three-Body Problem *318*

11.3.1	Eulerian Points	320
11.3.2	Lagrangian Points	322
11.4	Circular Restricted Three-Body Problem	323
11.4.1	Energy Conservation in the CR3BP	323
11.4.2	Jacobi's Integral	325
11.4.3	Orbits in the Synodic System	326
11.4.4	Stability and Dynamics at Eulerian Points	328
11.4.5	Stability and Dynamics at Lagrangian Points	334
	Problems	339

12	**Orbit Perturbations**	*341*
12.1	Problem Setting	341
12.1.1	General Considerations	341
12.1.2	Gaussian Variational Equations	343
12.2	Gravitational Perturbations	345
12.2.1	Geoid	345
12.2.2	Gravitational Potential	346
12.3	Numerical Perturbation Calculation	350
12.3.1	Cowell's Method by Recurrence Iteration	350
12.3.2	Encke's method	353
12.4	Analytical Perturbation Calculation	354
12.4.1	Lagrange's Planetary Equations	354
12.4.2	Gravitational Perturbations of First Order	354
12.4.3	Higher-Order Gravitational Perturbations – Triaxiality	357
12.4.4	Lunisolar Perturbations in GEO	362
12.5	Solar Radiation Pressure	365
12.5.1	Effect of Solar Radiation	366
12.5.2	Temporal Evolution of the Orbit	370
12.5.3	Correction Maneuvers	371
12.6	Drag	374
12.6.1	General Considerations	374
12.6.2	Elliptical Orbits	377
12.6.3	Circularization	380
12.6.4	Circular Orbits	381
12.6.5	Orbit Lifetime	383
	Problems	387

13	**Coordinate Systems**	*393*
13.1	Space Coordinate Systems	393
13.2	Time Coordinates	399

14 Orbit Determination *403*

14.1   Orbit Measurements   *403*
14.1.1   Radar tracking   *403*
14.1.2   Other Tracking Systems   *405*
14.2   Method of Orbit Determination   *407*
14.3   Orbit Estimation   *408*
14.3.1   Simple Orbit Estimation   *408*
14.3.2   Lambert's Method   *409*
14.4   Conversion of Orbital Elements   *412*
14.4.1   Transformation $\boldsymbol{r}, \boldsymbol{v} \to a, e, i, \omega, \Omega, \theta$ *412*
14.4.2   Transformation $a, e, i, \omega, \Omega, \theta \to \boldsymbol{r}, \boldsymbol{v}$ *415*
14.5   State Vector Propagation   *415*
14.5.1   Propagation $r, v, \gamma \to r', v', \gamma'$ *416*
14.5.2   Propagation $\boldsymbol{r}, \boldsymbol{v} \to \boldsymbol{r'}, \boldsymbol{v'}$ *418*
14.5.3   Universal Propagator   *419*
         Problems   *420*

15 Rigid Body Dynamics *421*

15.1   Fundamental Physics of Rotation   *421*
15.1.1   Physical Basics   *421*
15.1.2   Equations of Rotational Motion   *428*
15.1.3   Reference Frames   *429*
15.1.4   Translation vs. Rotation   *432*
15.2   Torque-free Motion   *433*
15.2.1   Basic Considerations   *433*
15.2.2   Stability and Nutation   *434*
15.2.3   Nutation of a Torque-Free Symmetrical Gyro   *437*
15.2.4   Nutation and Energy Dissipation   *440*
15.2.5   General Torque-Free Motion   *443*
15.3   Gyro under External Torque   *444*
15.4   Gravity-Gradient Stabilization   *446*
15.4.1   Gravity-Gradient Torque   *447*
15.4.2   Gravity-Gradient Induced Attitude Changes   *448*
15.4.3   Stability of Gravity-Gradient Oscillations   *449*
15.4.4   Pitch Oscillation   *450*
15.4.5   Coupled Roll-Yaw Oscillation   *452*

Appendix

A	**Astrodynamic Parameters**	*455*
A.1	Mean Orbit Radius	*455*
A.1.1	Titius–Bode Law	*455*
A.1.2	Average over True Anomaly	*456*
A.1.3	Time Average	*456*
A.2	Mean Orbital Velocity	*457*
A.2.1	Average over True Anomaly	*457*
A.2.2	Time Average	*458*
B	**Approximate Analytical Solution for Uneven Staging**	*459*

Color Plates *463*

References *473*

Index *475*

Preface

"There is no substitute for true understanding"

Kai Lai Chung

If you want to cope with science, you have to understand it – truly understand it. This holds in particular for astronautics. "To understand" means that you have a network of relationships in your mind, which permits you to deduce an unknown fact from a well-known fact. The evolution of a human being from birth to adulthood and beyond consists of building up a comprehensive knowledge network of the world, which makes it possible to cope with it. That you are intelligent just means that you are able to do that – sometimes you can do it better, and sometimes worse.

True understanding is the basis of everything. There is nothing that would be able to substitute true understanding. Computers do not understand – they merely carry out programmed deterministic orders. They do not have any understanding of the world. This is why even a large language computer will always render a false translation of the phrase: "It is the horse which rides the child." It won't be able to understand what riding means, and thus not know who is riding on what or whom. Most probably, and according to the word sequence, it would translate it as if the horse is riding the child. No computer program in the world is able to substitute understanding. You have to understand yourself. Only when you understand are you able to solve problems by designing excellent computer programs. Nowadays, real problems are only solved on computers – written by bright engineers and scientists.

The goal of this book is to build up a network of astronautic relationships in the mind of the reader. If you don't understand something while reading this book, I made a mistake. The problem of a relational network, though, is that the underlying logic can be very complex, and sometimes it seems that our brains are not suitable for even the simplest logic. If I asked you, "You are not stupid, are you?", you would normally answer, "No!" From a logical point of view, a double negation of an attribute is the attribute itself. So your "No!" means that you consider yourself stupid. You, and also we scientists

Astronautics. Ulrich Walter
Copyright © 2008 WILEY-VCH Verlag GmbH & Co. KGaA, Weinheim
ISBN: 978-3-527-40685-2

and engineers, do not want this embarrassing mistake to happen time and time again, and so we use mathematics. Mathematical logic is the guardrail of human thinking. Physics, on the other hand, is the art of applying this logic consistently to nature in order to be able to understand how it works. So it comes as no surprise to find a huge amount of formulas and a lot of physics in this book.

Some might think this is sheer horror. But now comes the good news. You don't need to remember most of the formulas – neither for exams nor for later. To understand astronautics, you just need to know the formulas shaded gray and to remember those bordered black. They are all you require to tell you the essential story. There you should pause and try to understand their meaning and lift the secrets of nature. You don't need to remember all the other formulas, but you should be able to derive these stepping stones for yourself. Thereby you will always be able to link nodes in your relational network whenever you deem it necessary. To treat formulas requires knowing a lot of tricks. You will learn them only by watching others doing such "manipulation" and, most importantly, by doing it yourself. Sometimes you will see the word "exercise" in brackets. This indicates that the said calculation would be a good exercise for you to prove to yourself that you know the tricks. Sometimes it might denote that there is not the space to fully lay out the needed calculation because it is too lengthy or quite tricky. So, you have to guess for yourself whether or not you should do the exercise. Nonetheless, only very few of you will have to derive formulas professionally later. For the rest of you: just try to follow the story and understand how consistent and wonderful nature is. Those who succeed will understand the words of Richard Feynman, the great physicist, who once expressed his joy about this by saying: "The pleasure of finding things out."

Take the pleasure to find out about astronautics.

April 2007

Ulrich Walter

Acknowledgments

I am grateful to Olivier L. de Weck, Bernd Häusler, and Hans-Joachim Blome for carefully reading the manuscript and for many fruitful suggestions. My special thanks go to Winfried Hofstetter, who contributed Sections 9.6.2 and 9.7, and to Julia Bruder for her tedious work of translating the original German manuscript into English. Many expounding passages of this book wouldn't be in place without the bright questions of my students, who reminded me of the fact that a lot of implicitness scientists got used to is not that trivial as its seems to be.

Many figures in this book were drawn by the interactive plotting program *gnuplot* v4.0. My sincere thanks to its authors Thomas Williams, Colin Kelley, Hans-Bernhard Bröker, and many others for establishing and maintaining this versatile and very useful tool for free public use. The author is grateful to the GeoForschungsZentrum Potsdam for providing the geoid views and the visualization of the spherical harmonics in the color tables.

Writing this book was only possible by sacrificing most of the weekends and vacation over the past four years. Needless to say that I deeply regret that I couldn't spent more of this time with my beloved wife Beate and my wonderful kids Natalie and Angela. My deepest gratitude goes to you for accepting this without any saying.

List of Symbols

Indices

0	inner, at the beginning (zero)
a	with respect to the atmosphere
air	atmosphere
apo	apoapsis
B	body system
c	combustion
cm	center of mass
col	collision
crit	critical
D	aerodynamic drag force
e	at exit, exhaust; or at reentry
eff	effective
esc	escape (velocity)
ext	external
f	final (mass)
F	force
GG	gravity gradient
H	Hohmann
i	initial (mass)
I	inertial reference system
ion	ionic
in	initial, at entry, incoming

int	internal
kin	kinetic (energy)
L	aerodynamic lift; or
	payload
LVLH	local vertical local horizontal reference system
max	maximum
min	minimum
micro	microscopic
n	nozzle; or
	normal (vertically to ...)
opt	optimum (value)
osc	oscillation
out	final, at exit, outgoing
p	propellant; or
	planet
P	principal axes system
per	periapsis
ϕ	diameter
pot	potential (energy)
r	radial; or
	reflection; or
	radiation
rms	root mean square (= quadratic mean)
rot	rotation
S/C	spacecraft
SOI	sphere of influence
syn	synodic
t	tangential
T	transfer orbit
θ	vertically to radial
trans	translation
tot	total
s	structural
t	throat (of thruster)
∞	external, at infinity

$*$	effective (thrust), total
$\oplus$	Earth
$\odot$	Sun
∇	spacecraft
$\bullet$	inner (orbit)
$\circ$	outer (orbit)
$\perp$	vertical to beam direction, (effectively wetted surface)
$\times$	at orbit crossing

Latin symbols

a	semi-major axis (of a Keplerian orbit); or
	speed of sound; or
	acceleration
A	area
b	semi-minor axis (of a Keplerian orbit); or $b := L \cdot \tan \gamma_e / (2D)$
B	ballistic coefficient (without index: for drag)
c	$c := L \cdot \cot \gamma_e / (2D)$; or
	speed of light
c^*	characteristic velocity, $c^* := p_0 A_t / m_p$
c_p	specific heat capacity at constant pressure
c_V	specific heat capacity at constant volume
C_3	characteristic energy, $C_3 := v_\infty^2$
C_∞	infinite-expansion coefficient
C_D	drag coefficient
C_f	thrust coefficient
C_L	lift coefficient
C_n	nozzle coefficient (nozzle efficiency)
CM	center of mass
δx	variation (small changes) of x
$\delta v_\parallel$	differential increase in orbital velocity due to kick-burn in flight direction
$\delta v_{\perp O}$	differential increase in orbital velocity due to kick-burn vertical to flight direction, within orbital plane, outbound
$\delta v_{\perp \perp}$	differential increase in orbital velocity due to kick-burn vertical to flight direction and vertical to orbital plane, parallel to angular momentum
Δv	delta-v budget

D	drag force; or diameter
$diag(\ldots)$	diagonal matrix with elements $(\ldots)$
e	eccentricity; or electric charge unit; or Eulerian number $e = 2.718\,281\,828\ldots$
E	energy; or elliptic eccentric anomaly
f^x	f function, see definition Eq. (10.4.2)
F	force; or hyperbolic eccentric anomaly
g	Earth's gravitational constant (mean value at sea level: $g_0 = 9.798\,28$ m s^{-2})
G	gravitational constant, $G = 6.672\,59 \times 10^{-11}$ m^3 kg^{-1} s^{-2}
h	specific angular momentum (i.e., per mass unit); or molar enthalpy; or height (above sea level), altitude
H	enthalpy; or scale height
i	inclination
$\mathbf{I}$	inertial tensor
I_{sp}	weight-specific impulse
j	charge flow density
JD	Julian Date
k	orbit number: $k := rv^2/\mu$
k_B	Boltzmann constant, $k_B = 1.380\,650 \times 10^{23}$ J K^{-1}
L	lift force; or angular momentum
m	orbiting mass (without index: of a space craft)
$\dot{m}$	mass flow rate (without index: of a space craft)
M	central mass (central body); or total mass of a system of bodies; or mean anomaly; or molar mass
Ma	Mach number $Ma := v/a$

MJD	Modified Julian Date
n	rocket stage number; or mean motion; or mean number of excited degrees of freedom of gas molecules
p	pressure; or propellant; or linear momentum, $p = mv$; or semi-latus rectum $p = h^2/\mu = a(1-e^2)$; or $p := H/(\varepsilon_e R)$
P	power; or perturbative function
q	electric charge; or $q := H \cot^2 \gamma_e / (\varepsilon_e R)$
$\dot{q}$	heat flux
$\dot{Q}$	heat flow rate
r	orbital radius
R	radius (of a celestial body, in particular the Earth); or universal gas constant $R = 8.314$ J K^{-1} mol^{-1}
$\boldsymbol{R}$	Rotation (matrix)
R_s	specific gas constant of air, $R_s = 286.91$ J K^{-1} kg^{-1}
Re	Reynolds number
S/C	spacecraft
sgn(x)	sign function: sgn(x) = sign of x
SOI	sphere of influence
St	Stanton number, $St \approx 0.1\%$
t	time
t_0	time at passage through periapsis, epoch (see Eq. (7.3.17))
T	temperature; or orbital period
$\boldsymbol{u}$	unit vector, e.g., $\boldsymbol{u}_r \equiv \hat{\boldsymbol{r}} = \boldsymbol{r}/r$
U	internal energy of a gas; or total electrical voltage; or gravitational potential
v	velocity (orbital, of the spacecraft; or drift velocity of propellant gas)
v_h	$v_h := \mu/h$
V	volume; or electrical potential

Greek symbols

α	thrust angle; or
	angle of attack; or
	proper acceleration; or
	mass-specific power output of an electrical plant
β	$\beta := v/c$
δ	Dirac's delta function; or
	turn angle
δ_{nm}	Kronecker symbol
$\delta(\lambda, \varepsilon_e)$	see definition Eq. (10.4.3)
Δ	aiming radius
ε	structural factor; or
	specific orbital energy (a.k.a. total mechanical energy); or
	expansion ratio; or
	$\varepsilon := v^2/v_0^2$ (see definition Eq. (10.2.12))
ε_0	absolute dielectric constant
γ	flight path angle, i.e. entry angle for reentry; or
	$\gamma := 1/\sqrt{1-\beta^2} = 1/\sqrt{1-v^2/c^2}$
κ	adiabatic index; or
	dimensionless drag, κ_D, or lift, κ_L, coefficient
λ	payload ratio; or
	dimensionless altitude variable (see definition Eq. (10.2.13))
η	efficiency, in particular thermal efficiency
ρ	(atmospheric) mass density; or
	surface reflectivity; or
	invers radius: $\rho := 1/r$
μ	standard gravitational parameter: $\mu := GM$; or
	reduced mass (and variations of it); or
	mass flux density: $\mu := \dot{m}_p/A$ (see Section 4.1.3); or
	bank angle (roll angle)
μ_i	mass ratio (see Eqs. 2.8.1 and (3.1.7))
σ	Stefan–Boltzmann constant, $\sigma = 5.6704 \times 10^{-8}$ W m^{-2} K^{-4}; or
	proper speed: $\sigma = c \cdot \text{arctanh}\,\beta$
τ	eigentime

θ	true anomaly, orbit angle; or pitch angle: $\theta := \alpha + \gamma$
ω	angular velocity (frequency), $\omega = d\theta/dt$; or argument of periapsis
Ω	right ascension of ascending node (RAAN)

Others

$:=$	definition equation; the symbol preceding the colon is defined by the expression following the equal sign
$= const$	the expression preceding the equal sign is constant (invariant) with respect to a given variable
$\Rightarrow$	from this follows ...
@	the equation preceding this symbol holds at the condition following it
$O(\varepsilon^n)$	Landau notation (a.k.a. big O notation); $O(\varepsilon^n)$ is the magnitude (order) of the residual power (here ε^n) of a series expansion, and means that the residual is of order ε^n
$\dot{x}$	derivative with respect to time, $\dot{x} := dx/dt$
$\bar{x}$	geometric mean
$\mathbf{x}$	vector x
$\mathbf{X}$	matrix X
$\hat{r}$	unit vector along direction r, $\hat{r} = r/r = u_r$
$\angle(\mathbf{a}, \mathbf{b})$	angle between vector $(\mathbf{a})$ and vector $(\mathbf{b})$
$\langle y \rangle_x$	average of y with respect to x over interval $[a, b]$, $\langle y \rangle_x := \frac{1}{b-a} \int_a^b y \cdot dx$
$\langle y \rangle$	time average, $\langle y \rangle = \langle y \rangle_t$
$\uparrow$	increasing
$\uparrow\uparrow$	strongly increasing
$\downarrow$	decreasing
$\downarrow\downarrow$	strongly decreasing

Abbreviations

AOA	angle-of-attack
CFPAR	constant flight path angle rate
CM	center of mass
CPR	constant-pitch-rate
CR3BP	circular restricted three-body problem
ET	external tank
FPA	flight path angle
GEODSS	ground-based electro-optical deep space surveillance
GEO	geosynchronous Earth orbit
GG	gravity-gradient force
GMT	greenwich mean time
GSO	geosynchronous orbit
GTO	GSO transfer orbit
GVE	Gaussian variational equation
IAU	international astronomical union
ICRF	international celestial reference frame
ISS	international space station
JD	Julian date
LEO	low Earth orbit
LPE	Lagrange's planetary equation
LVLH frame	local vertical, local horizontal frame
MJD	modified Julian date
OMS	orbital maneuvering system
R3BP	restricted three-body problem
RAAN	right ascension of ascending node
RTG	thermoelectric generator
SOI	sphere of influence
SRB	solid rocket booster
SSME	Space Shuttle main engine
SSTO	single stage to orbit
TAEM	terminal area energy management
TDRS	tracking and data relay satellite
UT	universal time
WSB	weak stability boundary

1
Rocket Fundamentals

Many people have had, and still have, misconceptions about the basic principle of a rocket. Here is a comment of the publisher of the renowned *New York Times* from 1921 about the pioneer of US astronautics, Robert Goddard, who at that time was carrying out the first experiments with liquid propulsion engines:

> "Professor Goddard ... does not know the relation of action to reaction, and of the need to have something better than a vacuum against which to react – to say that would be absurd. Of course he only seems to lack the knowledge ladled out daily in high schools."

The publisher's doubts whether rocket propulsion in vacuum could work is based on our daily experience that you can only move forwards by pushing backwards against an object or medium. Rowing is based on the same principle. You use the blades of the oars to push against the water. But this example already shows that the medium you push against, which is water, does not have to be at rest, it may move backwards. So basically it would suffice to fill a blade with water and push against it by very quickly guiding the water backwards with the movement of the oars. Of course, the forward thrust of the boat gained thereby is much lower compared with rowing with the oars in the water, as the large displacement resistance in the water means that you push against a far bigger mass of water. But the principle is the same. Instead of pushing water backwards with a blade, you could also use a pile of stones in the rear of your boat, and hurl them backwards as fast as possible. With this you would push ahead against the accelerating stone. And this is the basis of the propulsion principle of a rocket: it pushes against the gases it ejects backwards with full brunt. So, with the propellant, the rocket carries the mass against which it pushes to move forwards, and this is why it also works in vacuum.

This repulsion principle, which is called the "rocket principle" in astronautics, is based on the physical principle of conservation of momentum. It states that the total (linear) momentum of a system remains constant with time. If, at initial time t_0 the boat (rocket) with mass m_1 plus stone (propellant) with mass m_2 had velocity v_0, implying that the initial total momentum was $p_0(t_0) = (m_1 + m_2)v_0$, then, at some time $t_+ > t_0$, on hurling the stone (propellant) away with velocity v_2 the boat will have velocity v_1 (neglecting water friction) and the total momentum $p(t_+) = m_1 v_1 + m_2 v_2$ must be the same. That is

$$p(t_0) = p(t_+) \quad \text{principle of the conservation of (linear) momentum}$$

from which follows

$$(m_1 + m_2) \cdot v_0 = m_1 v_1 + m_2 v_2$$

Note: *The principle of conservation of momentum is only valid for the vector form of the momentum equation, which is quite often ignored. A bomb that is ignited generates a huge amount of momentum out of nothing, which apparently would invalidate an absolute value form of the momentum equation. But if you add up the vectorial momenta of the bomb's fragments, it becomes obvious that the vectorial linear momentum has been conserved.*

Given m_1, m_2, v_0 and velocity v_2 of the stone (propellant) expelled, one is able to calculate from this equation the increased boat (rocket) velocity v_1. Doing so, this equation affirms our daily experience that hurling the stone backwards increases the speed of the boat, while doing it forwards decreases its speed.

1.1
The Rocket Principle

With a rocket, the situation is a bit more complicated, as it does not eject one stone after another, but it emits a continuous gas jet. It can be shown (see Ruppe (1966, p. 24ff)) that ejecting the same amount of mass continuously rather than in chunks maximizes the achievable thrust. In order to describe the gain of rocket speed by the continuous mass ejection stream adequately in mathematical and physical terms, we have to consider the ejected mass and time steps as infinitesimally small and in an external rest frame, a so-called inertial (unaccelerated, see Section 13.1) reference system. This is depicted in Fig. 1.1, where in an inertial reference system with its origin at the center of the Earth a rocket with mass m in space experiences no external forces.

At a given time t the rocket may have velocity v and momentum $p(t) = mv$. By ejecting the propellant mass $dm_p > 0$ with **effective exhaust velocity** v_* – the meaning of which will become clear in the next section – and hence with propellant momentum $p_p(t + dt) = (v + v_*) \cdot dm_p$, it will lose part of its mass $dm = -dm_p < 0$ and hence gain rocket speed dv by acquiring momentum $p_r(t + dt) = (m + dm)(v + dv)$.

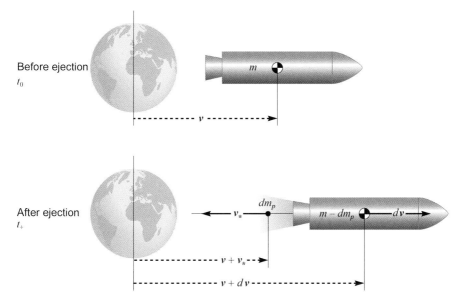

Figure 1.1 A rocket in force-free space before (above) and after (below) it ejected a propellant mass dm_p with effective exhaust velocity v_*, thereby gaining speed dv. Velocities relative to the external inertial system (Earth) are dashed, and those with regard to the rocket are solid.

Note: *In the literature $dm > 0$ often denotes the positive mass flow rate of the propellant, and m the mass of the rocket. This is inconsistent, and leads to an erroneous mathematical description of the relationships, because if m is the mass of the rocket, logically dm has to be the mass change of the rocket, and thus it has to be negative. This is why in this book we will always discriminate between rocket mass and propulsion mass using the consistent description $dm = -dm_p < 0$ implying $\dot{m} = -\dot{m}_p < 0$ for their flows.*

For this line of events we can apply the principle of conservation of momentum as follows:

$$p(t) = p(t+dt) = p_p(t+dt) + p_r(t+dt)$$

From this follows

$$mv = -dm\,(v+v_*) + (m+dm)\,(v+dv)$$
$$= mv - dm \cdot v_* + m \cdot dv + dm \cdot dv$$

As the double differential $dm \cdot dv$ mathematically vanishes with respect to the single differentials dm and dv, we get with division by dt:

$$m\dot{v} = \dot{m}v_* \tag{1.1.1}$$

According to Newton's second law (Eq. (7.1.12)), $F = m\dot{v}$, the term on the left side corresponds to a force due to the repulsion of the propellant, which we correspondingly indicate by

$$F_* = \dot{m}v_* \tag{1.1.2}$$

with $\dot{m} = -\dot{m}_p < 0$. This means that the thrust of a rocket is determined by the product of propellant mass flow rate and exhaust velocity. Observe that due to $\dot{m} < 0$ F_* is exactly in opposite direction to the exhaust velocity v_* (But depending on the steering angle of the engine, v_* and hence F_* does not necessarily have to be in line of the flight direction v.). Therefore with regard to the absolute values we can write

$$\boxed{F_* = -\dot{m}v_* = \dot{m}_p v_*} \quad \text{propellant force (thrust)} \tag{1.1.3}$$

Equation (1.1.2), or (1.1.3) respectively, is of vital importance for astronautics, as it describes basic physical facts, just like every other physical relationship, relating just three parameters, such as $W = F \cdot s$ or $U = R \cdot I$. This is its statement: thrust is the product of exhaust velocity times mass flow rate. Only the two properties together make up a powerful thruster. The crux of the propellant is not its "energy content" (in fact the energy to accelerate the propellant might be provided externally, which is the case with ion propulsions), but the fact that it possesses mass, which is ejected backwards, and thus accelerates the rocket forwards by means of conservation of momentum. The higher the mass flow rate, the larger the thrust. If "a lot of thrust" is an issue, for instance during launch, when the thrust has to overcome the gravitational pull of the Earth, and since the exhaust speed of engines is limited, you need thrusters with a huge mass flow rate. The more the better. Each of the five first-stage engines of a Saturn V rocket had a mass flow rate of about 2.5 metric tons per second, in total 12.5 tons per second, to achieve the required thrust of 33 000 N (corresponds to 3400 tons of thrust). This tremendous mass flow rate is exactly why, for launch, chemical thrusters are matchless up to now, and they will certainly continue to be so for quite some time.

1.2
Rocket Thrust

1.2.1
Pressure Becomes Thrust

If the masses dm_p were stones, and if we hurled them backwards, then the thrust would just be the repulsion of the stones. But generally we hurl gases with the engine. Gases are a loose accumulation of molecules, which, depending on temperature, display internal molecular motion, and thus generate pressure. On the other hand, the rocket at launch moves in an atmosphere whose gas molecules exert an external pressure. In order to understand the impact of the propellant gas pressure and external ambient pressure on the engine's thrust, let's have a look at the pressure conditions in an engine (see Fig. 1.2).

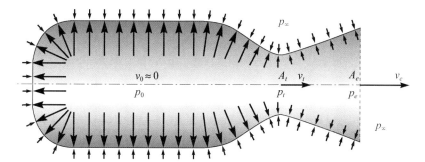

Figure 1.2 Pressure conditions inside and outside an engine chamber.

Inside the combustion chamber, and depending on the location within the chamber, we assume a variable pressure p_{int}, which exerts the force $dF_{int} = p_{int} \cdot dA$ on a wall segment dA. In the area surrounding the chamber we assume an equal external ambient pressure p_∞. The propellant force F_* generated by the chamber must be the sum of all effective forces acting on the entire engine wall with surface S

$$F_* = \iint_S dF_{eff} = \iint_S (p_{int} - p_\infty) \cdot dA \tag{1.2.1}$$

The surface vector can be split into two components: a radial component u_r and an axial component u_x (Fig. 1.3),

$$dA = dA_r + dA_x = (\sin\theta \cdot u_r + \cos\theta \cdot u_x) \cdot dA$$

where the wall angle θ is the angle between surface normal and chamber axis. If the *combustion chamber is axially symmetric*, then we have

$$\iint_S (p_{int} - p_\infty) \cdot dA_r = 0$$

and therefore we only get axial contributions

$$F_* = \iint_S (p_{int} - p_\infty) \cdot dA_x = u_x \iint_S (p_{int} - p_\infty)\cos\theta \cdot dA \tag{1.2.2}$$

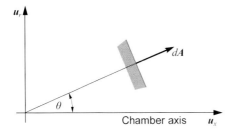

Figure 1.3 Definition of the wall angle with regard to the chamber axis.

Maintaining the internal pressure conditions, and thus without a change in thrust, we now deform the combustion chamber, so that we get a rectangular combustion chamber (see Fig. 1.4). Now that all wall angles are only $\theta = 0°$, $90°$, $180°$, $270°$ the following is valid

$$F_* = -\iint_{A_\phi} (p_{int} - p_\infty)(-1) \cdot dA - \iint_{A_\phi - A_t} (p_{int} - p_\infty) \cdot dA \tag{1.2.3}$$

where F_* now expresses the propellant force of the combustion chamber in forward direction, the direction in which the total force is effectively pushing.

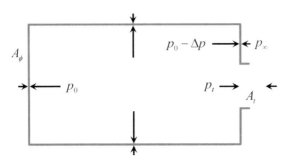

Figure 1.4 Pressure conditions in the idealized combustion chamber.

As there is no wall at the throat with the surface A_t, no force can be exerted on it, and thus on the chamber's back side the integral is limited to the surface $A_\phi - A_t$. The maximum combustion chamber pressure $p_{int} = p_0$ is on the front side of the chamber, where the gas is about at rest. Because the gas flow increases in the direction of the throat where it exits the chamber, the pressure at the rear of the chamber is reduced by a certain amount Δp: $p_{int} = p_0 - \Delta p(r)$, and due to the axial symmetry of the chamber this pressure drop is also axially symmetrical, so that at the throat $p_{int} = p_0 - \Delta p(r) = p_t$ applies.

So Eq. (1.2.3) reads as follows:

$$F_* = (p_0 - p_\infty) A_\phi - \iint_{A_\phi - A_t} (p_0 - p_\infty) \cdot dA + \iint_{A_\phi - A_t} \Delta p \cdot dA$$

As

$$\iint_{A_\phi - A_t} (p_0 - p_\infty) \cdot dA = (p_0 - p_\infty)(A_\phi - A_t)$$

and

$$\iint_{A_\phi - A_t} \Delta p \cdot dA = \iint_{A_\phi} \Delta p \cdot dA - \iint_{A_t} \Delta p \cdot dA = \iint_{A_\phi} \Delta p \cdot dA - (p_0 - p_t) A_t$$

we get

$$F_* = (p_t - p_\infty) A_t + \iint_{A_\phi} \Delta p \cdot dA \tag{1.2.4}$$

Let's have a closer look at the integral of the last equation. It describes a force which results from the pressure reduction along the rear combustion chamber wall. This pressure reduction is due to the propellant flow through the throat. This mass flow, of course, does not generate a sudden pressure drop at the rear wall, but rather a pressure gradient along the chamber axis, i.e.

$$\iint_{A_\phi} \Delta p \cdot d\mathbf{A} \rightarrow - \iiint_{chamber} \nabla p \cdot dV$$

The pressure gradient corresponds to an acceleration field dv/dt of the mass flow. According to the Euler equation of hydrodynamics, they are intimately connected with each other via the mass density ρ:

$$-\nabla p = \rho \frac{dv}{dt} \qquad \text{Euler equation}$$

This equation expresses Newton's law in hydrodynamics. If we apply the Euler equation to the volume integral, we obtain

$$\iiint_{chamber} \nabla p \cdot dV = - \iiint_{chamber} \frac{dv}{dt} \frac{dm_p}{dV} dV = - \int_0^{v_t} \dot{m}_p \cdot dv$$

The velocity integral now ranges from the velocity at the front part of the chamber, where the pressure gradient (and hence the drift velocity of the propellant) is zero, to its throat, where the velocity takes on the exit value v_t.

According to the continuity equation (Eq. (1.2.9)), the mass flow rate $\dot{m}_p$ is invariant along the combustion chamber and also in the subsequent nozzle, and thus it is constant. So we find

$$\iint_{A_\phi} \Delta p \cdot d\mathbf{A} = -\iiint_{chamber} \nabla p \cdot dV = \dot{m}_p \int_0^{v_t} dv = \dot{m}_p v_t \tag{1.2.5}$$

If we apply this result to Eq. (1.2.4), we get

$$F_* = \dot{m}_p v_t + (p_t - p_\infty) A_t$$

So far our considerations have been independent of the exact form of the combustion chamber, as long as it is axially symmetric. So we can consider the nozzle to be also a part of the combustion chamber. Then all the parameters considered so far at the throat of the combustion chamber are also valid for the nozzle exit, i.e.

$$\boxed{F_* = \dot{m}_p v_e + (p_e - p_\infty) A_e =: F_e + F_p} \tag{1.2.6}$$

We recover its vectorial form by the direction information in Eq. (1.2.2)

$$\mathbf{F}_* = -\mathbf{u}_e \left[\dot{m}_p v_e + (p_e - p_\infty) A_e\right] \tag{1.2.7}$$

where $\mathbf{u}_e$ is the unit vector of the exit surface in the direction of the exhaust jet and v_e the **exhaust velocity**. The first term on the right side of Eq. (1.2.6) is called **momentum thrust** F_e, and the second term is called **pressure thrust** F_p. The first name is well chosen, because if you integrate expression $\dot{m}_p v_e$ with regard to time, you get the momentum $m_p v_e$, which is merely the recoil momentum of the ejected propellant. The second term is formally not quite correct, as according to Eq. (1.2.5), the momentum thrust is also generated by a pressure on the chamber because of its internal pressure gradient. At the end it's all pressure which accelerates the engine, and with it the rocket.

Effective exhaust velocity

If we compare Eq. (1.2.6) with Eq. (1.1.3), we can see that the effective exhaust velocity is made up of two contributions:

$$v_* = v_e + (p_e - p_\infty) \frac{A_e}{\dot{m}_p} \quad \text{effective exhaust velocity} \tag{1.2.8}$$

The expression "effective exhaust velocity" makes it clear that it is not only about exhaust velocity v_e, but modified by the pressure thrust. However, for a real thruster the pressure thrust indeed is only a small contribution. For an ideally adapted nozzle with $p_e \approx p_\infty$ (Section 4.1.6) it even is negligibly small.

1.2.2
Momentum Thrust and Pressure Thrust

Ultimately, if it is only pressure that drives a rocket, how does this fit together with the rocket principle discussed in Section 1.1, which was based on repulsion and not on pressure? And what is the physical meaning of "pressure thrust"? You often find the statement that pressure thrust occurs when the pressure at the exit (be it nozzle or chamber exit) hits the external pressure. The pressure difference at this point times the surface is supposed to be the pressure thrust. Though the result is right, the explanation is not. First, the exit pressure does not abruptly meet the external pressure. There is rather a smooth pressure transition from the exit pressure to the external pressure covering in principle an infinite volume behind the engine. Second, even if such a pressure difference could be traced back mathematically to a specific surface, this would not cause a thrust, because, as we will see later, the gas in the nozzle expands backwards with supersonic speed, and such a gas cannot have a causal effect on the engine to exert a thrust on it.

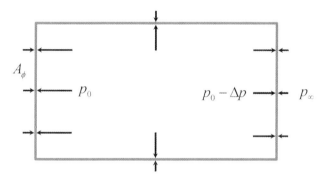

Figure 1.5 Pressure conditions of the idealized combustion chamber if it could be, hypothetically, fully closed.

For a true explanation let's imagine for a moment, and purely hypothetically, a fully closed combustion chamber (see Fig. 1.5) with the same pressure conditions as in the idealized combustion chamber with mass flow rate (see Fig. 1.4). The surface force on the front side would be $F_{front} = (p_0 - p_\infty)A_\phi$ on the front side, and $F_{rear} = (p_0 - \Delta p - p_\infty)A_\phi$ on the rear side. Hence the net forward thrust would be $F_* = F_{front} - F_{rear} = \Delta p \cdot A_\phi$. Because the wall angle on the rear side is 0° and because of Eq. (1.2.5), this translates into $F_* = \Delta p \cdot A_\phi = \dot{m}_p v_t$. Therefore, we can say the following:

> The **momentum thrust** F_e physically results from the fact that, in a hypothetically closed engine chamber, due to the mass flow rate $\dot{m}_p$ there is a bigger chamber pressure on the front side compared to the by Δp smaller pressure on the back side. This causes a net pressure force $\Delta p \cdot A_\phi$.

Ultimately it is the Euler equation, which relates the mass flow rate $\dot{m}_p$ with the pressure differences in the pressure chamber. In order to have the hypothetical gas flow indeed flowing, we need to make a hole with area A_t into the rear side (see Fig. 1.4). Once this is done, the counterthrust at the rear side decreases by

$$\Delta F_{rear} = -(p_0 - \Delta p - p_\infty)A_t = -(p_t - p_\infty)A_t$$

which in turn increases the forward thrust by the same amount. But this is just the pressure thrust. Therefore:

> The **pressure thrust** F_p is the additional thrust which originates from the absence of the counter-pressure force at the exit opening of the engine.

If the exit pressure happens to be equal to the external pressure, then the external pressure behaves like a wall, the pressure thrust vanishes, and we have an ideally adapted nozzle (see Section 4.1.6).

1.2.3 Continuity Equation

The momentum thrust can also be described in a different mathematical form. Let's have a general look at the behavior of propellant gas perfusing an engine. A propellant mass dm_p perfuses a given cross section of the engine with area A with velocity v (see Fig. 1.6). During the time interval dt, the volume of amount $dV = A \cdot ds = Av \cdot dt$ will have passed through it. Therefore

$$dm_p = \rho \cdot dV = \rho A v \cdot dt$$

where ρ is the mass density, which we assume to be constant. From this we derive the mass flow rate equation

$$\boxed{\dot{m}_p = \rho v A} \qquad \text{continuity equation (conservation of mass)} \qquad (1.2.9)$$

A constant mass density simply means that nowhere within the volume new mass is generated or disappears. This is exactly what the word "continuity"

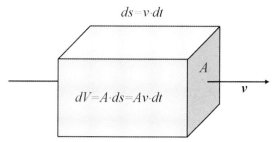

Figure 1.6 The volume dV which a mass flow with velocity v passes in time dt.

means. We could also call it "conservation of mass". So the conservation of mass directly implies Eq. (1.2.9).

At the engine exit, the continuity equation reads $\dot{m}_p = \rho_e v_e A_e$. Applying this to Eq. (1.2.6) yields

$$F_e = \dot{m}_p v_e = \rho_e A_e v_e^2 \tag{1.2.10}$$

This equation begs the question whether the momentum thrust is linearly or quadratically dependent on v_e. The answer depends on the engine in question. Depending on the type (e.g. electric or chemical engine) of engine, a change of its design in general will vary all the parameters v_e and $\dot{m}_p, \rho_e, A_e$ in a specific way. This is why the demanding goal of engine design is to tune all the engine parameters, including v_e, such that the total thrust is maximized. Hence it is not only v_e alone which is decisive for the momentum thrust of an engine, but it is necessary to adjust all the relevant engine parameters in a coordinated way.

1.3
Rocket Performance

The mechanical power of an exhaust jet, the so called **jet power**, is defined as the change of the kinetic energy of the ejected gas (jet energy) per time unit, i.e.

$$P_e := \frac{dE_e}{dt} = \frac{d}{dt}\left(\frac{1}{2}m_p v_e^2\right) = \frac{1}{2}\dot{m}_p v_e^2 = \frac{1}{2}F_e v_e \quad \text{jet power} \tag{1.3.1}$$

It describes the time rate of expenditure of the jet energy. The thrust power of an engine is the thrust energy generated per time unit, i.e.

$$P_* := \frac{dE_*}{dt} = \frac{d}{dt}\left(\frac{1}{2}m_p v_*^2\right) = \frac{1}{2}\dot{m}_p v_*^2 = \frac{1}{2}F_* v_* \quad \text{thrust power} \tag{1.3.2}$$

where the latter parts in both equations occur because of Eqs. (1.1.3) and (1.2.6). The power transmitted to a spacecraft (S/C) with velocity v is simply calculated according to classic physics by the product of force times velocity, i.e.

$$P_{S/C} = F_* \cdot v \quad \textbf{transmitted spacecraft power} \quad (1.3.3)$$

Note that the forces (here F_e and F_*) are independent of the chosen reference system, whereas the velocities v_e and v_* are only meant with respect to the rocket. So *jet and thrust power are properties with respect to the rocket*, while the *transmitted spacecraft power is valid in the rocket system and the external inertial reference system* because v is the same in both of them. Note, however, that v depends on the chosen external reference system.

The so-called <u>total</u> impulse I_{tot} of an engine is the integral product of total thrust and propulsion duration

$$I_{tot} := \int_0^t F_* dt = v_* \int_0^t \dot{m}_p \cdot dt \quad (1.3.4)$$

$$= m_p v_* \quad @\ v_*(t) = const \quad \textbf{total impulse}$$

The latter is only valid as long as the effective exhaust velocity is constant. This is, in its strict sense, not the case during launch, where the external pressure (and hence the effective exhaust velocity) varies due to the pressure thrust.

The total impulse can be used to define the very important (weight-)specific impulse which characterizes the general performance and therefore is a figure of merit of an engine. The **weight-specific impulse** is defined as "*the achievable total impulse with respect to a given propellant weight $m_p g_0$*", i.e. with Eq. (1.3.4)

$$\boxed{I_{sp} := \frac{I_{tot}}{m_p g_0} = \frac{v_*}{g_0}} \quad @\ v_*(t) = const \quad \textbf{(weight-)specific impulse} \quad (1.3.5)$$

By this definition the specific impulse has the curious, but simple, dimension "second." Typical values are 300–400 seconds for chemical propulsions, 300–1500 seconds for electrothermal propulsions (Resistojet, Arcjet), and approximately 2000–6000 seconds for electrostatic (ion engines) and electromagnetic engines (see Fig. 1.7).

In Europe, in particular at ESA, the **mass-specific impulse** with definition "I_{sp} = *the achievable total impulse with respect to a given propellant mass m_p*" is more common. This leads to the simple identity $I_{sp} = v_*$. However, the definition "I_{sp} = weight-specific impulse" is more established worldwide, which is why we also will use it throughout this book. In either case you should keep in mind that quite generally:

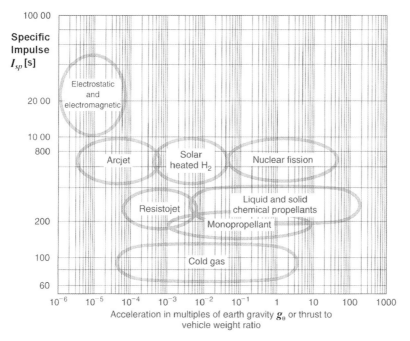

Figure 1.7 Specific impulse and specific thrust of different propulsion systems.

> The specific impulse is an important figure of merit of an engine, and is in essence the effective exhaust velocity.

1.4 Rocket Equation of Motion

Apart from its own thrust, also external forces determine the trajectory of a rocket. They are typically summarized to one external force F_{ext}

$$F_{ext} := F_G + F_D + F_L \ldots \quad (1.4.1)$$

with F_G = gravitational force, F_D = aerodynamic drag, and F_L = aerodynamic lift (see Fig. 1.8). For each of these external forces, the rocket can be considered as a point on which the external force acts. This point has a unique location with regard to the geometry of the rocket, and it is in general different for every type of force. The masses of the rocket can be treated as lumped together at the center of mass where the gravitational force applies. The aerodynamic drag and lift forces virtually apply at the so-called center of pressure. And possible magnetic fields have still another imaginary point of impact. If the

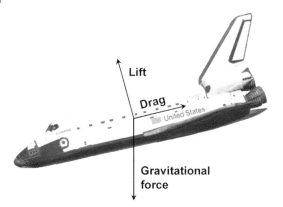

Figure 1.8 Effects of different external forces on a spacecraft.

latter do not coincide with the center of mass, which in general is the case, the distance in between results in torques due to the inertial forces acting effectively at the center of mass. Here, we disregard the resulting complex rotational movements, and we just assume that all the points of impact coincide, or that the torques are compensated by thrusters.

Newton's second law, Eq. (7.1.12), gives us an answer to the question of how the rocket will move under the influence of all the forces F_i including the propellant force:

$$m\dot{v} = \sum_{\text{all } i} F_i$$

We therefore find the following equation of motion for the rocket:

$$m\dot{v} = F_* + F_{ext}$$

and with Eq. (1.1.2), we finally obtain

$$\boxed{m\dot{v} = \dot{m}v_* + F_{ext}} \qquad \textbf{rocket equation of motion} \qquad (1.4.2)$$

This is the key differential equation for the motion of the rocket. In principle the speed can be obtained by a single integration step and its position by a double integration. Note that this equation applies not only to rockets but also to any type of spacecraft during launch, reentry or when flying in space with or without propulsion.

Problems

Problem 1.1 Balloon Propulsion

Consider a balloon, which is propelled by exhausting its air with density $\rho = 1.29 \text{ g dm}^{-3}$. The balloon has a volume of 2 dm^3, the exit (throat) diameter is $A_t = 0.5$ cm^2. Let's assume the balloon exhausts the gas with constant mass flow rate within 2 s. Show that the momentum thrust is $F_e = 0.026$ N and the pressure thrust is $F_p = 0.013$ N and hence that the momentum thrust is roughly twice as big as the pressure thrust.

Hint: Observe that the exhaust velocity at the throat does not reach the speed of sound. Make use of Bernoulli's equation $p + \frac{1}{2}\rho v^2 = const$.

Problem 1.2 Nozzle Exit Area of an SSME

The thrust of a Space Shuttle main engine (SSME) at 100% power level is 1.817×10^6 N at sea level and 2.278×10^6 N in vacuum. By using only this information, derive that the nozzle exit area is $A_e = 4.55$ m^2.

2
Rocket Flight

We now want to tackle the problem of solving the equation of motion (1.4.2). As will be seen in Section 2.1, even for many simple cases this equation can be solved only by numerical methods. Since this is not the objective of this book, we will treat only those important cases which can be analyzed analytically. This will give rise to an important characteristic quantity, the so-called "delta-v". Its relevance will be explored in Section 2.5.

Before turning to this, we will introduce some essential rocket mass definitions that we will use in this chapter:

$m =$ instantaneous total mass of the rocket,

$m_0 =$ total launch mass of the rocket,

$m_f =$ mass of the rocket at burnout (final mass),

$m_p =$ propellant mass of the rocket before launch or before a maneuver,

$m_s =$ structural mass of the rocket,

$m_L =$ payload mass.

From this it is obvious that

$$m_0 = m_p + m_s + m_L$$
$$m_f = m_0 - m_p = m_s + m_L$$
(2.0.1)

In the following calculations, the variable parameter m can often be interpreted as the instantaneous mass as well as the mass at burnout, so in most cases $m = m_f$ is valid.

2.1
General Considerations

Before we come to the few very important cases which can be examined analytically, let's have a look at the general solution of a flight in an external field.

Astronautics. Ulrich Walter
Copyright © 2008 WILEY-VCH Verlag GmbH & Co. KGaA, Weinheim
ISBN: 978-3-527-40685-2

This may be any relevant external field, but it is almost always the gravitational field. To do this, we separate the variables on the left side of the motion equation (1.4.2), and we get

$$dv = \frac{F_* + F_{ext}}{m} dt \qquad (2.1.1)$$

where F_* and F_{ext} are generally dependent on time. For example, during ascent in the atmosphere, the pressure thrust continually varies according to Eq. (1.2.6) because of the changing atmospheric pressure and/or the mass flow rate, which in particular holds for solid propellant rockets. In addition the thrust direction changes because of the so-called gimbaling, i.e. the steering of the nozzle to change flight direction. In all these cases, with a given $F_*(t)$, $F_{ext}(t)$, $\dot{m}_p(t)$, or

$$m(t) = m_0 - \int_0^t \dot{m}_p(t') \, dt' \qquad (2.1.2)$$

respectively, one can calculate the resultant velocity change by explicit integration:

$$\Delta v(t) = \int_0^t \frac{F_* + F_{ext}}{m} dt' = \int_{m_0}^m \frac{v_*}{m} dm + \int_0^t \frac{F_{ext}}{m} dt' \qquad (2.1.3)$$

where we have assumed $t_0 = 0$ for the sake of simplicity. We have written the left side of the equation in terms of the new and characteristic quantity "delta-v"

$$\Delta v(t) := \int_{v_0}^v dv = v(t) - v_0 \qquad (2.1.4)$$

which will turn out to be quite handy to describe spacecraft maneuvers in space (see Section 2.5). It describes the total change of the rocket's velocity vector due to all forces acting on the S/C over time t.

In order to determine the position of the S/C as a function of time, Eq. (2.1.3) needs to be integrated once more. For nearly every practical case these integrations need to be done by numerical methods. There is only one important case where both integrations can be performed fully analytically for an external force: the constant tangential thrust transfer under a gravitational force. This important case will be covered separately in Section 8.2. We now consider some other important specific limiting cases.

2.2
Rocket in Free Space

A limiting case occurs in free space when there are no external forces, $F_{ext} = 0$. In addition, in free space thrust maneuvers typically take place with a v_* which is constant both in absolute value and in direction. This special but most common situation simplifies Eq. (2.1.3) considerably to

$$\Delta v = v_* \int_{m_0}^{m} \frac{dm}{m}$$

so that it can be integrated straight away

$$\Delta v = -v_* \cdot \ln \frac{m_0}{m} \quad @ \ v_*(t) = const \tag{2.2.1}$$

with (v_0, m_0) the initial, (v, m) the final state of the S/C, and $\Delta v = v - v_0$ (Fig. 2.1). Note that the velocity change is independent of thrust level, of the duration, or of any time dependence of the thrust, and hence via Eq. (1.1.3) of any variation of the mass flow rate. So in free space, for "delta-v", the specific engine characteristics does not matter, nor whether the velocity-change boost is carried out over a short or a long time period. All what matters is initial and final mass and a constant exhaust speed.

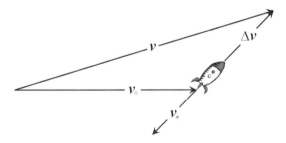

Figure 2.1 Direction of "delta-v" for a maneuver in free space.

Keeping in mind that Δv is always strictly antiparallel to v_* (see Fig. 2.1), Eq. (2.2.1) can be rewritten as an absolute value equation

$$\boxed{\Delta v = v_* \ln \frac{m_0}{m}} \quad @ \ v_*(t) = const \quad \text{rocket equation (one stage)} \tag{2.2.2}$$

which is also known as the **Ziolkowski equation**. Next to Eq. (1.1.3) this rocket equation is the most important equation in rocket flight. It can be written alternatively as

$$\frac{m}{m_0} = e^{-\Delta v/v_*} \tag{2.2.3}$$

or applying $m = m_f = m_0 - m_p$ at burnout

$$\frac{m_p}{m_0} = 1 - e^{-\Delta v / v_*} \tag{2.2.4}$$

Note that during ascent from Earth the condition $v_*(t) = const$ is not strictly fulfilled, as there are pressure thrust changes.

2.3
Impulsive Maneuvers

Another limiting case is where maneuvers are impulsive. Formally this means that for $t \to 0$ the mass specific total thrust $\int_0^t \frac{F_*}{m} dt'$ of the maneuver takes on a finite value. But since $\lim_{t \to 0} \int_0^t \frac{F_{ext}}{m} dt' = 0$ this indirectly implies

$$F_* \gg F_{ext},$$

i.e. a minute but powerful thrust maneuver. From Eq. (2.1.3) we therefore find

$$\Delta v = \int_0^t \frac{F_*}{m} dt' = \int_{m_0}^{m_f} \frac{v_*}{m} dm. \tag{2.3.1}$$

> For an impulsive maneuver external fields can be neglected and the so-called "delta-v" is determined solely by the thrust characteristics.

Note, that if the propellant mass expelled in an impulsive maneuver is not negligibly small compared to the total rocket mass, the "delta-v" of the maneuver is <u>not</u> $\Delta v \approx F_*/m \int dt' = F_* t / m$ as assumed quite frequently. Rather for a constant thrust (equals mass flow rate at $v_* = const$)

$$\Delta v = \int_{m_0}^{m_f} \frac{v_*}{m} dm = v_* \int_{m_0}^{m_f} \frac{dm}{m} = -v_* \cdot \ln \frac{m_0}{m}$$

i.e. Eq. (2.2.1). Impulsive maneuvers are of high relevance for orbit transfers. This is why we will investigate their effects on orbits in more detail in Chapter 8.

2.4
Rocket in a Gravitational Field

Because the rocket equation of motion (1.4.2) is universal it also applies for a rocket without propulsion in a gravitational field: $m\dot{v} = m\ddot{r} = F_G = -\mu r/r^3$.

2.4 Rocket in a Gravitational Field

This situation is more frequent with satellites or space probes. So on one hand it is an extremely important case, whereas on the other hand it is somewhat difficult to treat, which is why we explore it separately in Chapter 7.

2.4.1
Brief Thrust

If the thrust $F_* = -\dot{m}_p v_*$ is constant, i.e. $\dot{m}_p = const$ and $v_* = const$, but if $F_* \gg F_{ext}$ does not apply, we have to take into account the effect of the gravitational field during the maneuver. From Eq. (2.1.3) we find

$$\Delta v(t) = v_* \int_{m_0}^{m} \frac{dm}{m} + \int_{0}^{m} \frac{F_{ext}}{\dot{m}} \frac{dm}{m}$$

Often the thrust maneuver is short compared to any variation of the external field, which implies $F_{ext} = const$. In this case

$$\Delta v = -\left(v_* - \frac{F_{ext}}{\dot{m}_p}\right) \cdot \ln \frac{m_0}{m_f}$$
$$= \frac{F_* + F_{ext}}{\dot{m}_p} \cdot \ln \frac{m_0}{m_f} \qquad @ \; v_*, F_{ext} = const \qquad (2.4.1)$$

This is quite an interesting result. It claims that for a brief thrust maneuver we only need to substitute $v_* \to v_* - F_{ext}/\dot{m}_p$ in the familiar rocket equation (2.2.2). In closing we mention that the impulsive maneuver considered above is just a limiting case of Eq. (2.4.1) for $F_* = -\dot{m}_p v_* \gg F_{ext}$.

2.4.2
Gravitational Loss

We now consider the ascent of a rocket from a celestial body (cf. Section 6.4) with the gravitational force F_G as the only external force acting on the rocket. From Eq. (2.1.1) we find for the instantaneous speed gain

$$dv = \frac{v_*}{m} dm + \frac{F_G}{m} dt$$

We assume that in course of the ascent the gravitational field does not change significantly and therefore $F_G = mg_0 \approx const$. To find the absolute value of the instantaneous speed gain, we multiply this equation by the instantaneous unit speed vector $\hat{v}(t)$, finding

$$dv = \frac{\hat{v} v_*}{m} dm + g_0 \cos(\gamma + 90°) \cdot dt = \frac{\hat{v} v_*}{m} dm - g_0 \sin\gamma \cdot dt$$

where $\gamma = \angle(v, F_G) - 90°$ is the so-called **flight path angle**, which is the angle between the flight path and the local horizon (see Fig. 6.4). For a steering-free ascent $\hat{v}v_* = -v_* \approx const$. In general $\hat{v}\, v_*/m$ is the mass-specific speed gain in a gravitational field-free environment, the integral of which is not of specific interest for us here, so we will denote it quite generally as Δv_0. Therefore we find for the velocity after time t

$$\Delta v = \Delta v_0 - \underbrace{g_0 \int_0^t \sin\gamma \cdot dt'}_{gravitational\ loss} \qquad (2.4.2)$$

So, in contrast to a rocket in free space, the achieved velocity for ascent depends on the flight direction relative to the local horizon and the time t to engine shutdown. For ascent $\gamma > 0°$ and therefore the integral term in Eq. (2.4.2) is positive. (For a reentering S/C with $\gamma < 0°$ it would be negative.) Therefore an ascending rocket does not achieve the same velocity increase as for a propulsion maneuver in free space, which is why the entire integral term is called **gravitational loss** term. Gravitational loss is particularly striking for vertical ascent, when $\gamma = 90°$. In this case

$$\Delta v = \Delta v_0 - g_0 t \quad @\ \text{vertical ascent}$$

In the extreme case when the thrust at ascent just balances the gravitational force, then $g_0 t = \Delta v_0$ and therefore $\Delta v = v_0 = 0$: The rocket barely hovers above the launch-pad until the fuel is used up. We therefore derive the following general rule:

> The longer the ascent time and the larger the angle between the flight path and the gravitational force, the smaller is the final speed at engine shutdown.

Therefore, for vertical ascent, not only the specific impulse $I_{sp} \propto v_*$ is a figure of merit of an engine, but also a high thrust, which reduces ascent time and thus gravitational losses. Consequently, powerful but admittedly low-efficient chemical boosters are regularly used during vertical ascent, while for the upper stages when the rocket turns horizontally higher-efficient but lower-power liquid Hydrogen/Oxygen thrusters take over.

Though we have found a clue how to get into orbit efficiently we are still far from answering the question: What is the optimum ascent trajectory? We will investigate this problem in more detail in Section 6.4.

2.5 Delta-v Budget

The "delta-v" figure Δv appeared in the above equations for the first time. It has a special, double relevance in astronautics. On the one hand, it represents the mass-specific momentum change of a rocket: $\Delta v = \Delta p/m$. Momentum changes are necessary to change from a given Keplerian orbit to another Keplerian orbit, or from a Hohmann transfer orbit into a planetary orbit, or vice versa. We know from conservation laws that momentum is a basic physical parameter. Another important basic parameter is energy. To track energy changes is very important, as the initial increase in kinetic energy $\Delta E = mv \cdot \Delta v$ generated by a small Δv may transfer into different forms of energy by means of the energy conservation law, e.g. into potential energy, and in lower Earth orbits unfortunately also into frictional energy. This is why it should come as no surprise that a rocket, which formally gains velocity through a kick-burn Δv, may actually decrease its total velocity v when for instance a rocket fires in a gravitational potential. Then more kinetic energy is transferred into potential energy than kinetic energy is produced by the kick-burn. Overall, due to a higher final orbit and in accordance with $v = \sqrt{\mu/r}$ (see Eq. (7.4.3)) the orbit velocity is paradoxically reduced, even though the spacecraft initially received a velocity increase Δv.

The nice feature with Δv is that it measures all the possible energy demands of a mission. Since it measures also momentum and angular momentum demand, it is a perfect measure for the total thrust demand for an entire mission.

Remark: *Changes of angular momentum caused by thrust maneuvers are related to Δv demand in a quite complicated way, as because of $h = r \times p = mrv \cdot \sin(\angle(r,v)) = mrv \cdot \cos \gamma$, every change of angular momentum depends strongly on the thrust direction and the so-called flight path angle γ. For a circular orbit and for a small tangential thrust $\Delta v_\parallel$, from the above it is easy to show that $\Delta h = mr \cdot \Delta v_\parallel$.*

Even better, according to Eq. (2.2.4), the all important propellant demand m_p has a one-to-one correspondence to Δv. This is a very handy relationship. Although the required propellant is actually the determining factor in space flight, later on it does not shows up in orbit calculations. The orbit equations, such as the vis-viva equation (7.2.10), however, mostly deal with the Δv parameter. Equation (2.2.4) now links the two parameters in a very convenient way. Even more conveniently, the propellant demand of two successive Δv maneuvers corresponds to the sum $\Delta v_1 + \Delta v_2$ of the individual maneuvers according to

$$\Delta v = v_* \ln \frac{m_0}{m_2} = v_* \ln \frac{m_0 m_1}{m_1 m_2} = v_* \ln \frac{m_0}{m_1} + v_* \ln \frac{m_1}{m_2} = \Delta v_1 + \Delta v_2$$

since the propulsion effort is independent on the sign of Δv, we derive for the effort

$$\Delta v = \sum_{\text{all maneuvers}} |\Delta v_i| \qquad (2.5.1)$$

So we can make the following comment:

> The figure Δv is a perfect measure for the total propulsion demand of a mission due to the additive property together with its direct relation to momentum, angular momentum, and total energy. This is why it is called "delta-v budget" (a.k.a. "propulsion demand") per se.

2.6
Fuel Demand – Star Trek Plugged

Keep in mind that, according to Eq. (2.2.4), the propellant demand and the propulsion demand have a non-linear relationship: propellant demand strongly grows for increasing propulsion demand. For practical purposes, the relative change of the launching mass of a rocket $\Delta m_0/m_0$ for a given change in propulsion demand $\Delta(\Delta v)$ is a very interesting relation. This relation of relative change can shown (exercise, Problem 2.1) to be

$$\frac{\Delta m_0}{m_0} = \exp\left[\frac{\Delta(\Delta v)}{v_*}\right] - 1 \approx \frac{\Delta(\Delta v)}{v_*} \quad @\ \Delta(\Delta v) \ll v_* \qquad (2.6.1)$$

Example
An interplanetary probe is to be accelerated to sufficient escape velocity to leave the solar system. A flyby maneuver via Jupiter requires $\Delta v_2 = 6.33\ \text{km s}^{-1}$. Direct escape from Earth orbit requires $\Delta v_1 = 8.82\ \text{km s}^{-1}$. How much more launching mass do you need for direct escape, if the chemical propulsions have an I_{sp} of 306 s? Answer:

$$v_* = g_0 \cdot I_{sp} = 9.81 \cdot 306\ \text{m s}^{-1} = 3.00\ \text{km s}^{-1}$$

$$\frac{\Delta m_0}{m_0} = \exp\left[\frac{2.49}{3.00}\right] - 1 = 1.29$$

The additional propulsion demand for direct escape is only 39%; the increase of the lauch mass due to the additional propellant demand, however, is 129%!

Let's have a closer look at Eq. (2.2.4). It refutes what many science fiction fans believe: that good classical propulsion just needs a lot of energy.

Remark: *By "good classical propulsion" we refer to classical recoil propulsion not to exotic propulsion like "warp" propulsion. When you see on a cinema screen a spacecraft accelerating with a thundering roar (which of course does not make sense at all in a vacuum as outer space!) during a spacecraft battle, this obviously is recoil propulsion.*

The truth however is this: What a flight maneuver needs more than anything else is propellant mass. A lot of it. As for large maneuvers, such as an inversion of the flight direction, Δv gets very large, the exponent tends to zero, and the used propellant mass tends to 100% of the spacecraft mass, which is an extremely uncomfortable perspective for the passengers. You could object, arguing that "Star Trek" et al. have engines providing unlimited exhaust velocity v_*, which would reduce the propellant demand in line with Eq. (2.2.4). But that's not possible. Because, from Einstein, we know that the maximum possible exhaust velocity is the velocity of light c. Assuming that "Star Trek" relativistic rocket engines (of course) have $v_* = c$, one can prove (cf. Section 2.9) that, for big maneuvers, Eq. (2.2.4) takes on the form

$$\frac{m_p}{m_0} = 1 - \frac{1}{2}\sqrt{1 - \left(\frac{\Delta v}{c}\right)^2} \quad @ \ v_* = c, \ \Delta v \to c$$

If Capt. Kirk now wants to carry out a reversion maneuver, he has to decelerate first, i.e. $\Delta v = v \approx c$, then he has to accelerate again in the opposite direction, i.e. again $\Delta v = v \approx c$. This relativistic equation has to be used for both maneuvers one after the other, leading to

$$\frac{m_p}{m_0} = 1 - \frac{1}{4}\left[1 - \left(\frac{v}{c}\right)^2\right] \quad @ \ v_* = c, \ \Delta v = 2v \to c$$

In other words, if, in a galactic fight with an enemy, Kirk only flew with 90% of the velocity of light (which would indeed be below his dignity), he would need 95.25% of the spacecraft's mass as propellant for one single reversion maneuver. If he flew with 99% velocity of light, he would need 99.5% of the spacecraft's mass. It is quite strange that you never see any of the necessary huge propellant tanks in the movies!

But propellant shortage would be the least of Kirk's problems. The energy required for the reversion maneuver would be more of a problem. A relativistic calculation for the total energy demand of a full slam on the brakes would be

$$E = \frac{2m_0 c^2}{\sqrt{1 - v^2/c^2}}$$

If the spacecraft does a reversion maneuver using up the double amount of energy, and let's assume the ideal case that Kirk gets his energy from annihilation of matter and antimatter – the very best that could be imagined – then he would need the energy mass-equivalent

$$m = \frac{4m_0}{\sqrt{1 - v^2/c^2}}$$

of matter and antimatter, half the amount of each. So, if Kirk flies with 90% velocity of light, he would need the mass equivalent of 9.18 spacecraft masses, and with 99% of the velocity of light he would need 28.4 spacecraft masses. But from a logical point of view, this is not possible at all, as the spacecraft only has one spacecraft mass.

2.7
Rocket Efficiency

The principle of rocket propulsion is that a certain amount of energy is utilized to accelerate propulsion mass in order to gain rocket speed via repulsion and hence rocket kinetic energy. Of course, it is the goal to design a rocket which gets out as much kinetic energy as possible from a given amount of energy spent. The quantity to measure this is the *total rocket efficiency* η_{tot}. It is defined as

$$\eta_{tot} := \frac{\text{rocket kinetic energy at burnout}}{\text{utilized energy}}$$

$$= \frac{\frac{1}{2}m_f v_f^2}{E_0} \quad \text{total rocket efficiency} \quad (2.7.1)$$

The utilized energy is converted into rocket kinetic energy in two steps. First the engine converts the utilized energy into thrust with *internal efficiency* η_{int}

$$\eta_{int} = \frac{\text{generated thrust energy}}{\text{utilized energy}} = \frac{\frac{1}{2}m_p v_*^2}{E_0} \quad \text{internal efficiency} \quad (2.7.2)$$

The internal efficiency is independent from the motion state of the rocket. It is therefore characteristic for an engine and has to be evaluated separately for different kinds of engines (see for instance Eq. (4.1.24)). In a second (propulsion) step the thrust energy is converted into kinetic energy of the rocket based on the conservation of momentum. The efficiency of this second conversion step is called *external efficiency* – a.k.a. integral or mechanical efficiency – of a rocket. It is defined as

$$\eta_{ext} := \frac{\text{rocket kinetic energy at burnout}}{\text{generated thrust energy}}$$

(2.7.3)

$$= \frac{\frac{1}{2} m_f v_f^2}{\frac{1}{2} m_p v_*^2} \qquad \textbf{external efficiency}$$

In total we have

$$\eta_{tot} = \eta_{ext} \cdot \eta_{int} \qquad (2.7.4)$$

How is the external efficiency determined? The key point is that velocity is a property relative to a reference system. Velocity changes when a different reference system is used.

Although never mentioned explicitly in the literature, *the reference system here is the one in which the rocket has zero velocity at the beginning of the propulsion phase*. In this case $v_f = v_0 + \Delta v = \Delta v$. Applying this relation to Eq. (2.7.3) yields

$$\eta_{ext} = \frac{m_f v_f^2}{m_p v_*^2} = \frac{m_f}{m_0 - m_f} \frac{\Delta v^2}{v_*^2} = \frac{(\Delta v / v_*)^2}{m_0 / m_f - 1}$$

With Eq. (2.2.3) we finally obtain

$$\eta_{ext} = \frac{(\Delta v / v_*)^2}{\exp(\Delta v / v_*) - 1} \qquad (2.7.5)$$

This function is displayed in Figure 2.2. It has a maximum at $\Delta v / v_* = 1.59362...$, which, according to Eq. (2.2.3), corresponds to $m_f / m_0 = 0.203188...$. Usually it is the case that the optimum operating point is around this maximum and an acceptable economic limit is usually reached at about $\Delta v \approx 3v_*$, when the payload portion is only 5.0%. In this way it is argued that a rocket can be operated efficiently only for $\Delta v < 3v_*$.

Some words of caution are in place. The only reasonable reference system relative to which the rocket has zero speed is the Earth's surface at ascent. However, for ascend from the launch-pad to a target orbit with final velocity $v_f = \Delta v$, it is not just the rocket equation (2.2.2) alone, but also gravitational losses (see Section 2.4.2), as well as drag and steering losses (see Section 6.4.1), that need to be considered. This reduces the external efficiency significantly, and the expression for it can no longer be given explicitly. So, as long as the rocket equation (2.2.2) is applied to derive Eq. (2.7.5), the only case for which it applies is the impulsive maneuver (see Section 2.3), because then all external

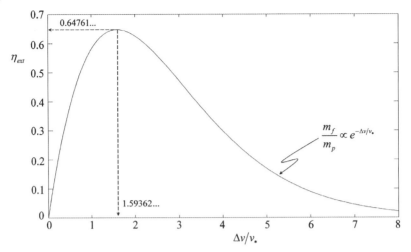

Figure 2.2 External efficiency of a rocket as a function of the propulsion demand.

fields are negligible. But even in this case the external efficiency is meaningless, because the objective of a thrust maneuver is to achieve a given delta-v. The kinetic energy gained by the maneuver is irrelevant in contrast. The only thing that matters is this: How much propellant is expended to achieve the given delta-v? The answer is provided by the rocket equation or Eq. (2.2.4), respectively. Only the rocket equation is able to tell whether an impulsive maneuver is efficient or not – apart from the fact that efficiency is a discretionary notion. The external efficiency therefore is of no practical relevance, which is why it is rarely used. In contrast, the internal efficiency is a valuable figure of merit of an engine.

2.8
Payload Considerations

When looking at Eq. (2.2.3), at first glance one might think that the burnout mass is the payload mass, $m = m_f = m_L$. That would mean that at a given Δv one would be able to get a payload of any size into space, if you only choose the launch mass big enough. However, this thought discards the structural mass m_s of the rocket, which includes the mass of the outer and inner mechanical structure of the rocket in particular the tank mass, the mass of the propulsion engines including propellant supply (pumps), avionics including cable harness, energy support systems, emergency systems, etc. Structural mass trades directly with payload mass, and hence

$$m_f = m_s + m_L$$

2.8 Payload Considerations

In practice structural mass limits the payload mass to such an extent that later on we will have to look for alternative propulsion concepts, the so-called staging concepts, to reduce m_s. For further considerations, in particular for the later stage optimization, we define the following mass ratios:

$$\mu := \frac{m_f}{m_0} = \frac{m_s + m_L}{m_0} \qquad \text{mass ratio} \qquad (2.8.1)$$

$$\varepsilon := \frac{m_s}{m_0 - m_L} = \frac{m_s}{m_s + m_p} \qquad \text{structural ratio} \qquad (2.8.2)$$

$$\lambda := \frac{m_L}{m_0 - m_L} = \frac{m_L}{m_s + m_p} \qquad \text{payload ratio} \qquad (2.8.3)$$

From this it follows

$$\mu(1+\lambda) = \frac{m_s + m_L}{m_0} \frac{m_s + m_p + m_L}{m_s + m_p} = \frac{m_s + m_L}{m_s + m_p} = \varepsilon + \lambda$$

hence

$$\mu = \frac{m_f}{m_0} = \frac{\varepsilon + \lambda}{1 + \lambda} \qquad (2.8.4)$$

So the rocket equation (2.2.3) can be written as

$$\frac{\Delta v}{v_*} = -\ln \frac{\varepsilon + \lambda}{1 + \lambda}$$

or

$$\lambda = \frac{e^{-\Delta v/v_*} - \varepsilon}{1 - e^{-\Delta v/v_*}} \qquad (2.8.5)$$

This equation is represented in Fig. 2.3. It directly relates the payload ratio to the achievable propulsion demand at a given structural ratio of the rocket and effective exhaust velocity of the engine. So, because the structural mass is not negligibly small, it is no longer possible to achieve any propulsion demands you like. In numbers this says that:

> As $\varepsilon = 0.05$ represents the lower limit of the structural mass of a rocket, the obtainable propulsion demand is limited to $\Delta v < 2.5 \cdot v_*$ at $\lambda = 3\,\%$.

The effective velocity of chemical rockets ascending through Earth's atmosphere is limited to $v_* \leq 4 \text{ km s}^{-1}$, limiting the available propulsion demand to $\Delta v < 10 \text{ km s}^{-1}$. If, for instance, the goal is to get in a single stage into orbit

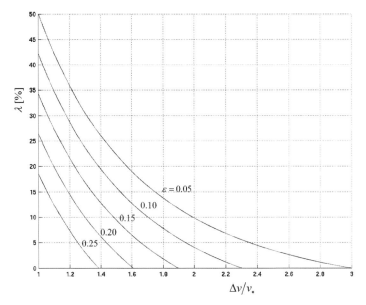

Figure 2.3 Obtainable payload ratios at a given propulsion demand for different structural factors.

(so-called SSTO), for which in practice $\Delta v = 9.2 \text{ km s}^{-1}$ is required, then even with an optimum $v_* = 4 \text{ km s}^{-1}$ for a LH2/LOX engine the achievable payload ratio is a mere $\lambda = 3.4\,\%$. Even if the structural ratio would be an ideal $\varepsilon = 5\,\%$ we would arrive at only $\lambda = 5.6\,\%$. So, in principle, an SSTO rocket is possible, but only at the expense of an unacceptable low payload mass. This is why there is no way around a staged rocket (cf. calculation following Eq. (3.3.5)) to which we come in the next chapter.

2.9
Relativistic Rocket[1]

All what was said to here was based on Newton's classical mechanics. It holds as long as the speed of the rocket v is well below the speed of light c. We know from the theory of special relativity, which Einstein developed at the beginning of the last century, that physics behaves differently if $v \approx c$. May rockets eventually fly close to the speed of light? In order to find out, we need to know what is needed to get it close to the speed of light and how it performs there. But note that the need to apply relativistic physics depends on the precision which is needed to describe a given situation. A satellite navigation system in earth orbit, for instance, needs a high-precision time-

1) Section 2.9 is partly adopted from Walter (2006).

keeping system on-board with a stability of less than $\Delta t/t \approx 10^{-12}$ which allows a position on Earth to be determined with roughly 10 cm accuracy. At an orbital speed of 3.9 km s^{-1}, relativity contributes to the time deviation with $\Delta t/t = v^2/2c^2 \approx 8.5\ 10^{-11}$, which is not negligible. Therefore, also at much lower speeds relativity must be taken into account if the accuracy of the description is high.

Our goal here is to understand how relativity works for a spacecraft close to the speed of light and how this relates to classical mechanics at lower speeds. We start out by assuming a one-dimensional motion of the rocket, with thrust direction and hence acceleration along the x axis. The main ingredients of relativity will not be touched by this restriction. This implies that the position of a rocket in time can be appropriately described by the 2-vector (x, t). We define two reference frames: the "primed" reference frame of an external inertial observer $O'(x', t')$ and the "unprimed" reference frame of the rocket under consideration $R(x, t)$, which is supposed to have an instantaneous velocity v relative to O'.

2.9.1
Space Flight Dynamics

For relativistic physics it is important to note that among all existing reference frames, there is one preferred frame: the rest frame. This is the frame of the object under consideration in which it is at rest. Any other external observer having velocity v relative to this rest frame observes the properties of the object such as length, time, speed, and acceleration differently to the object itself. Since there may be an infinite number of observers and therefore many different views of the object properties, relativistic physics holds that only one has a proper view of the object: the object itself. In this sense relativity is an absolute concept.

Therefore relativistic physics introduces the notion of "proper." In general, a "proper" measure of a quantity is that taken in the relevant instantaneous rest frame, therefore also called the proper reference frame. So "proper" is everything an astronaut experiences in his rocket. This is why we will not put a prime on such quantities, and those as observed from outside will carry a prime. In general the observed values are dependent on the reference frame, with of course one exception: $v = v'$. Adopting this notion, what is of relevance first is how the proper measures relate to the measures of external observers.

Proper time (also called eigentime) τ is the time the watch of an astronaut in a rocket shows. Special relativity holds that τ is related to the time t' of the external observer O by

$$d\tau = dt = \sqrt{1 - \beta^2} \cdot dt' \qquad (2.9.1)$$

where we adopt the convenient relativistic notation $\beta := v/c$ and $\gamma := 1/\sqrt{1-\beta^2}$. We will sometimes denote dt by $d\tau$ in order to point out that the proper time is meant. It should be noted that in special relativity relation (2.9.1) holds for any condition of the rest frame even if it is accelerated, because, and contrary to common misjudgement, special relativity is not restricted to constant relative velocities or inertial reference frames.

Einstein pointed out that acceleration is an absolute concept: An astronaut does not experience rocket velocity in his rest frame, but he does experience acceleration. Let us assume that the astronaut experiences an acceleration α. Then special relativity tells us that this is related to the acceleration a' as seen by an external observer through

$$\alpha = \gamma^3 a' \quad \textbf{proper acceleration} \tag{2.9.2}$$

Because acceleration is an absolute concept we are able to define

$$d\sigma := \alpha(t) \cdot dt \tag{2.9.3}$$

where $d\sigma$ is an increase in speed as measured in the instantaneous rest frame. We integrate to get

$$\sigma(\tau) = \int_0^\tau \alpha(t) \cdot dt \tag{2.9.4}$$

This equation tells us that σ is the integral of the acceleration as experienced in the proper reference frame and hence is the speed as experienced by an astronaut, who sees the outer world going by. Since this is the true meaning of "proper," σ is a proper speed. In order to find the relation of this proper speed to the relative speed v, we apply Eqs. (2.9.1) and (2.9.2) to Eq. (2.9.4) and get

$$\sigma = \int_0^{t'} \gamma^2 a' \cdot dt'$$

As $a' = dv/dt' = c \cdot d\beta/dt'$ we find

$$\sigma = \int_0^v \frac{dv}{1 - v^2/c^2} = c \int_0^\beta \frac{d\zeta}{1 - \zeta^2} = c \cdot \operatorname{arctanh} \beta \quad \textbf{proper speed}$$

or

$$\beta = \frac{v}{c} = \tanh\left(\frac{\sigma}{c}\right) \tag{2.9.5}$$

2.9 Relativistic Rocket

It is now shown that the proper speed is proper in a more general sense. Let us consider a second rocket or any other object in space having the known speed u relative to the astronaut's R system. We want to know what its speed u' is as measured by O'. Special relativity tells us that

$$u' = \frac{u+v}{1+uv/c^2} = c\frac{\beta_u + \beta_v}{1+\beta_u\beta_v} \tag{2.9.6}$$

The problem with this transformation equation is that it is not linear as in classical physics, where the Galileo transformation $u' = u + v$ holds. In addition, Eq. (2.9.6) limits u' to the range $0 \leq u' \leq c$ if v starts out from below c. This can be seen immediately if one inserts even limiting velocities $u = c$. This is Einstein's famous law that nothing goes faster than the speed of light. It is exactly this non-linearity and limited range of values which causes problems when treating special relativity mathematically. We now apply Eq. (2.9.5) to Eq. (2.9.6) and find

$$\tanh\left(\sigma'_u/c\right) = \tanh(\sigma_u/c + \sigma_v/c)$$

where we have used the algebraic equation for any two values x and y

$$\frac{\tanh x + \tanh y}{1 + \tanh x \cdot \tanh y} = \tanh(x+y)$$

As this must hold for any proper speed values we find

$$\sigma'_u = \sigma_u + \sigma_v \tag{2.9.7}$$

i.e., proper speed recovers the linearity of speed transformation in special relativity.

According to Eq. (2.9.5) the proper speed goes to infinity if the externally observed speed goes to the speed of light. This is to say that, from an astronaut's point of view, there is no speed limit. He actually can travel much faster than the speed of light. But of course he cannot travel faster than infinitely fast. This is the reason why the observer also sees a speed limit: the speed of light. So the ultimate reason why nothing can ever go faster than the speed of light is that nothing can ever go faster than infinitely fast. Note that from this point of view photons always travel infinitely fast. They experience that any distance in the universe is zero: for them the universe is one point. Because their proper time is zero, one would say that they don't even exist. But this would be wrong. They come into existence at one point in our universe, they transfer energy, momentum, angular momentum, and information to any other point in proper zero time, thereby causally linking any two parts in our universe, and at the instance their work is done they are gone. This is why causality is the basic conservation law and hence the cement of our universe, and not the speed of light. The speed of light c may vary throughout our universe, but the fact that the proper time at $v = c$ is always zero and cannot become negative – implying that no inverse causality is possible – is firm.

In order to show that the concept of proper speed has relevance to the concept of classical speed, we finally show that for small speeds the proper speed turns over into the classical concept of speed v for $v \to 0$:

$$\sigma = c \cdot \operatorname{arctanh} \beta = c \left(\beta + \tfrac{1}{3} \beta^3 + \cdots \right) \approx c \beta = v \qquad (2.9.8)$$

We summarize by noting that the proper speed exhibits four important properties: It is proper, it transforms linearly, it takes on real numbers, and it turns over into the classical concept of velocity at low speeds. This implies that it is a natural extension of the classical speed into special relativity and is mathematically integrable.

2.9.2
Relativistic Rocket Equation

With the concept of proper speed at hand we start out to derive the relativistic rocket equation. We want to do this in its most general form. The two physically distinct rocket propulsion systems are mass propulsion and photon propulsion. We take both into account, and assume that upon combustion a portion ε of the propellant mass will be converted into energy with a certain efficiency η and that a portion δ of it expels the exhaust mass with velocity v_e, while the other portion $(1 - \delta)$ is expelled as exhaust photons, and the rest is lost. Therefore the overall energy scheme looks like Fig. 2.4. In the rest frame R of the rocket, momentum conservation holds. Taking the momenta of both exhaust components and that of the rocket into account we can write

$$(1 - \varepsilon) \, dm \cdot v_e + (1 - \delta) \, \eta \varepsilon \cdot dm \cdot c + (m + dm) \, dv = 0$$

where again we count dm negatively since m is the mass of the rocket. From the above equation we find

$$dv = -v_* \frac{dm}{m} \qquad (2.9.9)$$

with the effective exhaust velocity

$$v_* := (1 - \varepsilon) \, v_e + (1 - \delta) \, \eta \varepsilon c$$

Note that all terms in Eq. (2.9.9) are unprimed and are therefore terms measured in the proper reference frame, including dv. Now, in classical physics, the relation $dv = dv'$ holds, and hence the equation can be readily integrated

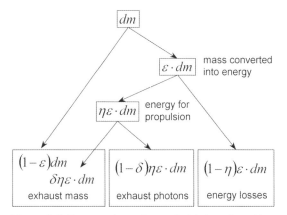

Figure 2.4 Energy scheme for a relativistic rocket with energy losses and expelled propulsion mass and photons.

to yield the classical rocket equation $\Delta v = v_* \ln(m_0/m)$, see Eq. (2.2.2). But $dv = dv'$ is no longer valid for relativistic speeds. However, if we identify $dv = d\sigma$ we again can directly integrate to obtain

$$\Delta\sigma = v_* \ln \frac{m_0}{m} \qquad \text{\textbf{relativistic rocket equation}} \qquad (2.9.10)$$

So the relativistic rocket equation is to the utmost extend complementary to the classical rocket equation (2.2.2). In order to show that Eq. (2.9.10) is in accordance with today's more convenient form of the relativistic rocket equation, we apply Eq. (2.9.5) and the algebraic equation for the free variable x

$$\operatorname{arctanh} x = \ln \sqrt{\frac{1+x}{1-x}}$$

to Eq. (2.9.10) and with Eq. (2.9.6) we derive the common form of the relativistic rocket equation

$$\frac{m_0}{m} = \left(\frac{1+\Delta\beta}{1-\Delta\beta}\right)^{\frac{1}{2\beta_*}} \qquad (2.9.11)$$

with

$$\beta_* = \frac{v_*}{c}$$

From Eqs. (2.9.3) and (2.9.9) we can also derive the thrust F_* of the relativistic rocket in its rest frame

$$F_* = m\alpha = m\frac{d\sigma}{d\tau} = -v_*\frac{dm}{d\tau} = -\dot{m}v_* \qquad \text{\textbf{relativistic rocket thrust}} \qquad (2.9.12)$$

which is identical to the classical equation (1.1.3).

2.9.3
Exhaust Considerations

Because a portion of the converted energy propels the exhaust mass, the energy obtained from the propellant $dE_m = \delta\eta\varepsilon \cdot dm \cdot c^2$ has to equal the relativistic energy of the propelled mass $dm_e = (1-\varepsilon)\,dm$, i.e.

$$dE_m = \delta\eta\varepsilon \cdot dm \cdot c^2 = \gamma_e dm_e \cdot c^2 - dm_e \cdot c^2 = (\gamma_e - 1)(1-\varepsilon)\,dm \cdot c^2 \quad (2.9.13)$$

This implies that for a given δ, η the two terms γ_e (or β_e) and ε are interrelated, namely

$$\varepsilon = \frac{\gamma_e - 1}{\delta\eta + \gamma_e - 1} \quad \text{or} \quad 1 - \varepsilon = \frac{\delta\eta}{\delta\eta + \gamma_e - 1} \quad (2.9.14)$$

or (the other way around)

$$\beta_e = \sqrt{1 - \left[\frac{1-\varepsilon}{1-\varepsilon(1-\delta\eta)}\right]^2} \quad (2.9.15)$$

We summarize by saying that internal energy considerations define the relativistic exhaust mass speed. For the effective exhaust velocity from Eq. (2.9.9) this implies that

$$\beta_* = (1-\varepsilon)\beta_e + (1-\delta)\eta\varepsilon$$

$$= (1-\varepsilon)\sqrt{1 - \left[\frac{1-\varepsilon}{1-\varepsilon(1-\delta\eta)}\right]^2} + (1-\delta)\eta\varepsilon \quad (2.9.16)$$

For a rocket, which exhausts just mass, $\delta = 1$, we find

$$\beta_* = (1-\varepsilon)\sqrt{1 - \left[\frac{1-\varepsilon}{1-\varepsilon(1-\eta)}\right]^2} \quad \text{effective exhaust-mass velocity} \quad (2.9.17)$$

In case the rocket has 100% efficiency, $\eta = 1$, we find the expression

$$\beta_* = \sqrt{2\varepsilon - \varepsilon^2} \quad @ \; \delta, \eta = 1$$

For a photon rocket we get

$$\beta_* = \eta \quad @ \; \varepsilon = 1, \delta = 0 \quad (2.9.18)$$

Matter–Antimatter Annihilation Drive

As an example, let us assume a matter–antimatter annihilation rocket (rather like those in Figs. 2.5 and 2.6). We assume that our rocket annihilates H_2 and anti-H_2 ($\overline{H}_2$) molecules stored as solid pellets in a storage tank below 14 K (the freezing temperature of hydrogen and hence also anti-hydrogen), typically at 1–2 K to avoid sublimation. In order to confine the neutral antimatter, either their diamagnetism would hold them together in a strong external magnetic field or they would be electrically charged and suspended in an array of electrostatic traps. Otherwise we neglect all the technical obstacles which come along with such storing devices.

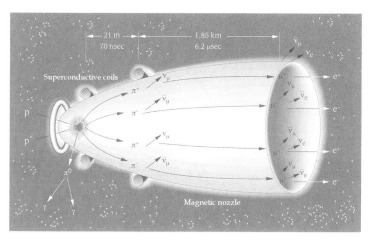

Figure 2.5 Working schema of a matter-antimatter annihilation drive. See also color figure on page 464.

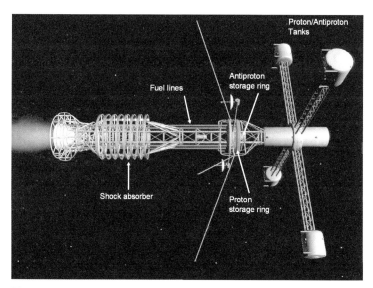

Figure 2.6 Artist view of the ICAN-II relativistic proton-antiproton annihilation drive rocket. See also color figure on page 464.

Upon annihilation of an H and $\overline{H}$ atom with a total rest mass of 1877.6 MeV, 22.30% of them are converted into charged pions, 14.38% into neutral pions and the electron and positron into two γ-rays. The charged pions can be deflected backwards by a magnetic field to provide propulsion force. Let us assume that this can be done with 100% efficiency. The neutral pions are lost because after a 0.06 μm travel distance they decay into 709 MeV γ-rays, which has to be considered as a major hazard to the crew. So, effectively, we have 418.8 MeV of pion rest mass as propulsion mass while the remainder is converted into energy, i.e. $\varepsilon = 0.7769$. Of this energy, 748.6 MeV goes into the kinetic energy of the pions, $\eta = 0.5132$, and the rest is lost.

As long as the γ-rays cannot be directed backward as well (there seems to be no way of doing that), thus adding to the thrust via photonic propulsion, this drive is a purely mass-exhaust drive. From Eq. (2.9.17) we then find an *ultimate effective exhaust-mass velocity* of

$$\beta_* = 0.2082 \quad @ \ H\text{-}\overline{H} \text{ annihilation} \quad (2.9.19)$$

For a given total rocket mass at a given time, this can be used to calculate the travel speed at this instance from rocket equation (2.9.10) or (2.9.11).

2.9.4
External Efficiency

As for a classical rocket in Section 2.7 we want to derive the external rocket efficiency η_{ext} of a relativistic rocket, which was defined by

$$\eta_{ext} := \frac{\text{rocket kinetic energy at burnout}}{\text{generated thrust energy}} =: \frac{E_{kin}}{E_*}$$

From Eq. (2.9.13) plus the photon energy we have

$$E_* = (\gamma_e - 1)(1-\varepsilon) m_p c^2 + (1-\delta) \eta \varepsilon m_p c^2$$

$$= \left[(\gamma_e - 1)(1-\varepsilon) + (1-\delta) \eta \varepsilon\right] \left(\frac{m_i}{m} - 1\right) mc^2$$

$$= \frac{\eta(\gamma_e - 1)}{\delta\eta + \gamma_e - 1} \left(e^{\sigma/v_e} - 1\right) mc^2 = \eta\varepsilon \left(e^{\sigma/v_e} - 1\right) mc^2$$

In the second line we have applied Eq. (2.9.14) and the relativistic rocket equation (2.9.10). And trivially

$$E_{kin} = \gamma mc^2 - mc^2 = (\gamma - 1) mc^2$$

Employing the definition we derive

$$\eta_{ext} = \frac{\gamma - 1}{\gamma_e - 1} \frac{\delta\eta + \gamma_e - 1}{\eta \left(e^{\sigma/v_*} - 1\right)} \quad (2.9.20)$$

$$= \frac{1}{\eta\varepsilon} \frac{\gamma - 1}{e^{\sigma/v_*} - 1} \quad \textbf{relativistic external efficiency}$$

For non-relativistic speeds, i.e. $\gamma \to 1 - v^2/2c^2$, $\sigma \to v$, and for $\delta = 1$, we recover the classical external rocket efficiency (see Eq. (2.7.5))

$$\eta_{ext} = \frac{(v/v_*)^2}{e^{v/v_*} - 1}$$

Note that while the external efficiency in the relativistic regime is dependent on the internal efficiency η, this does not hold for classical speeds.

2.9.5
Space-Time Transformations

It is an important and well-known feature of special relativity that observed values for space and time intervals depend on the reference frame of the external observer. This is what the word "relativity" actually refers to. With the concept of proper speed it is easy to derive the space-time transformation equations between the absolute (proper) reference frame of the spacecraft and that of an external observer, which we now denote (σ, τ) and (x', v', t'), respectively. From Eqs. (2.9.1) and (2.9.5) and denoting $\phi := \sigma/c$, the so-called **rapidity**, we find

$$c \cdot dt' = \cosh\phi \cdot c \cdot d\tau$$

and because

$$dx' = v \cdot dt' = c \cdot \tanh\phi \cdot dt' = c \cdot \tanh\phi \cdot \cosh\phi \cdot d\tau = \sinh\phi \cdot c \cdot d\tau$$

we can write in short-hand vector notation

$$\begin{bmatrix} dx' \\ c \cdot dt' \end{bmatrix} = \begin{bmatrix} \sinh\phi \\ \cosh\phi \end{bmatrix} \cdot c \cdot d\tau \quad (2.9.21)$$

Note that because $dx = 0$ for the rocket in the rest frame we do not have a transformation matrix as in the general case. In order to derive the space-time transformations for any rocket–observer relation, we have to determine the rapidity (proper speed) and then solve the two differential equations (2.9.21). This will be done now for the two most simple cases.

Cruising Rocket

For a cruising (non-accelerated) rocket $\cosh\phi = \cosh(\sigma/c) = \gamma = const$ and Eqs. (2.9.21) can easily be integrated to give the well-known space-time transformation between inertial observers

$$t' = \frac{1}{\gamma}\tau \tag{2.9.22a}$$

$$x' = \frac{v}{\gamma}\tau \tag{2.9.22b}$$

$$\frac{v}{c} = \tanh\frac{\sigma}{c} \tag{2.9.22c}$$

As an example: A rocket that travels 90% the speed of light as seen from an external observer or $\sigma = 1.47 \cdot c$ of proper speed would cross our milky way with diameter $d = 100\,000$ ly within $t' = d/v = 111\,000$ yr or $\tau = 48\,000$ yr in proper time.

Constant-Acceleration Rocket

If the acceleration α is constant, $\sigma = c\phi = \alpha\tau$. By integrating Eqs. (2.9.21) we find

$$\frac{\alpha t'}{c} = \sinh\frac{\alpha\tau}{c} = \sinh\frac{\sigma}{c} \tag{2.9.23a}$$

Also, and with Eq. (2.9.23a) and $\cosh x = \sqrt{1 + \sinh^2 x}$, we have

$$\frac{\alpha x'}{c^2} = \cosh\frac{\alpha\tau}{c} - 1 = \sqrt{1 + \left(\frac{\alpha t'}{c}\right)^2} - 1 \tag{2.9.23b}$$

Applying Eq. (2.9.23a) with $\tanh x = \sinh x / \sqrt{1 + \sinh^2 x}$ to Eq. (2.9.5) yields

$$\frac{v}{c} = \tanh\frac{\sigma}{c} = \frac{\alpha t'}{\sqrt{c^2 + (\alpha t')^2}} \tag{2.9.23c}$$

From Eq. (2.9.23b) we find

$$\left(x' + \frac{c^2}{\alpha}\right)^2 - (ct')^2 = \left(\frac{c^2}{\alpha}\right)^2$$

This denotes that the space-time trajectory of a rocket with constant acceleration is a hyperbola.

Let us reconsider the case of an ultimate manned H–$\overline{H}$ annihilation rocket ($\beta_{ex} = 0.2082$) which now crosses the Milky Way ($x'_f = 100\,000$ ly) with a comfortable acceleration of $\alpha = 1g$. According to Eq. (2.9.23b) this would take only $\tau_f = 11.9$ yr in proper time of an astronaut. His final proper speed would be $\sigma_f = \alpha \tau_f = 12.2\ c$ and the rocket's mass ratio can be calculated from the rocket equation (2.9.10) to be $m_i/m_f = 3.33 \cdot 10^{25}$. If the final spacecraft mass is, say, 100 metric tons (Space Shuttle), then the launch mass would need to be $m_i = 1.67\ m_\odot$, where $m_\odot$ is the mass of the Sun! That is not saying anything about the engines that would have to propel such a mass at 1 g.

Problems

Problem 2.1 Launch Mass Changes

Prove Eq. (2.6.1) by deriving the relative change of the launch mass of a rocket $\Delta m_0/m_0$ for a relative change of the propulsion demand "delta-v" $\Delta(\Delta v)/(\Delta v)$. Find the approximation for small $\Delta(\Delta v)$.

3
Rocket Staging

In Section 2.7 and 2.8, we found out that there are limits to the obtainable terminal velocity or to the payload mass because of the finite structural mass. This limit is crucial for the construction of a rocket. If a rocket is to reach the low Earth orbit, a propulsion demand of about 9.2 km/s has to be taken into account, which is at the limit of feasibility for today's chemical propulsions. If a higher payload ratio beyond 3% is required, or the S/C needs to leave the gravitational field of the Earth, one has to take measures to increase the rocket efficiency.

The best method to do this is rocket staging. "Staging" means to construct a rocket such that some tanks and/or engines are integrated into one stage that can be jettisoned after use, thereby reducing the mass to be further accelerated. Figure 3.1 depicts four types of rocket staging, where parallel staging and serial staging are the most common.

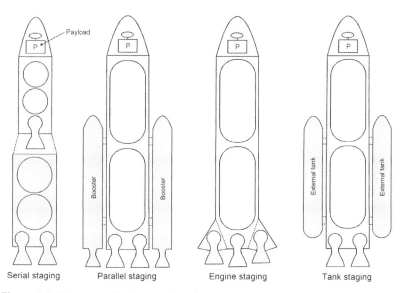

Figure 3.1 Four types of rocket staging. Stages which are jettisoned during ascent are marked in grey.

Astronautics. Ulrich Walter
Copyright © 2008 WILEY-VCH Verlag GmbH & Co. KGaA, Weinheim
ISBN: 978-3-527-40685-2

3.1
Serial Staging – General Considerations

3.1.1
Definitions

Serial staging (a.k.a. multistaging, multi-stepping, tandem staging) means that several rockets (n stages) sit on top of each other. One stage after the other is fired during operation, and the burnt-out stages are jettisoned. The advantages are:

1. The engines can be adapted to the changing environment upon ascent. The lowest stage can be chosen for a high thrust to quickly escape the Earth's gravitational potential, whereas the upper stage(s) in (almost) free space can be dimensioned for best efficiency (I_{sp}).

2. Jettisoning the lower engines, which are no longer necessary, and the tanks, decreases structural mass and hence increases payload mass.

The concept of serial and parallel staging dates back to the first military applications in 1529. Figure 3.2 depicts the sketch of the first staged rocket as introduced by the Austrian military technician Conrad Haas in his *Kunstbuch* (German, meaning *Art Book*). The concept of staging then passed via a publication of the German rocket pioneer Johannes Schmidlap to the Polish military engineer Kazimierz Siemienowicz, whose book *Artis Magnae Artilleriae pars prima* (*Great Art of Artillery, the First Part*) published in 1650 was translated into many languages in Europe, and became for two centuries the basic artillery manual. So rocket staging became a well-known technology in artillery rocketry and passed from there also to the rocket-hype decades 1920–1940. Already in those days the question was raised, whether ignition of the second stage should wait until its speed had reduced to zero, i.e. make use of the full impetus, or to ignite it immediately. The answer is given by Eq. (2.4.2). The gravitational loss demands for a preferably short ascent and hence a preferably short ignition sequence.

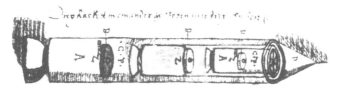

Figure 3.2 A three-stage rocket of Conrad Haas from 1529. The picture inscription (German) reads: "Three nested rockets with three shots". Note that each stage has already a bell-shaped nozzle. By courtesy of Barth (2005).

For a mathematical analysis of serial staging, we introduce the concept of a **partial rocket** *i*. Let's assume we have four separate propulsion units (see Fig. 3.3), which are set up in four stages on top of each other. The first partial rocket refers to the sum of all the four propulsion units plus the true payload. The second partial rocket refers to the sum of the three stages 2–4 arranged on top of that, plus the true payload. The third partial rocket refers to the sum of the third and fourth stage plus the true payload. Finally, the fourth partial rocket to the upper unit plus the true payload.

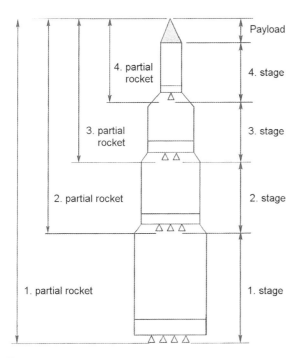

Figure 3.3 Definitions for a staged rocket.

We define the payload of a partial rocket as the mass of the next partial rocket, i.e.

$$m_{L,i} := m_{0,i+1} \tag{3.1.1}$$

The **true payload** is defined as the payload of the last partial rocket. In line with the single-stage rocket, $m_{0,i}$ and $m_{f,i}$ will be the initial and the final mass of the *i*th partial stage; therefore

$$m_{0,i} = m_{p,i} + m_{s,i} + m_{0,i+1}$$

$$m_{f,i} = m_{0,i} - m_{p,i} = m_{s,i} + m_{0,i+1} \tag{3.1.2}$$

We also define the following ratios (cf. Eqs. (2.8.1) to (2.8.3)) for each partial rocket i:

$$\mu_i := \frac{m_{f,i}}{m_{0,i}} = \frac{m_{0,i+1} + m_{s,i}}{m_{0,i}} = \frac{m_{L,i} + m_{s,i}}{m_{0,i}} \qquad \text{mass ratio} \qquad (3.1.3)$$

$$\varepsilon_i := \frac{m_{s,i}}{m_{0,i} - m_{0,i+1}} = \frac{m_{s,i}}{m_{s,i} + m_{p,i}} \qquad \text{structural ratio} \qquad (3.1.4)$$

$$\lambda_i := \frac{m_{L,i}}{m_{0,i} - m_{0,i+1}} = \frac{m_{L,i}}{m_{s,i} + m_{p,i}} = \frac{m_{0,i+1}}{m_{s,i} + m_{p,i}} \qquad \text{payload ratio} \qquad (3.1.5)$$

Observe that the structural mass and the payload mass of a partial stage are taken relative to the initial mass of its *stage* and not its partial rocket. This is reasonable because it will facilitate the optimization of stages, which will be the main objective of this chapter. Because

$$1 + \lambda_i = \frac{m_{p,i} + m_{s,i} + m_{0,i+1}}{m_{p,i} + m_{s,i}} = \frac{m_{0,i}}{m_{p,i} + m_{s,i}}$$

it follows that

$$\frac{\lambda_i}{1 + \lambda_i} = \frac{m_{0,i+1}}{m_{p,i} + m_{s,i}} \frac{m_{p,i} + m_{s,i}}{m_{0,1}} = \frac{m_{0,i+1}}{m_{0,1}}$$

and

$$\frac{\varepsilon_i}{1 + \lambda_i} = \frac{m_{s,i}}{m_{p,i} + m_{s,i}} \frac{m_{p,i} + m_{s,i}}{m_{0,1}} = \frac{m_{s,i}}{m_{0,1}}$$

We therefore obtain for the so-called total payload ratio (ratio of the true payload to the total launch mass)

$$\lambda_* := \frac{m_{L,n}}{m_{0,1}} = \frac{m_{L,i}}{m_{0,i}} \frac{m_{0,i}}{m_{0,i-1}} \cdots \frac{m_{0,3}}{m_{0,2}} \frac{m_{0,2}}{m_{0,1}} \qquad \text{total payload ratio}$$

from which follows

$$\lambda_* = \prod_{i=1}^{n} \frac{\lambda_i}{1 + \lambda_i} \qquad (3.1.6)$$

and

$$\mu_i = \frac{m_{L,i} + m_{s,i}}{m_{0,i}} = \frac{\varepsilon_i + \lambda_i}{1 + \lambda_i} \qquad (3.1.7)$$

Remark: *In order to be consistent to Eq. (3.1.5) λ_* should be defined as $\lambda_* := m_{L,n}/(m_{0,1} - m_{L,n})$. In this case in all the following equations $\lambda_* \to \lambda_*/(1+\lambda_*)$ should be replaced. In Eq. (3.2.2) the replacement should read $\lambda_* \to \lambda_*(1+\lambda_*)^2$. Because the inconsistent definition is used throughout literature and because it is insignificantly different to the consistent form we will adopt it here also.*

3.1.2
Rocket Equation

We are now looking for a serial-staged rocket equation equivalent to Eq. (2.2.2) for the single-staged rocket. According to Eq. (2.2.2) for each partial rocket i,

$$\Delta v_i = v_{*,i} \ln \frac{m_{0,i}}{m_{f,i}} = -v_{*,i} \ln \mu_i$$

holds. Serial staging with instant firing of the following stage means that the terminal velocity of one partial rocket is the initial velocity of the following partial rocket, i.e. $v_{f,i} = v_{0,i+1}$. So the following is valid:

$$\Delta v = v_{f,n} - v_{0,1} = \left(v_{f,n} - v_{0,n}\right) + \left(v_{f,n-1} - v_{0,n-1}\right) + \ldots + \left(v_{f,1} - v_{0,1}\right)$$

$$= \sum_{i=1}^{n} \Delta v_i$$

Therefore we get for the total velocity increase (propulsion demand)

$$\Delta v = -\sum_{i=1}^{n} v_{*,i} \ln \mu_i = \sum_{i=1}^{n} v_{*,i} \ln \frac{1+\lambda_i}{\varepsilon_i + \lambda_i} \qquad \text{serial-stage rocket equation} \qquad (3.1.8)$$

3.2
Stage Optimization

3.2.1
Road to Stage Optimization

Our goal now is to optimize the serial stages such that we get a maximum payload into orbit or to achieve a maximum velocity gain. In order to know how to perform the optimization, let's state the problem explicitly. For the optimal construction of a serial-staged rocket, the following quantities must be considered:

- Technically given quantities
    - exhaust velocities $v_{*,i}$
    - structural ratios ε_i
    - total launch mass $m_{0,1}$ of the rocket

- Technically variable parameters
    - number of stages n
    - payload ratios λ_i
- Target quantities
    - total payload ratio λ_*
    - propulsion demand Δv

The latter are determined by the technical parameters through Eqs. (3.1.6) and (3.1.8).

The objective of a stage optimization now is first to specify one target quantity and then to maximize the other by variation of n and λ_i. However, n is not a true variable: first, because n is an integer; and, second, because, with every additional stage, the rocket becomes more efficient (see Fig. 3.4) and therefore an optimum rocket would have infinitive many stages. So, optimizing n cannot be the true objective. Rather, the following holds. Because every stage adds to the propulsion demand by a summand (see Eq. (3.1.8)), in the following optimization procedure n is the smallest stage number for which at a given propulsion demand the optimized payload ratios can be just determined. We will see at the end of Section 3.2.2 precisely what this means, and in Section 3.4 we will learn how n is determined in general.

Therefore the payload ratios remain as the only variables, which have to be optimized by the following two different optimization approaches

1. max λ_*, i.e. maximize the total payload ratio λ_* at a given Δv
2. max Δv, i.e. maximize the obtainable Δv at a given λ_*

The first approach is taken for instance by Ruppe (1966) and the second for instance by Griffin and French (2004). So, with max λ_*, Eq. (3.1.6) is the target function to maximize, and Eq. (3.1.8) is the secondary condition; and for max Δv it is the other way round. As both proceedures are described in the literature and also used in practice, we also want to explore both, and will see that in principle they are equivalent.

3.2.2
General Optimization

Now λ_* and Δv are to be maximized with respect to λ_i under secondary conditions. This can be achieved with the so-called Lagrange multiplier method, whereby a secondary condition can be taken into account by adding it to the partial derivatives (to be set to zero) via a so-called Lagrangian Multiplier γ.

> **Remark:** *For a comprehensible description of the Lagrange multiplier method and why it works, see for instance Reif (1965, Appendix A10), or Gluss, David and Weisstein, Eric W. "Lagrange Multiplier." From MathWorld–A Wolfram Web Resource. http://mathworld.wolfram.com/LagrangeMultiplier.html.*

By this method we find for both max cases and for each partial rocket $j = 1, \ldots, n$ the condition equations

$$\frac{\partial \lambda_*}{\partial \lambda_j} + \gamma \frac{\partial (\Delta v)}{\partial \lambda_j} = 0 \quad @ \max \lambda_* \tag{3.2.1a}$$

$$\frac{\partial (\Delta v)}{\partial \lambda_j} + \gamma' \frac{\partial \lambda_*}{\partial \lambda_j} = 0 \quad @ \max \Delta v \tag{3.2.1b}$$

respectively. Obviously these two equations actually set the same mathematical problem if one identifies $\gamma = 1/\gamma'$. Applying the partial derivations

$$\frac{\partial (\Delta v)}{\partial \lambda_j} = v_{*,j} \frac{\varepsilon_j + \lambda_j}{1 + \lambda_j} \frac{\partial}{\partial \lambda_j} \left(\frac{1 + \lambda_j}{\varepsilon_j + \lambda_j} \right) = -v_{*,j} \frac{1 - \varepsilon_j}{1 + \lambda_j} \frac{1}{\varepsilon_j + \lambda_j}$$

$$\frac{\partial \lambda_*}{\partial \lambda_j} = \frac{\lambda_*}{\lambda_j / (1 + \lambda_j)} \frac{\partial}{\partial \lambda_j} \left(\frac{\lambda_j}{1 + \lambda_j} \right) = \frac{\lambda_*}{\lambda_j (1 + \lambda_j)}$$

to Eq. (3.2.1a) yields

$$\frac{\lambda_*}{\lambda_j (1 + \lambda_j)} = \gamma v_{*,j} \frac{1 - \varepsilon_j}{1 + \lambda_j} \frac{1}{\varepsilon_j + \lambda_j}$$

From this it follows that

$$v_{*,j} \lambda_j \frac{1 - \varepsilon_j}{\varepsilon_j + \lambda_j} = \frac{\lambda_*}{\gamma} =: \alpha = const \quad @ j = 1, \ldots, n \tag{3.2.2}$$

This is a key equation: first, because it claims that if $v_{*,i} = v_* = const$ and $\varepsilon_i = const$ then all payload ratios are equal; and, second, because, once the constant α is known, all optimized payload ratios follow from it immediately by solving for λ_j:

$$\lambda_{j,opt} = \frac{\alpha \varepsilon_j}{(1 - \varepsilon_j) v_{*,j} - \alpha} \quad @ j = 1, \ldots, n \tag{3.2.3}$$

Inserting this equation into Eqs. (3.1.8) and (3.1.6) one arrives at

$$\Delta v = \sum_{i=1}^{n} v_{*,i} \ln \frac{v_{*,i} - \alpha}{\varepsilon_i v_{*,i}} \tag{3.2.4}$$

and

$$\lambda_* = \prod_{i=1}^{n} \frac{\alpha \varepsilon_i}{(1 - \varepsilon_i)(v_{*,i} - \alpha)} \tag{3.2.5}$$

The two optimization methods now differ merely in that, in the

- max λ_* case, at a given Δv, the constant α is numerically determined from secondary condition (3.2.4), which inserted into Eq. (3.2.5) yields the maximized λ_*,

- max Δv case, at a given λ_*, the constant α is numerically determined from the secondary condition (3.2.5), which inserted into Eq. (3.2.4) yields the maximized Δv.

Quite generally α can be determined only numerically, which, however, is no problem if, for instance, Newton's method is applied. At this point the number of stages, which occurs in both equations, comes into play. In Eq. (3.2.5) it is not crucial, since for every given $\lambda_* < 0.5$ and stage number n one can find a λ_i and hence an α. This can best be seen from Eq. (3.1.6) from which it is derived. However, in Eq. (3.2.4) n is crucial, because in the max λ_* case n has to be raised such that, for a given propulsion demand, there exists an α, and in the max Δv case n has to be raised such that, for an α derived from Eq. (3.2.5), a required propulsion demand is achieved. Therefore we have that:

> The number of stages n is the smallest integer number for which there exists an α and hence a self-consistent solution $\Delta v \left(\lambda_* \right)$ or $\lambda_* \left(\Delta v \right)$ from Eqs. (3.2.4) and (3.2.5).

If we insert α so derived into Eq. (3.2.2) we also obtain the optimum payload ratios. With this we have finally achieved the stage optimization goal.

One could argue that the procedure laid out here will not work in the max λ_* case because according to Eq. (3.2.2) the constant α depends on λ_*, which in itself is a variable to be determined. Rather one would have to set up a system of $n+1$ equations from the corresponding secondary condition (3.2.4) or (3.2.5) plus the n equations (3.2.3) to find from this the $n+1$ quantities (λ_i, α) self-consistently. The point is that λ_* is not a variable to be optimized (which are the λ_i) but a target quantity, which is mathematically a (yet to be determined) constant. With the introduction of α in Eq. (3.2.2) we merely redefine the constant Lagrange multiplier γ with the help of the quasi-constant λ_*.

3.3
Analytical Solutions

We are now seeking analytical solutions to Eqs. (3.2.4) and (3.2.5). We start out with the most simple case of uniform staging and move towards more general cases. Whatsoever, the solution equation will be identical for both maximization cases, because mathematically we just merge the two equations by eliminating the constant α. We always obtain *one* equation which relates Δv and λ_* via ε_i and $v_{*,i}$. This is why we will disregard in the following the index *max* of the variable to be maximized.

3.3.1
Uniform Staging

The most simple case is the uniform staging where the structural factors and effective exhaust velocities are all the same:

$$v_{*,i} = v_* = const$$

$$\varepsilon_i = \varepsilon = const$$

Via Eq. (3.2.3) this implies that all payload ratios must be identical, i.e. $\lambda_i = \lambda = const$. The single optimum payload ratio for all stages can be derived from Eq. (3.1.6) as

$$\frac{1}{\lambda_{opt}} = \frac{1}{\lambda_*^{1/n}} - 1 \qquad (3.3.1)$$

The maximized velocity gain of an n-staged rocket is finally obtained by inserting Eq. (3.3.1) into the serial-stage rocket equation (3.1.8):

$$\Delta v = -v_* \sum_{i=1}^{n} \ln\left[\lambda_*^{1/n}(1-\varepsilon) + \varepsilon\right] = -nv_* \ln\left[\lambda_*^{1/n}(1-\varepsilon) + \varepsilon\right] \qquad (3.3.2)$$

From this follows

$$\lambda_*^{1/n} = \frac{e^{-\Delta v/nv_*} - \varepsilon}{1 - \varepsilon} \qquad (3.3.3)$$

In the max Δv case the maximized Δv is calculated with a given λ_* from Eq. (3.3.2), and with Eq. (3.3.3) it is just the other way around. In any case, the smaller the payload ratio λ_*, the bigger is Δv. The absolute maximum Δv is achieved when λ_* vanishes, in which case we find

$$\Delta v_{max} = -nv_* \ln \varepsilon \quad @ \; \lambda_* = 0 \qquad (3.3.4)$$

This sets an upper limit to what can be achieved. If we insert Eq. (3.3.3) into Eq. (3.3.1) we derive for the optimized uniform payload ratio

$$\lambda_{opt} = \frac{\lambda_*^{1/n}}{1 - \lambda_*^{1/n}} = \frac{e^{-\Delta v/nv_*} - \varepsilon}{1 - e^{-\Delta v/nv_*}} \qquad (3.3.5)$$

3 Rocket Staging

Having found an optimum solution to the serial-staged rocket we might ask: How does it compare with the single-stage rocket? In terms of achievable payload ratio, Eq. (3.3.3) compares with Eq. (2.8.4) for the single stage. For a single stage rocket (SSTO) with 7% structural ratio and $v_* = 4\text{ km s}^{-1}$, we are able to lift only 3.4% total payload ratio into low Earth orbit (LEO). With an optimized two-stage rocket we lift 7.0%, and hence twice as much under the same conditions; and with a three-stage rocket the total payload ratio we obtain is only a little more, viz. 7.6%. Obviously a two-stage rocket is a perfect vehicle to LEO. This is why all rockets into LEO have two stages.

This simple case of uniform staging enables us to calculate the ultimate rocket with an infinite number of stages. The transition to infinite stages can be performed in Eq. (3.3.2). The result is (exercise, Problem 3.1):

$$\Delta v = -v_* \left(1 - \varepsilon\right) \ln \lambda_* \quad @\ n \to \infty$$

From this follows

$$\lambda_* = \exp\left[-\frac{\Delta v}{v_* \left(1 - \varepsilon\right)}\right] \quad @\ n \to \infty \tag{3.3.6}$$

So, even in this ideal case the total payload ratio of an optimized rocket decreases exponentially with Δv budget. Figure 3.4 displays the functional de-

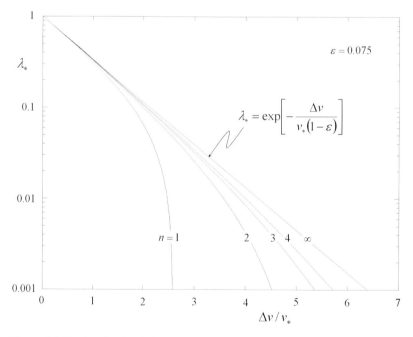

Figure 3.4 Total payload ratio λ_* as a function of the normalized terminal velocity $\Delta v/v_*$.

pendences of Eqs. (3.3.3) and (3.3.6). So, with a given structural ratio Δv and stage number n the achievable total payload ratio λ_* can be easily derived and vice versa. The result shows that already with a few stages it is possible to considerably increase the fraction of the payload mass. However, it also demonstrates that with more than 3 or 4 stages, the profit hardly outweighs the additional complexity of the rocket. Obviously the straight line Eq. (3.3.6) is an asymptote for the achievable λ_* and Δv, respectively. And, whatever the number of stages, the total payload ratio always decreases exponentially with an increasing terminal velocity.

3.3.2
Uniform Exhaust Velocities

We now relax the conditions to

$$v_{*,i} = v_* = const$$

ε_i arbitrary

From Eq. (3.2.4) it follows that

$$e^{\Delta v/v_*} = \prod_{i=1}^{n} \frac{v_* - \alpha}{\varepsilon_i v_*} = \prod_{i=1}^{n} \frac{1}{\varepsilon_i} \prod_{i=1}^{n} \left(1 - \frac{\alpha}{v_*}\right) = \frac{1}{\bar{\varepsilon}^n} \left(1 - \frac{\alpha}{v_*}\right)^n$$

with the geometric mean

$$\bar{x} = \left(\prod_{i=1}^{n} x_i\right)^{1/n}$$

Solving for α/v_* yields

$$\frac{\alpha}{v_*} = 1 - \bar{\varepsilon} \cdot e^{\Delta v/nv_*} \tag{3.3.7}$$

Inserting this into Eq. (3.2.5)

$$\lambda_* = \prod_{i=1}^{n} \frac{\alpha \varepsilon_i}{(1 - \varepsilon_i)(v_* - \alpha)} = \frac{\bar{\varepsilon}^n}{(1 - \bar{\varepsilon})^n} \cdot \frac{1}{(v_*/\alpha - 1)^n}$$

delivers

$$\lambda_*^{1/n} = \frac{e^{-\Delta v/nv_*} - \bar{\varepsilon}}{1 - \bar{\varepsilon}} \tag{3.3.8}$$

From this follows

$$\Delta v = -nv_* \ln\left[\lambda_*^{1/n}(1 - \bar{\varepsilon}) + \bar{\varepsilon}\right] \tag{3.3.9}$$

depending on which is the target quantity to be maximized. Inserting Eq. (3.3.7) into Eq. (3.2.3) one obtains for the optimized payload ratios

$$\lambda_{i,opt} = \frac{e^{-\Delta v/nv_*} - \bar{\varepsilon}}{\frac{\bar{\varepsilon}}{\varepsilon_i} - e^{-\Delta v/nv_*}} = \frac{\lambda_*^{1/n}}{\frac{\bar{\varepsilon}}{\varepsilon_i}\frac{1-\varepsilon_i}{1-\varepsilon} - \lambda_*^{1/n}} \tag{3.3.10}$$

These are the analytical results for a rocket with uniform exhaust velocities $v_{*,i} = v_*$.

3.3.3
Uneven Staging

We finally examine the most general case where both the exhaust velocities and the structural ratios are uneven, i.e.

$v_{*,i}$ arbitrary

ε_i arbitrary

Due to the complexity of Eqs. (3.2.4) and (3.2.5) no exact analytical solutions can be given. But in the following we will provide approximate solutions, which for a first rocket layout are sufficiently exact. Without loss of generality we first define

$$v_{*,i} = v_* + \Delta v_{*,i} \quad \text{and} \quad \varepsilon_i = \varepsilon + \Delta \varepsilon_i$$

where

$$v_* = \frac{1}{n}\sum_{i=1}^{n} v_{*,i} \quad \text{and} \quad \varepsilon = \frac{1}{n}\sum_{i=1}^{n} \varepsilon_i$$

are the arithmetic means of the exhaust velocities and structural ratios of all n stages. Assuming $\Delta v_{*,i} \ll v_*$ and $\Delta \varepsilon_i \ll \varepsilon$ and according to Appendix B it is possible to derive the following approximate solutions:

$$\lambda_*^{1/n} = \frac{e^{-\Delta v/nv_* - C} - \bar{\varepsilon}}{1-\varepsilon} \tag{3.3.11}$$

with

$$C = \frac{1}{n}\sum_i \frac{\Delta v_{*,i}}{v_*}\left\{\frac{\Delta \varepsilon_i}{\varepsilon_i} + \frac{1}{2}\left(1 - \frac{1}{\bar{\varepsilon}}e^{-\frac{\Delta v}{nv_*}}\right)^2 \frac{\Delta v_{*,i}}{v_*}\right.$$

$$\left. + O\left[\left(\frac{\Delta \varepsilon_i}{\varepsilon}\right)^2\right] + O\left[\left(\frac{\Delta v_{*,i}}{v_*}\right)^2\right]\right\}$$

From this follows

$$\Delta v = -n v_* \left\{ \ln \left[\lambda_*^{1/n} (1-\bar{\varepsilon}) + \bar{\varepsilon} \right] + C' \right\} \tag{3.3.12}$$

with

$$C' = \frac{1}{n} \sum_i \frac{\Delta v_{*,i}}{v_*} \left\{ \frac{\Delta \varepsilon_i}{\varepsilon_i} + \frac{\lambda_*^{2/n}}{2} \left(\frac{1-\bar{\varepsilon}}{\bar{\varepsilon}} \right)^2 \frac{\Delta v_{*,i}}{v_*} \right.$$

$$\left. + O \left[\left(\frac{\Delta \varepsilon_i}{\varepsilon} \right)^2 \right] + O \left[\left(\frac{\Delta v_{*,i}}{v_*} \right)^2 \right] \right\}$$

and

$$\lambda_{i,opt} = \frac{e^{-\Delta v/n v_* - C'} - \bar{\varepsilon}}{\frac{\bar{\varepsilon}}{\varepsilon_i} - e^{-\Delta v/n v_* - C'}} = \frac{\lambda_*^{1/n} + K}{\frac{\bar{\varepsilon}}{\varepsilon_i} \frac{1-\varepsilon_i}{1-\bar{\varepsilon}} - \lambda_*^{1/n} - K} \tag{3.3.13}$$

with

$$K = \left[\lambda_*^{1/n} + \frac{\bar{\varepsilon}}{1-\bar{\varepsilon}} \right] C'$$

Note that even for quite uneven exhaust velocities the solutions are applicable.

Example
Saturn V, for instance, had three stages with $v_{*,1} = 304$ s and $v_{*,2} = v_{*,3} = 421$ s in vacuum, therefore $v_* = 382$ s. A calculation of C via the above equations introduces a relative error of only

$$\frac{\Delta C}{C} \approx \sum_{i=1}^3 \left(\frac{\Delta v_{*,i}}{v_*} \right)^3 = 0.64 \%$$

Nevertheless, in specific cases Eqs. (3.2.4) and (3.2.5) should rather be solved numerically, for instance with Newton's method. Newton's iterative equation for solving Eq. (3.2.4) reads

$$\alpha' = \alpha - \frac{\Delta v - \sum_{i=1}^n v_{*,i} \ln \frac{v_{*,i} - \alpha}{\varepsilon_i v_{*,i}}}{\sum_{i=1}^n \frac{v_{*,i}}{v_{*,i} - \alpha}} \tag{3.3.14}$$

and that for Eq. (3.2.5) reads

$$\alpha' = \alpha \cdot \left\{ 1 + \frac{\ln\left[\lambda_* \cdot \prod_{i=1}^{n} \frac{1-\varepsilon_i}{\varepsilon_i}\left(\frac{v_{*,i}}{\alpha} - 1\right)\right]}{\sum_{i=1}^{n} \frac{v_{*,i}}{v_{*,i} - \alpha}} \right\} \tag{3.3.15}$$

3.4
Stage Number Determination

So far, the number of stages n unfolded from Eqs. (3.2.4) and (3.2.5) as the smallest number that permits their self-consistent solution. However, it would be more pleasing if n could be determined directly. Such a determination is only possible in the case of uniform staging, to which we turn now.

Remark: *We follow here the approach by Ruppe (1966, vol. 1). Note that Ruppe uses definitions of structural ratios and payload ratios different from those used here.*

A uniform staging might be considered unrealistic. But it is not if we are interested only in the stage number, because n is an integer and deviations of the true quantities $v_{*,i}$, ε_i from their assumed uniform means do not have a crucial impact on the integer n.

As in Section 3.3.1 for a uniform staging

$$v_{*,i} = v_*, \quad \varepsilon_i = \varepsilon, \quad \text{for all} \quad i = 1, \ldots, n$$

holds, from which according to Eq. (3.2.3) uniform payload ratios follow

$$\lambda_i = \lambda, \quad \text{for all} \quad i = 1, \ldots, n$$

With this assumption, we can simplify the total payload ratio Eq. (3.1.6) and the rocket equation (3.1.8) to

$$\lambda_* = \prod_{i=1}^{n} \frac{\lambda}{1+\lambda} = \left(\frac{\lambda}{1+\lambda}\right)^n$$

from which follows that

$$\ln \lambda_* = n \ln \frac{\lambda}{1+\lambda} \tag{3.4.1}$$

and

$$\Delta v = -nv_* \ln \mu = -nv_* \ln \frac{\varepsilon + \lambda}{1+\lambda} \tag{3.4.2}$$

We can separate n in Eq. (3.4.1) and insert it into Eq. (3.4.2), yielding

$$\frac{\ln \lambda_*}{\Delta v / v_*} = -\ln \frac{\lambda}{1+\lambda} \bigg/ \ln \frac{\varepsilon + \lambda}{1+\lambda} =: g(\lambda)$$

We now have one equation independent on n, which we have to optimize with regard to λ.

From here we proceed as follows: The function $g(\lambda)$ is positively curved with a maximum. We therefore can determine the maximum of λ_* by determining the root of $d\lambda_*/d\lambda$. Since $\ln v_*$ is a strictly monotonic function in v_*, $d \ln \lambda_*/d\lambda \propto dg/d\lambda$ will deliver the same result. By setting $dg/d\lambda = 0$ we will therefore find the optimum λ. Thereafter we have to options: either (i) Δv is given, then from Eq. (3.4.2) the wanted stage number n can be derived, and by inserting n into Eq. (3.4.1) also the maximum λ_* is found; or (ii) λ_* is a given, in which case we derive the required stage number from Eq. (3.4.1) and the achievable Δv from Eq. (3.4.2).

In order to facilitate the derivation of the maximum of $g(\lambda)$, we apply the following transformation to the new variables:

$$x := \frac{\lambda}{1+\lambda} \quad \text{and} \quad y := \frac{\varepsilon}{1+\lambda} \tag{3.4.3}$$

From this follows

$$g(x) = -\frac{\ln x}{\ln (x+y)} \tag{3.4.4}$$

We are now set to do the derivation:

$$\frac{dg(x)}{d\lambda} = \frac{dg(x)}{dx}\frac{dx}{d\lambda} = -\frac{(1/x)\ln(x+y) - [1/(x+y)]\ln x}{[\ln(x+y)]^2} \frac{1}{(1+\lambda)^2}$$

We derive the optimum x by setting the expression to zero:

$$x_{opt} \ln x_{opt} = (x_{opt} + y) \cdot \ln (x_{opt} + y)$$

or

$$x_{opt}{}^{x_{opt}} = (x_{opt} + y)^{x_{opt}+y} \tag{3.4.5}$$

Our task with Eq. (3.4.5) is to find the optimum payload ratio x_{opt} at a given y. To do this, we will first analyze the graphical representation of the left and right sides of the equation. Let's have a look at the function $f(x) = x^x$ in Fig. 3.5. It has a minimum at $x = 1/e$ ($e = 2.71828\ldots$). This can be easily verified by differentiating the function $\exp(x \ln x) = x^x$. For small ε and hence y we expect x_{opt} somewhere around the minimum of the function, and

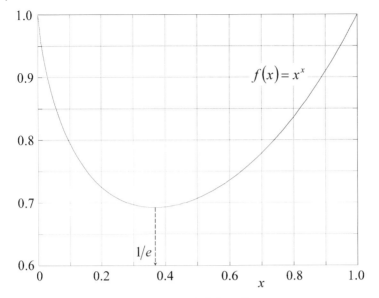

Figure 3.5 Graphical representation of $f(x) = x^x$.

therefore we approximate x^x as a parabola around its minimum. Whatever the parabola looks like, Eq. (3.4.5) tells us that we look for the intersection of two parabolas with the same shape, having minima at $1/e$ and $1/e - y$. Because of their symmetrical property, the two parabolas intersect exactly in the middle, which is the solution we are looking for:

$$x_{opt} = \frac{1}{e} - \frac{y}{2} \tag{3.4.6}$$

By means of Eq. (3.4.3) we transform back to the former variables λ and ε and obtain for the optimized payload ratio

$$\lambda_{opt} = \frac{1}{e-1}\left(1 - \frac{e\varepsilon}{2}\right) = 0.582 - 0.791\varepsilon \tag{3.4.7}$$

Predetermined Δv

If Δv is given, we insert Eq. (3.4.7) into Eq. (3.4.2), which for $\varepsilon \ll 1$ delivers the stage number required for this Δv:

$$n = -\frac{\Delta v}{v_* \ln\frac{1}{e}\left[1 + \frac{(e-1)\varepsilon}{2}\right]} \approx \frac{\Delta v}{v_*}\frac{1}{1 - \frac{(e-1)\varepsilon}{2}} \approx \frac{\Delta v}{v_*}\left(1 + \frac{e-1}{2}\varepsilon\right) \tag{3.4.8}$$

For common structural ratios $\varepsilon \approx 0.1$ we get

$$n \approx 1.09 \frac{\Delta v}{v_*} \approx \frac{\Delta v}{v_*} \quad @ \; \varepsilon \approx 0.1 \qquad (3.4.9)$$

For technical and financial reasons, usually rather fewer stages are taken into consideration. With Eq. (3.4.8) the maximized total payload ratio results from Eq. (3.4.1) as

$$\lambda_* = \exp\left\{-\frac{\Delta v}{v_*}[1 + (e-1)\varepsilon]\right\} \qquad (3.4.10)$$

Because from an economical perspective a rocket with $n > 3$ is pointless, we find from Eq. (3.4.9) that only $\Delta v \leq 3v_*$ is attainable. This limitation is in line with similar results we had obtained earlier from the examination of the payload ratio in Section 2.8.

> **Example**
> As an example to determine the number of stages via Eq. (3.4.9), we consider two important mission types:
>
> 1. **Ascent into low earth orbit**. As a rule of thumb the minimum propulsion demand is $\Delta v \approx 9.2 \text{ km s}^{-1}$ including ascent losses. Assuming $v_* \approx 3.9 \text{ km s}^{-1}$ as an average value for chemical thrusters, the optimum stage number is $n = 2.57 \to 2+$. In fact, today all launchers into LEO have two stages, where the first stage usually is parallel-staged.
> 2. **Interplanetary flight**. For interplanetary flights one has to leave the Earth's gravitational field. The propulsion demand for a mission to the Moon, for instance, is $\Delta v \approx 13.1 \text{ km s}^{-1}$, including ascent losses. Equation (3.4.3) tells us that an optimum rocket should have $n = 3.66 \to 3+$. Indeed, all launchers into geosynchronous Earth orbit (GEO), or to the Moon or beyond, have three or more stages.

Predetermined λ_*

In case λ_* is given, we insert Eq. (3.4.7) into Eq. (3.4.1), which by the same token as the case above yields

$$n \approx \ln\left(\frac{1}{\lambda_*}\right) \cdot \left(1 - \frac{e-1}{2}\varepsilon\right) \approx -\ln \lambda_* \qquad (3.4.11)$$

and from Eq. (3.4.1) we obtain for the achievable Δv budget

$$\frac{\Delta v}{v_*} = [1 - (e-1)\varepsilon] \cdot \ln \frac{1}{\lambda_*} \qquad (3.4.12)$$

With Eqs. (3.4.9) and (3.4.11) we have found very simple and handy expressions to determine the minimum required stage number. Usually a target trajectory is given for a mission, which implies a certain propulsion budget, and the total launch mass of the rocket is increased (most commonly by strap-on boosters) to match the required payload mass via $m_{0,1} = m_{L,n}/\lambda_*$. Therefore the "predetermined Δv case" is more relevant.

Comparing the optimization of structural ratios presented in this section (Eq. (3.4.7)) with that in Section 3.3 (Eq. (3.3.5)), we see that we have found here a maximum payload ratio independent from the stage number and propulsion demand. That these results are nevertheless consistent with each other can be recognized when the limiting case $\varepsilon = 0$ is considered. If we insert Eq. (3.4.8) into Eq. (3.3.5) we get $\lambda_{opt} = 1/(e-1)$ just as from Eq. (3.4.7).

3.5
Parallel Staging

Parallel staging means that several stages (here k stages) are mounted, and also activated in parallel (see Fig. 3.1). Let's determine the corresponding rocket equation. The total thrust of a parallel-staged rocket is the sum of the generally different thrusts of all stage engines

$$F_* = \sum_{i=1}^{k} F_{*,i} = \sum_{i=1}^{k} \dot{m}_{p,i} v_{*,i} \qquad (3.5.1)$$

Also the total mass flow rate is the sum of all stage engines

$$\dot{m}_p = \sum_{i=1}^{k} \dot{m}_{p,i} \qquad (3.5.2)$$

Analogously to Eq. (1.1.3) we can set up a thrust equation

$$F_* = \dot{m}_p \langle v_* \rangle \qquad (3.5.3)$$

whereby we have indirectly defined a mean effective exhaust velocity

$$\langle v_* \rangle := \frac{F_*}{\dot{m}_p} = \frac{\sum_{i=1}^{k} \dot{m}_{p,i} v_{*,i}}{\sum_{i=1}^{k} \dot{m}_{p,i}} \quad \text{mean effective exhaust velocity} \qquad (3.5.4)$$

In words we have the following:

> For parallel staging the effective exhaust velocity is calculated from the mean of exhaust velocities weighted by the respective mass flow rates of the stage thrusters.

If the exhaust velocities are all the same, $v_{*i} = v_{*j}$, Eq. (3.5.3) results in $\langle v_* \rangle = v_*$, as expected. With this definition the rocket equation reads analogously to Eq. (2.2.2)

$$\Delta v = \langle v_* \rangle \cdot \ln \frac{m_0}{m} \qquad \text{rocket equation, parallel staging} \qquad (3.5.5)$$

So, all results so far remain valid if one replaces v_* by $\langle v_* \rangle$. In addition, as the rocket equation for parallel staging Eq. (3.5.5) plus Eq. (3.5.4) is formally identical with that of a single-stage rocket, any combination of serial staging and parallel staging can be easily calculated.

Advantages and Disadvantages of Parallel Staging Compared with Serial Staging

Advantages

- The total engine weight is fully used during propulsion time. It does not have to be carried as "dead" payload of the following stage as in serial staging.
- Thereby in near-Earth space one achieves a higher acceleration and hence a lower gravitational loss (see Section (2.4.2)).
- Attaching additional boosters easily adapts a system to larger payload masses.
- Empty tank mass is minimized by propellant transfer with collecting pipes.
- Development costs are minimized by standardizing structure and engines.
- A smaller overall length of a rocket reduces bending moments and longitudinal and lateral oscillations.

Disadvantages

- Structural load is higher after launch.
- This implies higher dynamic drag losses.
- Due to the relatively long combustion time, the boosters have to be dimensioned to operate over a higher altitude range. Therefore the nozzle cross sections cannot be optimally dimensioned for all altitudes, implying less thrust. With serial staging, on the other hand, the engines can be far better adapted to the respective operative ranges to achieve higher effective exhaust velocities.

After Saturn V, which was a true serial three-stage rocket, NASA quickly passed over to two-stage rockets with parallel staging of the first stage by reflecting on the archetypal Russian R-7 rocket. This mixed type of staging is a good trade-off between the pros and cons of serial and parallel staging. Engines that are just flanged on to the first stage are called strap-on boosters or just boosters. Today virtually any launcher is built like that: Delta, Sojus, Ariane, and Shuttle. Parallel staging of only the first stage has the advantage that by using two, four or six boosters the thrust can be easily adapted to varying payloads without changing the rocket design.

3.6
Other Types of Staging

Instead of integrating a tank plus engine in one stage, it is also possible (see Fig. 3.1) to stage only the tank (**tank staging**) or the engines (**engine staging**). These staging types can be combined with serial staging and parallel staging. Tank staging is an interesting alternative, for instance when a propellant component has a very low density (such as hydrogen), leading to a large and hence heavy empty tank at the end of the launching phase.

A quite interesting option of tank staging has been discussed since the 1970s. Only one type of engine is used for two different propellants (dual-propellant propulsion). Initially, a propellant with a higher density (e.g. kerosene and liquid oxygen) is burnt because of the higher thrust. At burnout, one merely switches to a propellant with lower density, but higher specific impulse (e.g. liquid hydrogen and oxygen). This staging type would be very effective (just one engine mass) and reliable (no full-stage separation) if the empty propellant tanks were attached externally to the rocket to be jettisoned.

Problems

Problem 3.1 Infinite-Stages Rocket
Starting from Eq. (3.2.2) show that for infinite many stages equation $\Delta v = -v_* (1 - \varepsilon) \ln \lambda_*$ holds (see Eq. (3.3.6)).

Problem 3.2 Space Shuttle Propulsion
Consider a Space Shuttle with a total launch mass of 2.017 t (1 t = 1000 kg); orbiter mass of 111 t; external tank (ET) of total mass 738 t, of which 705 t is H2/O2 fuel. There are two external Solid Rocket Boosters (SRB) of total mass 584 t each, of which is 500 t solid fuel, firing at $I_{sp} = 300$ s. The three Space Shuttle Main Engines (SSME) have a $v_e = 4.3 \text{ km s}^{-1}$ and a fuel flow rate of 500 kg s^{-1} each. Finally the orbiter incorporates two thrusters called Orbital Maneuvering Systems (OMS) with $v_e = 3.0 \text{ km s}^{-1}$ and 11 t of UDMH / N$_2$O$_4$ fuel.

There are three propulsion phases: During the first 120 s of ascent the SRBs burn in parallel with the three SSMEs. Then the SRBs are jettisoned and the orbiter with the ET continues ascent during this second phase until the ET is empty and the SSMEs are cut off. In the third phase, which includes the kick-burn into the target orbit, orbit maintenance maneuvers and deorbit burn the orbiter fires its OMSs.

Assuming that the mass flow rate is constant during all these phases show that the maximum Δv achievable with this Shuttle system adds up to $\Delta v_{\max} = 9.63 \text{ km s}^{-1}$.

4
Thermal Propulsion

The thrust from thermal propulsion engines results from the exhaust of propellant gases, which is achieved by the rapid expansion of the heated gas. The heat usually comes from the combustion of chemical propellants – which we will assume in the following without loss of generality – or from the supply of external heat, or from both. A chemical propellant therefore serves two different purposes at the same time: it is a provider of mass for the required mass flow rate, and a provider of energy to accelerate itself to exhaust velocity.

We now want to know how an engine converts combustion heat into thrust, that is, how the expansion of propellant gases can be described in terms of thermodynamics, and how, with a given amount of energy in the combustion chamber, we can determine and maximize the thrust of the propulsion with an adequate combustion chamber and nozzle design.

In doing so we will first assume a propulsion engine with an arbitrarily formed combustion chamber and nozzle. It is not important for us how the propellant actually gets into the chamber, but we merely assume that it somehow appears there with a given mass rate $\dot{m}_p$, and that it receives a certain amount of energy per mole, which heats it up. Either the energy supply might occur by its combustion and/or it might be supplied externally. The total received thermal energy per mole of propellant mass is the molar enthalpy, which we label h_0. The conditions mentioned above generally apply to mono- and bi-propellant, hybrid, dual mode, and thermoelectric thrusters. Even cold gas propulsion is applicable if one considers the product of pressure times molar volume in the combustion chamber as a molar enthalpy. Actually, we have the following:

> **Minimum engine requirements:** A propellant gas with a known pressure p_0, a received molar enthalpy h_0, and excited internal degrees of freedom n (see Eq. (4.1.2)) flows with a certain mass flow rate $\dot{m}_p$ and with sound velocity (see Eq. (4.1.13)) through a narrow throat with a given cross section A_t, and escapes through a nozzle by means of controlled expansion.

Astronautics. Ulrich Walter
Copyright © 2008 WILEY-VCH Verlag GmbH & Co. KGaA, Weinheim
ISBN: 978-3-527-40685-2

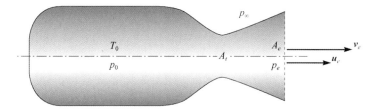

Figure 4.1 Determining parameters of a thermodynamic rocket.

This chapter deals with understanding the behavior of the propellant gas in such a chamber and nozzle, and their optimum design for maximum thrust.

The first part of this chapter will follow the gas flow from the combustion chamber (index 0), through the throat of the combustion chamber (index t), along the nozzle to the nozzle exit (index e), and outside (index ∞) (see Fig. 4.1). In the second part, we will analyze the optimum engine design.

4.1
Engine Thermodynamics

4.1.1
Physics of Propellant Gases

The properties of a thermal thruster are essentially determined by the properties of the propellant gas while flowing through the engine, from the combustion chamber right down to the nozzle exit. In order to understand the thrust characteristics, we therefore need to understand the basic behavior of a gas. Let's have a look at the physical and chemical characteristics of a propellant gas, before we apply this knowledge to optimize the design of the engine. The theory we will lay out below is based on the following assumptions for the gas flow in the engine:

- The gas is ideal (no viscous effects occur).
- The gas is homogeneous in the chamber and its composition does not change in space and time after combustion.
- The gas flow is continuous at the exit of the combustion chamber and beyond.
- Any change of gas state is adiabatic (heat losses are negligible).
- The acceleration of the rocket is negligible compared to acceleration of the gas.
- No shock waves occur within the nozzle.
- All conditions are the same for all points of a given cross section; in particular, the gas flow is axially symmetric.

Due to the last assumption, our theory is a one-dimensional theory of gas flow, which allows us to describe spatial conditions in the engine by only one variable, i.e. the distance traveled along the engine.

With these assumptions we are now going to exploit the behavior of the gas along the engine axis. Because the gas is ideal, the so-called intensive thermodynamic variables (variables that do not depend on the amount of gas) pressure p, gas density $\rho = m_p/V$, and gas temperature T are interdependent as described by the ideal gas law:

$$p = \rho \frac{R}{M_p} T \quad \text{ideal gas law} \tag{4.1.1}$$

with

$R =$ **universal gas constant** with value $R = 8.314 \text{ JK}^{-1}\text{mol}^{-1}$

$M_p =$ **molar mass** (mass of 1 mol $= 6.022 \times 10^{23}$ molecules)

of the propellant.

When the gas pressure forms due to combustion in the chamber, and when the gas thermally expands along the nozzle axis, internal energy, pressure, density, and temperature are constantly changing. We assume that the corresponding conversion processes are reversible and adiabatic. Physically speaking, such processes are called isentropic – they preserve entropy. The thermodynamic potential that adequately describes an isentropic process is the so-called enthalpy. It encompasses the internal energy (thermal energy) of the gas U, which corresponds to a microscopic, statistical motion of the molecules, plus the macroscopic displacement work pV of the gas, but not its macroscopic flow energy (kinetic energy). The enthalpy (amount of heat) H and the molar enthalpy h of an ideal gas with mass m_p are given by

$$H := U + pV = m_p c_p T \tag{4.1.2}$$

and

$$h = \frac{\kappa}{\kappa - 1} RT$$

with

$c_p = \frac{\kappa}{\kappa - 1} \frac{R}{M_p}$ **specific thermal heat capacity** at constant pressure

$\kappa = \frac{c_p}{c_V} = 1 + \frac{2}{n} \approx 1 + \frac{2}{8} = 1.25$ **adiabatic index**

c_V **specific thermal heat capacity** at constant volume

n average number of **excited degrees of freedom** of the gas molecules

The number of excited degrees of freedom n of the gas particles is an important characteristic of the gas, since it determines in how many micromechanical forms thermal energy is stored in the gas. As will be seen in a moment, amongst other things, this determines the temperature of the gas. The value of n depends on the specific type and composition of the gas. A propellant gas is usually composed of different types of molecules. Each gas component can move in all three directions in space, so it always has three translatory degrees of freedom $n_{trans} = 3$. If the gas component is monoatomic, the atom is not able to take up any more internal energies (gas ionization can be neglected in propellant engines), and $n = n_{trans} = 3$. For diatomic molecules there are two additional rotational degrees of freedom $n_{rot} = 2$ (two rotational axes perpendicular to the molecular axis – rotation around the molecular axis does not count, as quantum-mechanically there exists no corresponding moment of inertia), and one oscillatory (vibrational) degree of freedom along the molecular axis $n_{osc} = 1$, so in total $n = 6$. Polyatomic molecules mostly have a three-dimensional configuration, and thus three rotational axes and three oscillatory degrees of freedom, so $n = 9$. Two important exceptions, however, are: the linear CO_2 molecule with $n_{rot} = 2$ and $n_{osc} = 1$, therefore $n = 6$, and the important planar H_2O with $n_{rot} = 3$ and $n_{osc} = 2$, implying $n = 8$. For a propellant gas mixture the actual number of degrees of freedom is a stoichiometric average over all gas components, so in general it is not integer.

If you heat up a gas, the heat energy is distributed via impacts evenly between the molecules to cause translational, rotational and vibrational motion. But only translational motion determines the temperature of the gas. Physically speaking, the temperature of a gas is the kinetic energy of the average microscopic, translational motion, the so-called root-mean-square velocity v_{rms}, i.e. of the velocity of its molecules only. The corresponding relation is

$$\frac{n_{trans}}{2} k_B T = \frac{1}{2} m_p v_{rms}^2$$

where k_B is the Boltzmann constant. The interpretation of this equation is as follows: Each of the three translatory motions corresponds to the same amount of energy, $\frac{1}{2} k_B T$. If one were able to limit the motion of the gas to a line or a surface, it would therefore have the kinetic energy $\frac{1}{2} k_B T$ and $\frac{2}{2} k_B T$, respectively. For $\frac{3}{2} k_B T$ the gas can move freely in all three dimensions. Rotational and oscillatory motions physically do not contribute to temperature. Nevertheless, altogether they make up the internal energy U of the gas. Because there is no lower limit to translational energy and since the quantum energy of molecule rotation is very low, translational and rotational modes are always excited. Quantum vibrational energies are much higher and start to get excited at around 1000 K. Therefore at intermediate temperatures the number of excited degrees of freedom increases while κ decreases monotonically with increasing temperatures. At extremely high temperatures, around 10 000 K, which do not occur in thrusters, one also would have to consider ionizing degrees of freedom, i.e., gas plasma.

The number of theoretically accessible degrees of freedom of the molecules is between $n = 3$ for monoatomic noble gases and $n = 9$ for three-dimensional polyatomic molecules, or equivalently $1.22 \leq \kappa \leq 1.67$. Due to the high combustion chamber temperatures, almost all degrees of freedom of the mostly polyatomic molecules are excited, and as the gas is also a mixture of different components, $n \approx 8$ is a good average value for any rocket propellant with the corresponding adiabatic index $\kappa \approx 1.25$. Two examples are: $n(O_2/H_2) = 7.41$, $n(O_2/N_2H_4) = 8.70$, where X/Y denotes all reaction components of the oxidant X and the propellant Y. In the first case, the planarity of the resulting water molecule is obviously responsible for a relatively low number of degrees of freedom.

> **Example**
> What is the adiabatic index of dry air at standard conditions?
> The molar composition of dry air is 78% N_2, 21% O_2 and 1% Ar, which is 99% diatomic molecules with excited degrees of freedom $n = 5$ (at ambient temperature the one oscillation is not excited) and 1% atoms with $n = 3$. So we find $\kappa_{air} = 0.99 \times 1.400 + 0.01 \times 1.667 = 1.403$. This is exactly the value as given in relevant tables of thermodynamics.

On the path from the combustion chamber along the nozzle to the exit, the energy of the gas is continuously converted. Internal energy (heat) is converted into gas expansion work pV and macroscopic flow energy, and the other way round. Thereby the so-called thermodynamic state variables T, p, and ρ (the variables that characterize the gas state) change in line with the gas equation (4.1.1) and the macroscopic drift velocity v changes as well. In physics the so-called thermal efficiency η (a.k.a. *thermodynamic* or *cycle efficiency*) describes how efficiently internal energy (heat) is converted into work and gas flow in the course of these energy changes. As we assume only adiabatic (isentropic) processes, the following thermodynamic relations hold:

$$\eta = 1 - \frac{T}{T_0} = 1 - \left(\frac{p}{p_0}\right)^{\frac{2}{n+2}} = 1 - \left(\frac{\rho}{\rho_0}\right)^{\frac{2}{n}} \leq 1 \quad \text{thermal efficiency} \quad (4.1.3)$$

From this follows

$$\rho = \rho_0 (1-\eta)^{n/2} \qquad (4.1.4)$$

In the following it would be rather tedious to express the thermodynamic equations either by p (which is appropriate in most circumstances), or by T, or by ρ. We want to free ourselves from this ambivalence, so that:

> We use the thermal efficiency η as a substitute for p, T and ρ, which means that we can always immediately shift to any of these thermodynamic variables by the application of Eqs. (4.1.3).

In this way η is like a convenient exchange currency. This representation also has the very practical benefit of considerably simplifying the equations, though the equations derived will look quite different from those in the literature. Therefore for the most important equations we will also cite them in their familiar notation.

4.1.2
Flow Velocity

Let v be the gas flow velocity (drift velocity) and T its temperature at any point along the engine axis. According to the general law of the conservation of energy, the total energy, i.e. the kinetic energy (macroscopic flow energy) plus the enthalpy of the gas, expressed by the state variables v, T, has to be the same as in the combustion chamber (v_0, T_0)

$$\frac{1}{2}m_p v_0^2 + m_p c_p T_0 = \frac{1}{2}m_p v^2 + m_p c_p T \tag{4.1.5}$$

Assuming that the gas flow velocities in the combustion chamber at the location of heat generation are negligibly small, $v_0 \approx 0$, we derive from this the gas flow velocity at any position in the engine

$$v^2 = 2c_p T_0 \left(1 - \frac{T}{T_0}\right) = 2c_p T_0 \eta \tag{4.1.6}$$

By making use of Eq. (4.1.2) this equation leads to the important relationship

$$v = a_0 \sqrt{n\eta} \tag{4.1.7}$$

Remark 1: *Equation (4.1.7) is known in literature as Zeuner–Wantzel equation, reading*

$$v = \sqrt{\frac{2\kappa}{\kappa - 1} \frac{R}{M_p} T_0 \left[1 - (p/p_0)^{\frac{\kappa-1}{\kappa}}\right]}$$

Remark 2: *Although v_0 is not exactly zero, the neglected term $\frac{1}{2}m_p v_0^2$ formally can be ascribed in Eq. (4.1.6) to the chamber temperature via*

$$T_0 \to T_0' = T_0 + \frac{1}{2}\frac{v_0^2}{c_p}$$

The expression $m_p c_p T_0'$ is then the total combustion enthalpy.

In the above we have made use of the expression for the sound velocity

$$a = \sqrt{\frac{\kappa p}{\rho}} = \sqrt{\frac{\kappa RT}{M_p}} = \sqrt{\frac{\kappa RT_0}{M_p}}\sqrt{\frac{T}{T_0}} =: a_0\sqrt{1-\eta} \quad \text{sound velocity} \quad (4.1.8)$$

as given by physics. The latter is a result of Eq. (4.1.3). Note that Eq. (4.1.7) is valid for the gas velocity at any point along the flow right down to the nozzle exit. Also note that a decreasing pressure or temperature leads to an increase of ideal-cycle efficiency and also to an increasing flow velocity. We will have a closer look at this counter-intuitive behavior later on.

4.1.3
Mass Flow Density

We recall from Eq. (1.2.9) that the following continuity equation holds for the mass flow rate

$$\boxed{\dot{m}_p = \rho v A = const} \quad \text{continuity equation} \quad (4.1.9)$$

Since it is a result of mass conservation, the mass flow rate $\dot{m}_p$ must be constant over any cross section along the flow track.

Remark: *Strictly speaking we presume $\dot{m}_p = const$. By doing so, an areal cross section is defined via the continuity equation, on which ρ und v are constant. If the gas jet diverges, i.e. is no longer axial, such as in the nozzle, the area is no longer flat, but in lowest approximation a sphere segment. The flow vector is normal to the surface of this segment. The center of the sphere is the imaginary point where the gas flow lines converge. Depending on whether the exiting jet is under- or over-expanding (see later discussion following Eq. (4.1.19)) the exit area is a convex or concave sphere segment, respectively.*

Therefore, the important parameter **mass flow density** $\mu = \dot{m}_p/A$ is not constant. For μ we find from Eqs. (4.1.4), Eq. (4.1.7), and Eq. (4.1.8):

$$\mu := \frac{\dot{m}_p}{A} = a_0\rho_0\sqrt{n\eta(1-\eta)^n} = a\rho\sqrt{\frac{n\eta}{1-\eta}} \quad (4.1.10)$$

Remark: *Equation (4.1.10) is known in the literature as the Saint Venant equation, reading*

$$\frac{\dot{m}_p}{A} = \sqrt{\frac{2\kappa}{\kappa-1}p_0\rho_0(p/p_0)^{\frac{2}{\kappa}}\left[1-(p/p_0)^{\frac{\kappa-1}{\kappa}}\right]}$$

4.1.4
Flow at the Throat

As mass flow is constant and the nozzle cross section has its minimum at the throat, the mass flow density μ at a constant a_0 and ρ_0 has a maximum at the throat with

$$\eta_t = \frac{1}{n+1} \tag{4.1.11}$$

Equation (4.1.11) is easily derived by zeroing the first derivation of Eq. (4.1.10). By applying Eq. (4.1.11) to Eq. (4.1.10) we find for the mass flow density at the throat

$$\mu_t = \frac{\dot{m}_p}{A_t} = a_0 \rho_0 \sqrt{\left(\frac{n}{n+1}\right)^{n+1}} = a_t \rho_t \tag{4.1.12}$$

where the last term results from the last term of Eq. (4.1.10) due to $n\eta_t/(1-\eta_t) = 1$. In addition we get from Eq. (4.1.9), Eq. (4.1.12), and Eq. (4.1.8) the flow velocity at the throat

$$v_t = \frac{\dot{m}_p}{\rho_t A_t} = \frac{\rho_t a_t A_t}{\rho_t A_t} = a_t = a_0 \sqrt{\frac{n}{n+1}} \tag{4.1.13}$$

In words this is equivalent to the following:

> The flow velocity just reaches sound velocity at the throat. This is an important property of thermal propulsion engines.

Note: Sound velocity at the throat here is not the conventional $a_{air} \approx 343.4 \text{ m s}^{-1}$ at standard atmospheric pressure, but due to Eq. (4.1.8) $a_t = a_{air}\sqrt{T_t/T_{air}} \approx 1200 \text{ m s}^{-1}$. So a_t is much bigger and in addition dependent on T_t and hence on the temperature and pressure conditions in the pressure chamber.

Let's pause for a moment to ponder about what we have achieved so far, and what lies still ahead of us. Strictly speaking, the above maximization considerations at the throat, and with it this whole chapter, describe a key property of a thermal thruster. In view of the maximization principle, the physical principle of mass conservation (as expressed in Eqs. (4.1.9) and (4.1.10)), and the conservation of energy (whose result led to Eq. (4.1.7)), all the considerations that now follow are just more or less clever applications, new definitions, and analytical conversions of Eqs. (4.1.7), (4.1.9), (4.1.10), and (4.1.12).

4.1.5
Flow in the Nozzle

Behind the throat, the gas expands into the so-called Laval nozzle (a nozzle with a throat that widens in the flow direction) whereby its pressure decreases. According to Eqs. (4.1.3) and (4.1.7) the flow velocity therefore increases. So the thrust-determining exhaust velocity v_e is considerably larger at the exit of the nozzle than at the throat, and this is exactly what a nozzle is designed for. How big is this increase? The relation between v_e and v_t can be derived from Eqs. (4.1.7) and (4.1.13) to be

$$v_e = a_0\sqrt{n\eta_e} = v_t\sqrt{n+1}\sqrt{\eta_e} \tag{4.1.14}$$

In other words:

> The velocity gain factor of a nozzle is $\sqrt{n+1}\sqrt{\eta_e} \approx 3\sqrt{\eta_e}$, which in vacuum tends to the value of 3. So a nozzle increases momentum thrust by 200%, but because the expansion at the same time reduces the pressure thrust, the gain in total thrust is less than 67% (see Section 4.2.5).

It might be surprising to learn that the flow velocity increases with decreasing flow pressure. Intuitively, one would expect the contrary. Let's see why this seemingly paradoxical behavior occurs.

Hydrodynamics

This weird behavior is due to the hydrodynamic nature of the gas flow. The continuity equation (4.1.9)

$$\rho v A = const$$

hereby plays a key role. It states that, if the cross section A of the nozzle increases, the product ρv has to decrease. It now crucially depends on the relation $\rho \propto 1/v^\alpha$, or its differential $d\rho/\rho = -\alpha \cdot dv/v$, how v behaves. If $\alpha > 1$, then v increases with decreasing A, otherwise it decreases. So we have to determine α to solve the paradox. We start out by examining the gas equation (4.1.1), $p = \rho RT/M_p$, and the equation of energy conservation (4.1.6), $v^2 = 2c_p(T_0 - T)$. To arrive at one-to-one relations between the intensive thermodynamic variables p, ρ, T, v, which we are looking for, we need an additional relation between any two of them. This is furnished by the fact that, according to thermodynamics, for adiabatic processes $pV^\kappa = const$ holds. Because in general $\rho V = const$, we find $p \propto \rho^\kappa$ and hence

$$\frac{dp}{p} = \kappa \frac{d\rho}{\rho} \tag{4.1.15}$$

So p and ρ change in the same way: increasing p implies increasing ρ and vice versa (which we write in short as $p \uparrow \leftrightarrow \rho \uparrow$). In order to apply this differential equation, we have to differentiate $p = \rho R T / M_p$ and $v^2 = 2c_p (T_0 - T)$, which yields

$$\frac{dp}{p} = \frac{d\rho}{\rho} + \frac{dT}{T}$$

and

$$v\,dv = -c_p dT$$

With Eq. (4.1.15) we find the one-to-one relations

$$\frac{d\rho}{\rho}(\kappa - 1) = \frac{dT}{T} = -\frac{v^2}{c_p T}\frac{dv}{v} \tag{4.1.16}$$

Because $\kappa > 1$, they state: $\rho \uparrow \leftrightarrow T \uparrow \leftrightarrow v \downarrow$ and vice versa. The latter inverse behavior, which seems quite strange, is due to the law of energy conservation, to which we will turn later. Because of Eqs. (4.1.2) and (4.1.8) it follows that $c_p (\kappa - 1) T = \kappa R T / M_p = a^2$ and hence

$$\frac{d\rho}{\rho} = -\frac{v^2}{a^2}\frac{dv}{v} = -(Ma)^2 \frac{dv}{v} \tag{4.1.17}$$

where $Ma := v/a$ is the so-called **Mach number**. It is dimensionless by relating the flow velocity to the instantaneous sound velocity. With Eq. (4.1.17) we have identified $\alpha = (Ma)^2$. To find the explicit dependences $v \leftrightarrow A$ and $\rho \leftrightarrow A$, we differentiate $\rho v A = const$ and get

$$\frac{d\rho}{\rho} + \frac{dv}{v} + \frac{dA}{A} = 0$$

Inserting Eq. (4.1.17) into this, we finally obtain

$$\frac{dv}{v} = \frac{1}{(Ma)^2 - 1}\frac{dA}{A} \tag{4.1.18}$$

or

$$\frac{d\rho}{\rho} = -\frac{(Ma)^2}{(Ma)^2 - 1}\frac{dA}{A}$$

From the above equations we can read off the change of flow velocity as a function of change of cross section. We have to discern two cases:

1. subsonic case ($Ma < 1$): $A \uparrow \to \rho \uparrow$ and $v \downarrow \leftrightarrow T \uparrow$
2. supersonic case ($Ma > 1$): $A \uparrow \to \rho \downarrow$ and $v \uparrow \leftrightarrow T \downarrow$

In the subsonic case the flow velocity declines along the nozzle and its density increases, while for the supersonic case things are reversed. In Fig. 4.2 the physical causal chain is depicted according to the equations derived above, and in Fig. 4.3 quantitative results are shown.

Note: *This implies that, if the flow would not reach sound velocity at the throat, the flow in the nozzle would stay subsonic and even decrease. The condition $v = a$ at the throat therefore is a critical condition for a thermal propulsion engine.*

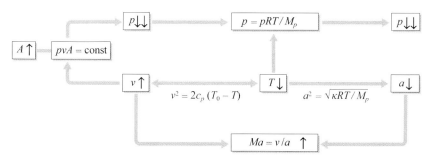

Figure 4.2 Qualitative dependence of the thermodynamic variables along the nozzle. The arrows ↑ and ↑↑ mean increase and strong increase, respectively, and ↓ and ↓↓ the corresponding decreases.

Physical Interpretation

This leaves open the question of why the flow behaves so differently at subsonic and supersonic speeds. This is due to the kinetic energy of the flow. Let's take a look at the law of energy conservation of the gas as given by Eq. (4.1.5):

$$m_p c_p T + \tfrac{1}{2} m_p v^2 = const$$

It shows that the kinetic energy increases quadratically with flow velocity. So any change in gas temperature implies decreasing flow speed changes for increasing flow speeds. This is expressed explicitly in Eq. (4.1.16): relative density and temperature changes are rigidly coupled, while the coupling between temperature and speed changes is quadratically in v. Therefore, for a given $\Delta \rho / \rho$ and with increasing v, the absolute value of $\Delta v / v$ decreases.

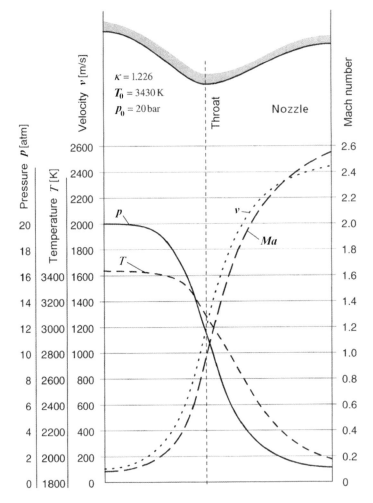

Figure 4.3 Course of thermodynamic variables along a nozzle.

So, overall the following happens. The space available for a given amount of gas enlarges with increasing cross section along the nozzle, $A \uparrow$. Thereby the density and the speed of the gas, flowing into the enlarging space ahead of it, change such that the amount of gas remains constant (conservation of mass, $\rho v A = const$). If the density would increase, $\rho \uparrow$, then, due to $\rho v A = const$, the flow speed v would have to strongly decrease, $v \downarrow\downarrow$. But at supersonic speeds this is not possible, because due to the gas equation $\rho \uparrow \leftrightarrow T \uparrow$ and, because of the equation of energy conservation, v decreases only little. Only at subsonic flow speeds the flow speed reduction would be big enough to compensate. Therefore at supersonic speeds the density has to drop, $\rho \downarrow$, causing the gas temperature via the gas equation to drop as well, which in turn raises the flow speed only little, $v \uparrow$. But this is all what is necessary according to $\rho v A = const$ because the density reduction already counteracts the increase of cross section to a large amount.

From the above equation of energy conservation with implication $T \uparrow \leftrightarrow v \downarrow$ one can read off another important feature: the first term is the enthalpy, which is the internal microscopic kinetic energy plus displacement work pV, and the second term is the kinetic flow energy of the gas. But because temperature is a direct measure of the translational microscopic motion of the gas molecules, as seen early on in this chapter, the inverse dependence can be interpreted in the following way:

> Supersonic expansion converts part of the microscopically disordered motion of the molecules (microscopic translations) into an increasingly macroscopically directed motion (flow velocity) of the gas.

Upon temperature reduction, fewer rotational and oscillatory modes of the gas molecules are excited. By means of impact processes they decay into microscopic translations, thus adding slightly to the flow velocity of the gas. However, this additional conversion effect, which lowers the excited degrees of freedom n and adds to the nozzle efficiency (see Section 4.2.5), is not considered in our derivations. We assume that n or κ is constant with temperature.

4.1.6
Ideal Nozzle Adaptation

A nozzle is used to increase exhaust velocity. However, our goal is not to maximize exhaust velocity, but thrust. In order to do so, let's have a look at the expression for total thrust. If the exit surface normal u_e and v_e go into the same direction, which usually is the case, the following holds for the amount of total thrust according to Eq. (1.2.6):

$$F_* = \dot{m}_p v_e + (p_e - p_\infty) A_e \tag{1.2.6}$$

Now consider the right side of Eq. (1.2.6) above. Expanding the gas through the nozzle increases v_e, but also reduces p_e. The first term tells us that this implies an increase of the momentum thrust, but the second term leads to a reduction of the pressure thrust. So, what we really want to know is how to choose p_e or v_e to maximize the thrust as the combined force of momentum thrust and pressure thrust.

Our optimization problem is even a bit more complicated. According to Eq. (1.2.6), thrust depends on three parameters, p_e, v_e, and A_e, if the *mass flow* rate is considered to be constant. The latter assumption is generally admissible, as supersonic speed in the nozzle has the positive effect that erratic flow variations cannot expand backwards into the combustion chamber. The Laval nozzle acts like a barrier, and the mass flow rate of the propulsion is only determined by the pressure in the combustion chamber and the diameter of the exit (see Eq. (4.1.26)).

It is our long term goal to find that optimum combination of $p_{e,opt}$, $v_{e,opt}$, and $A_{e,opt}$ for which thrust is maximum. Mathematically speaking, optimization means that we are looking for that combination where any variations δA_e, δv_e, and δp_e do not lead to any further increase of F_*, that is $\delta F_* = 0$. If we consider infinitesimally small variations, this can be expressed mathematically by the total differential as follows:

$$dF_* = dF_A + dF_e + dF_p = \frac{\partial F_*}{\partial A_e} dA_e + \frac{\partial F_*}{\partial v_e} dv_e + \frac{\partial F_*}{\partial p_e} dp_e = 0 \qquad (4.1.19)$$

Deriving the partial derivatives from Eq. (1.2.6) we find

$$dF_* = (p_{e,opt} - p_\infty) dA_e + \dot{m}_p dv_e + A_{e,opt} dp_e = 0$$

We now consider the fact that the thermodynamic variables v_e and p_e are directly dependent on each other via Eqs. (4.1.14) and Eq. (4.1.3). With these relationships and with the application of Eq. (4.1.10) we can calculate (exercise, Problem 4.1) the derivative dv/dp straightforwardly. Alternatively, Eq. (4.1.15) can be inserted into Eq. (4.1.16) and Eq. (4.1.9) applied. In both cases one obtains

$$dv = -\frac{A}{\dot{m}_p} dp \qquad (4.1.20)$$

This relation describes the change of the flow velocity with the pressure at any cross section A along the nozzle. (Observe that the mass flow rate $\dot{m}_p$ along the nozzle is always constant.) Applying this equation to the nozzle exit and inserting it into the above equation, the last two terms cancel each other out. So one arrives at

$$dF_* = (p_{e,opt} - p_\infty) dA_e = 0$$

This immediately brings us to the required p_e condition for maximum thrust:

> The thrust (propellant force) achieves its maximum for $p_e = p_\infty$, i.e. when the pressure thrust vanishes.

Remark: *We actually only showed that F_* is optimum at $p_e = p_\infty$. See Fig. 4.7 in Section 4.2.4 for the proof that it really is a maximum.*

For $p_e = p_\infty$ the corresponding **optimum exhaust velocity** can be derived from Eq. (4.1.14) as

$$v_{e,opt} = a_0 \sqrt{n \eta_\infty} \qquad (4.1.21)$$

The optimum exit area $A_{e,opt}$ will be derived later in Section 4.2.2.

Flow Expansion

A nozzle that achieves $p_e = p_\infty$ is called an **ideally adapted nozzle**. In a nozzle where the optimum is not achieved (see Fig. 4.4), be it $p_e < p_\infty$, which is called *over-expansion* because the jet is over expanded within the nozzle, or $p_e > p_\infty$, which is called *under-expansion* because the jet was not able to properly expand within the nozzle, divergences arise behind the nozzle exit. The jet direction then is not longer parallel to the engine axis. This leads to a loss of thrust, as there are thrust components perpendicular to the engine axis, which are then irreversibly lost. As pressure adiabatically lowers with increasing volume, thrust reductions due to an over- or under-expanding jet can be counteracted by an increase or decrease of the exit surface, which is equivalent to extending or reducing the length of the Laval nozzle.

Flow separation caused by overexpansion

Nozzle flows full

Under-expansion

Figure 4.4 Flow conditions of an over-expanding (top), ideally expanding (middle), and under-expanding (bottom) nozzle.

Shock Attenuation

For $p_e = p_\infty$, $dF_A = (p_e - p_\infty) dA_e = 0$. Therefore, from Eq. (4.1.19) it follows that

$$dF_p = -dF_e$$

In other words, every tiny pressure thrust variation is counterbalanced by a repulsion thrust variation. This is quite a remarkable effect, as pressure variations occur within every combustion chamber in the form of shock waves, which travel from the combustion chamber right beyond the nozzle exit. The inverse interdependence $dF_p = -dF_e$ ensures that none of these shock waves has any effect on the total thrust.

The situation is different with pogo vibrations, which are the nightmare of thruster manufacturers – and astronauts. Pogos are thrust vibrations along the engine axis, which may put the rocket under enormous stress even beyond its structural limits. They result from a positive feedback between thrust and mass flow rate. Assume, for instance, that at a given time there is a positive thrust variation. The increased acceleration force then fuels more propellant into the combustion chamber, leading to an increased mass flow rate, in turn giving rise again to a higher thrust, and so on. Pogos are not taken care of in our above considerations, as we assumed a constant mass flow rate. A small side chamber connected to the fuel line usually damps them out, because it gives and takes additional propellant depending on the acceleration on the propellant and hence on the fuel pressure in the fuel line.

4.1.7
Engine Performance

According to Section 1.3 engine performance and hence rocket performance is measured very generally by the engine's figure of merit, the specific impulse $I_{sp} = v_*/g_0$. Now that we have derived an expression for the exhaust velocity of thermal engines, we are ready to determine their I_{sp} explicitly. Usually I_{sp} is cited in literature for an ideally adapted nozzle in vacuum. At these conditions we get a *maximum obtainable specific impulse*, which can be calculated from Eq. (4.1.14) for $p_e = p_\infty = 0$. This yields with Eq. (4.1.8)

$$g_0 I_{sp,max} = v_{e,max} = a_0 \sqrt{n} = \sqrt{(n+2)\frac{RT_0}{M_p}} = \sqrt{\frac{2h_0}{M_p}} \qquad (4.1.22)$$

where $h_0 = M_p H_0/m_p$ is the molar form of the available enthalpy H_0, which comprises the molar combustion enthalpy H_p (heat of reaction) of the propellant with **combustion efficiency** η_c, that is $\eta_c H_p$, and the externally supplied energy E_{ext}

$$H_0 = \eta_c H_p + E_{ext} \qquad (4.1.23)$$

η_c is determined by the heat losses of the engine, which are quite considerable (see Fig. 4.5). From Eqs. (4.1.22) and (4.1.23) we get the remarkably simple result:

For an ideally adapted engine in vacuum, the maximum specific impulse and thus also the engine figure of merit does not depend on the design of the engine or the chamber pressure, but it depends exclusively on the propellant properties and the combustion efficiency of the engine. The best propellant is a propellant with the highest combustion molar enthalpy h_p and lowest molar mass M_p.

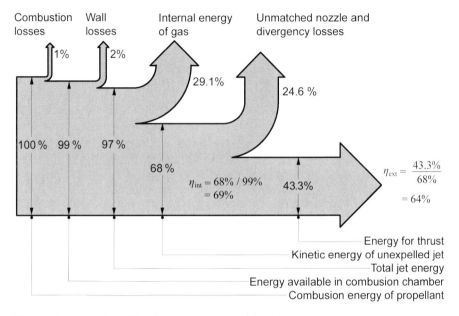

Figure 4.5 Energy flow of the third-stage engine of Ariane 1.

Because H_2/O_2 displays one of the highest molar enthalpies and hydrogen has the lowest molar mass, it fuels chemical thrusters with the highest efficiency available. Apart from employing a better propellant, the engine figure of merit can only be further increased at a given mass flow rate by utilizing the combustion enthalpy as much as possible in the combustion chamber (e.g. by a more efficient precombustion), or by injecting additional external energy (e.g. external nuclear–thermal energy), or by reducing heat losses.

Internal Efficiency

The internal efficiency of a propulsion engine is a measure for the effectiveness of converting the energy released into the combustion chamber into exhaust energy (kinetic energy of the exhausted jet). We recall from Eq. (2.7.2) that it is given as $\eta_{int} = \frac{1}{2} m_p v_*^2 / E_0$. For an ideally adapted engine without the supply of external energy, one then obtains

$$\eta_{int} = \frac{m_p v_*^2}{2\eta_c H_p} = \left(\frac{v_*}{v_{e,max}}\right)^2 \quad @ \ E_{ext} = 0 \tag{4.1.24}$$

> **Example**
> The SSME cryogenic engine of the Space Shuttle burns H_2/O_2 with $\eta_c H_p/m_p = 13.4$ MJ kg^{-1}, i.e. $v_{e,max} = 5.18$ km s^{-1}, and exhausts the propellant with $v_e = 4.44$ km s^{-1}. The internal efficiency for an ideally adapted nozzle $v_* = v_e$ therefore is $\eta_{int} = 0.735$, which is currently the best value for chemical engines.

4.1.8
Engine Thrust

We saw above that the figure of merit of an engine, $I_{sp,max}$, does not depend on the engine design, the chamber pressure in particular. The all-important thrust on the other hand does. How does the thrust depend on the engine design? We recall from Eq. (1.2.6) that the thrust is generally given by

$$F_* = \dot{m}_p v_e + (p_e - p_\infty) A_e$$

Inserting $v_e = a_0 \sqrt{n\eta_e}$ from Eq. (4.1.14) we get for the thrust

$$F_* = \dot{m}_p a_0 \sqrt{n\eta_e} + (p_e - p_\infty) A_e \tag{4.1.25}$$

How is the mass flow rate $\dot{m}_p$ related to chamber pressure p_0? To figure this out, we apply Eq. (4.1.12)

$$\dot{m}_p a_0 \sqrt{n} = a_0^2 \rho_0 A_t n \sqrt{\frac{n^n}{(n+1)^{n+1}}}$$

Since $a_0^2 \rho_0 = \kappa p_0$ (see Eq. (4.1.8)) and because $\kappa n = n+2$ we get the important relation

$$\dot{m}_p = \frac{A_t C_\infty}{a_0 \sqrt{n}} p_0 \tag{4.1.26}$$

with

$$C_\infty := (n+2) \sqrt{\frac{n^n}{(n+1)^{n+1}}} = \sqrt{\frac{2\kappa^2}{\kappa - 1} \left(\frac{2}{\kappa + 1}\right)^{\frac{\kappa+1}{\kappa-1}}} \quad \text{infinite-expansion coefficient}$$

Remark: *In the literature the quantity* $\Gamma = C_\infty/\sqrt{n+2}$ *is sometimes called the Vandenkerckhove function.*

We call C_∞ the infinite-expansion coefficient for reasons that will become clear in Section 4.2.4. For $n = 8$ ($\kappa = 1.25$) it has the value $C_\infty = 2.0810$. Because of Eq. (4.1.26) we can now rewrite the thrust more conveniently for engineering purposes as

$$F_* = p_0 A_t C_\infty \sqrt{\eta_e} + (p_e - p_\infty) A_e \tag{4.1.27}$$

Hence thrust increases roughly proportionally with chamber pressure. Roughly, because η_e is also weakly dependent on p_0. We will go into details in a moment. Apparently thrust also increases with the cross section of the throat. However, thrust cannot be increased infinitely by simply increasing the cross section of the throat: according to Eq. (4.1.26) a higher A_t lowers p_0 proportionally. You cannot beat physics!

From Eq. (4.1.27), with $p_e = p_\infty \to \eta_e = \eta_\infty$ for the thrust of an ideally adapted nozzle, we find

$$F_* = p_0 A_t C_\infty \sqrt{\eta_\infty} \quad @ \; p_e = p_\infty \tag{4.1.28}$$

In vacuum the thrust of an ideally adapted engine achieves its maximum value

$$F_* = p_0 A_t C_\infty \quad @ \; p_e = p_\infty = 0 \tag{4.1.29}$$

This is merely a theoretical value, as the nozzle exit cross section A_e would then grow infinitely big, as we will see in a moment (Eq. (4.2.3)).

While the maximum specific impulse is independent of the chamber pressure (see Eq. (4.1.22)) the maximum achievable thrust is nearly proportional to p_0. The reason for this is that for a given chamber due to Eq. (4.1.26) $\dot{m}_p \propto p_0$ and therefore $F_* = \dot{m}_p I_{sp,max} \propto p_0 \cdot const$.

Variation of Thrust with Chamber Pressure

How does the thrust and the specific impulse depend on the chamber pressure in general? From a practical point of view the answer is of significant importance, since the chamber pressure, which via Eq. (4.1.26) depends solely on the fuel flow, is the only continuously variable parameter in flight. The answer follows from Eq. (4.1.27) with $\eta_e = 1 - (p_e/p_0)^{\frac{2}{n+2}}$

$$F_* = p_0 A_t C_\infty \sqrt{1 - \left(\frac{p_e}{p_0}\right)^{\frac{2}{n+2}}} + (p_e - p_\infty) A_e \tag{4.1.30}$$

When calculating F_* from Eq. (4.1.30) for a given $p_0 \to p_0 + \Delta p_0$ one has to determine whether p_e also varies with p_0. In the following we assume that the engine nozzle is always ideally adapted, i.e. $p_e \approx p_\infty = const$. We therefore derive from Eq. (4.1.27) the relative thrust variation very generally as

$$\frac{dF_*}{F_*} = \frac{dp_0}{p_0} + \frac{1}{2}\frac{d\eta_e}{\eta_e} = \left(1 + \frac{p_0}{2\eta_e}\frac{d\eta_e}{dp_0}\right)\frac{dp_0}{p_0}$$

From $\eta_e = 1 - (p_e/p_0)^{\frac{2}{n+2}}$ it follows that

$$\frac{p_0}{2\eta_e}\frac{d\eta_e}{dp_0} = \frac{1}{n+2}\frac{1-\eta_e}{\eta_e} = \frac{1}{n+2}\frac{1}{(p_0/p_e)^{\frac{2}{n+2}} - 1}$$

and therefore

$$\frac{dF_*}{F_*} = \left(1 + \frac{1}{n+2}\frac{1-\eta_e}{\eta_e}\right)\frac{dp_0}{p_0} \quad \text{relative thrust variation} \quad (4.1.31)$$

Observe that due to Eq. (4.1.26) $dp_0/p_0 = d\dot{m}_p/\dot{m}_p$ and therefore any change in thrust is due to a change in mass flow rate. Hence

$$\frac{dF_*}{F_*} = \left(1 + \frac{1}{n+2}\frac{1-\eta_e}{\eta_e}\right)\frac{d\dot{m}_p}{\dot{m}_p}$$

Because of $F_* \propto p_0 I_{sp}$ we finally obtain for the relative specific impulse variation

$$\frac{dI_{sp}}{I_{sp}} = \frac{dF_*}{F_*} - \frac{dp_0}{p_0} = \frac{1}{n+2}\frac{1-\eta_e}{\eta_e}\frac{dp_0}{p_0} \quad (4.1.32)$$

It can be shown (exercise, Problem 4.2) that, if the chamber pressure varies, i.e. if for instance p_0 is continuously adjusted in flight, and the shape of the nozzle remains unchanged, such that $p_e \neq p_\infty = const$, then $p_e \propto p_0$ and Eqs. (4.1.31) and (4.1.32) remain unchanged. Therefore, at any rate ($p_e \propto p_0$ or $p_e = const$) we find for typical pressure ratios $p_e/p_0 \approx 1/70$ and for typical $n = 8$

$$\frac{dF_*}{F_*} = 1.075\frac{dp_0}{p_0} \quad \text{and} \quad \frac{dI_{sp}}{I_{sp}} \approx 0.075\frac{dp_0}{p_0} \quad (4.1.33)$$

Moreover, in the limiting case $\eta_e \to 1$, i.e. in vacuum, we derive from the above equations the following dependences for their maximum values:

$$\frac{dF_*}{F_*} = \frac{dp_0}{p_0} \quad \to \quad F_* \propto p_0$$

$$\frac{dI_{sp}}{dp_0} = 0 \quad \to \quad I_{sp} = const$$

both of which are in agreement with our results in Eqs. (4.1.29) and (4.1.22).

4.2
Thermal Engine Design

Engine design deals with the layout of the engine to achieve maximum total thrust. We already know two engine parameters which determine geometry: the cross-sectional area of the combustion chamber throat, A_t, and the cross-sectional area of the nozzle exit, A_e. In order to optimize the engine layout, we need to design both combustion chamber and nozzle. Let's go into details.

4.2.1
Combustion Chamber Design

Chamber pressure, propellant mass flow rate, and cross section of the throat need to be in balance to each other. With a larger cross section of the throat, the mass flow rate has to rise, in order to maintain the chamber pressure. Equation (4.1.26) describes the interplay between these three parameters. Their ratio determines the so-called characteristic velocity

$$c^* := \frac{p_0 A_t}{\dot{m}_p} = \frac{a_0 \sqrt{n}}{C_\infty} = \frac{1}{C_\infty}\sqrt{\frac{2h_0}{M_p}} = \frac{I_{sp,\max}}{g_0 C_\infty} = const \quad \text{characteristic velocity} \quad (4.2.1)$$

Because both C_∞ and $I_{sp,\max}$ depend only on propellant properties, c^* is constant and an alternative figure of merit for the engine. Equation (4.2.1) determines the design of the combustion chamber. On the left side of the equation we have the technically variable parameters, on the right side we have only propellant-specific parameters. For instance, with a given (turbo pumps, combustion rate) mass flow rate $\dot{m}_p$, and a maximum admissible pressure of the combustion chamber p_0, we can determine the necessary cross section A_t of the throat

$$A_t = \frac{\dot{m}_p}{C_\infty p_0}\sqrt{\frac{2h_0}{M_p}} = const$$

The chamber volume V_0 remains undefined by these considerations. It only comes into play when internal combustion processes are analyzed. Then the so-called characteristic length $L_c = V_0/A_t$ plays an important role for resonances in the chamber. A chamber stability analysis, however, is beyond the scope of this book.

4.2.2
Nozzle Design

The nozzle expands the propellant gas from the throat to the nozzle exit such that ideally $p_e = p_\infty$. For a given thermal efficiency η_e the exit cross section and the cross section of the throat need to be in a specific ratio. Applying Eq. (4.1.12) to the nozzle exit and from Eq. (4.1.10) we find the following for this expansion ratio:

$$\varepsilon := \frac{A_e}{A_t} = \frac{\dot{m}_p\, A_e}{A_t\, \dot{m}_p} = \frac{a_0 \rho_0 \sqrt{\left(\frac{n}{n+1}\right)^{n+1}}}{a_0 \rho_0 \sqrt{n\eta_e (1-\eta_e)^n}}$$

So we have

$$\varepsilon = \frac{C_\infty}{n+2} \frac{1}{\sqrt{(1-\eta_e)^n\, \eta_e}} \quad \text{expansion ratio} \qquad (4.2.2)$$

with

$$\eta_e = 1 - \left(\frac{p_e}{p_0}\right)^{\frac{2}{n+2}}$$

Remark: *Equation (4.2.2) is expressed in the literature as*

$$\varepsilon = \frac{A_e}{A_t} = \sqrt{\frac{\kappa-1}{2}\left[\frac{2}{(\kappa+1)}\right]^{\frac{\kappa+1}{\kappa-1}}} \cdot \left(\frac{p_0}{p_e}\right)^{\frac{1}{\kappa}} \left[1 - \left(\frac{p_e}{p_0}\right)^{\frac{\kappa-1}{\kappa}}\right]^{-1/2}$$

In order to find the optimum expansion ratio ε to be manufactured, $p_e = p_\infty$ and hence $\eta_e \to \eta_\infty$. With $\eta_\infty = 1 - (p_\infty/p_0)^{\frac{2}{n+2}}$ and A_t given, we hence can calculate the optimum expansion ratio to be

$$\frac{A_{e,opt}}{A_t} = \frac{C_\infty}{n+2} \frac{1}{\sqrt{(1-\eta_\infty)^n\, \eta_\infty}} \quad \text{optimum expansion ratio} \qquad (4.2.3)$$

With this we have achieved our long-term goal (see Section 4.1.6) to find the optimum parameters $p_{e,opt}$, $v_{e,opt}$, $A_{e,opt}$. The engine design results of Sections 4.2.1 and 4.2.2 following from that are summarized in Figure 4.6, which displays a workflow to determine the optimum engine design.

4.2.3
Thrust Coefficient C_f

Equation (4.1.27) gives rise to the definition of the so-called thrust coefficient C_f, which is of practical importance for engine design:

$$C_f := \frac{F_*}{p_0 A_t} \qquad (4.2.4)$$

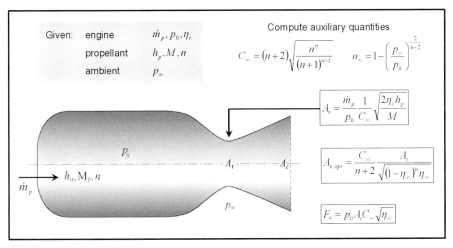

Figure 4.6 Workflow to determine the optimum engine design.

The rationale is that the thrust $F_* = C_f \cdot p_0 A_t$ now is simply a product of chamber-specific parameters $p_0 A_t$ and one characteristic parameter for the nozzle, namely C_f. Comparing Eq. (4.1.27) with Eq. (4.2.4) and applying Eq. (4.2.1), we derive the following expression for C_f:

$$C_f = \frac{v_*}{c^*} = C_\infty \sqrt{\eta_e} + \frac{p_e - p_\infty}{p_0} \frac{A_e}{A_t} \quad \text{thrust coefficient} \quad (4.2.5)$$

As we will see in Section 4.2.5, the thrust coefficient itself does not describe the nozzle efficiency, but has a one-to-one correspondence with it. So C_f just has engineering and not physical relevance.

4.2.4
Thrust Performance

In order to see how thrust depends on the expansion ratio and hence on the exit cross section, we have to determine the function $F_* = F_*(A_e)$ or $F_* = F_*(\varepsilon)$, respectively, by solving Eq. (4.2.2) with regard to $\eta_e = \eta_e(\varepsilon)$, and plug the result into Eq. (4.2.5). However, this can only be done numerically, and the result of which, $C_f = C_f(\varepsilon)$, is depicted in Fig. 4.7. The horizontal curves display the change of the thrust coefficient with the expansion ratio ε at a given chamber pressure ratio p_0/p_∞. The curves confirm our theoretical conclusion from Section 4.1.6 that maximum thrust occurs at $p_e = p_\infty$ indicated by the line crossing all others. The region around the trust maximum, in particular for high combustion chamber pressures, is so flat that a slightly suboptimally adapted nozzle does not gravely reduce thrust. Nevertheless, in course of the ascent through the atmosphere into space, over-expansion and under-expansion losses add up to typically about 10%.

Figure 4.7 shows that the thrust coefficient C_f steadily increases until for $A_e/A_t \to \infty$, i.e. in vacuum ($\eta_e \to \eta_\infty = 1$), $C_f = C_\infty = 2.0810$. In this case the infinitely sized nozzle allows the exhaust jet to expand infinitely to zero ambient pressure, which is why we call C_∞ infinite-expansion coefficient. Likewise the thrust tends to its limiting value of Eq. (4.1.29).

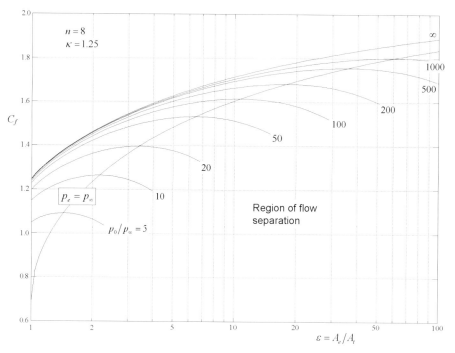

Figure 4.7 The dependence of the thrust coefficient C_f on the expansion ratio ε at $n = 8$ ($\kappa = 1.25$).

4.2.5
Nozzle Efficiency C_n

The thrust coefficient might be of engineering interest, but it does not provide an answer to the ultimate question: What is the thrust that a nozzle adds to the engine performance? We therefore define the nozzle coefficient (nozzle efficiency) as follows:

$$C_n := \frac{\text{thrust of an engine with nozzle}}{\text{thrust of an engine without nozzle}} = \frac{F_*}{F_{*,plain}} \quad (4.2.6)$$

According to Eq. (1.2.6), an engine without nozzle would provide the thrust

$$F_{*,plain} = \dot{m}_p v_t + (p_t - p_\infty) A_t$$

From Eqs. (4.1.11), (4.1.3), and (4.1.26) we find for the pressure at the throat

$$p_t = p_0 \left(\frac{n}{n+1}\right)^{\frac{n+2}{2}} = p_0 \frac{n}{\sqrt{n+1}} \frac{C_\infty}{n+2} \approx 0.55 \cdot p_0$$

With the relation $v_t = a_0 \sqrt{n}/\sqrt{n+1}$ from Eq. (4.1.13) and with $\dot{m}_p a_0 \sqrt{n} = p_0 A_t C_\infty$ from Eq. (4.1.25) we can rewrite this to

$$F_{*,plain} = p_0 A_t C_\infty \left(\frac{1}{\sqrt{n+1}} + \frac{n}{\sqrt{n+1}} \frac{1}{n+2} - \frac{p_\infty}{C_\infty p_0}\right)$$

$$= p_0 A_t \left(\frac{2 C_\infty \sqrt{n+1}}{n+2} - \frac{p_\infty}{p_0}\right) \quad (4.2.7)$$

We investigate the first term in the brackets and find with the definition Eq. (4.1.26) of C_∞ that

$$\frac{n+2}{2 C_\infty \sqrt{n+1}} = \frac{1}{2}\left(1 + \frac{1}{n}\right)^{n/2} \approx \frac{\sqrt{e}}{2}\left(1 - \frac{1}{4(n+1)}\right)$$

$$\approx \frac{\sqrt{e}}{2} \frac{35}{36}\left(1 + \frac{n-8}{315}\right) \approx 0.8015 \approx const$$

So, at any practical rate, this term is independent of n. Therefore we arrive at the interesting result that an engine without nozzle provides the thrust

$$F_{*,plain} = p_0 A_t \left(1.248 - \frac{p_\infty}{p_0}\right) \approx 1.248 \cdot p_0 A_t \quad (4.2.8)$$

The latter is due to $p_\infty/p_0 \approx 0.01$. For the thrust of an engine with a nozzle, Eq. (4.2.4) states that $F_* = p_0 A_t C_f$. If we insert this equation and Eq. (4.2.7) into Eq. (4.2.6) we obtain

$$C_n = \frac{C_f}{\frac{2 C_\infty \sqrt{n+1}}{n+2} - p_\infty/p_0} \approx \frac{C_f}{1.248 - p_\infty/p_0} \quad (4.2.9)$$

with

$$C_f = \frac{v_*}{c^*} = C_\infty \sqrt{\eta_e} + \frac{p_e - p_\infty}{p_0} \frac{A_e}{A_t}$$

We therefore arrive at the approximation adequate at any practical rate

$$\boxed{C_n \approx 0.81 \cdot C_f} \quad \text{nozzle efficiency} \quad (4.2.10)$$

The nozzle efficiency thus can be estimated to be 81% of the thrust coefficient. Because $F_* = C_n \cdot F_{*,plain}$ and in comparison to $F_* = C_f \cdot p_0 A_t$, the nozzle coefficient C_n is a more vivid substitute for C_f. The two are virtually in direct relation to each other.

If the nozzle is ideally adapted, then $C_f = C_\infty \sqrt{\eta_\infty}$, and we get

$$C_n = \frac{C_\infty \sqrt{\eta_\infty}}{1.248 - p_\infty/p_0} \quad @\ p_e = p_\infty \quad (4.2.11)$$

For a nozzle ideally adapted to outer space, one gets from Eq. (4.2.9) with $C_f = C_\infty$

$$C_n = \frac{n+2}{2\sqrt{n+1}} \approx \frac{5}{3} = 1.667 \quad @\ p_e = p_\infty = 0 \quad (4.2.12)$$

So, in space a nozzle ideally adapted to vacuum always increases the thrust by approximately 67% independent of the pressure in its combustion chamber. So nozzle efficiency is always lower than 67%. However, this is an ideal case because a nozzle adapted to vacuum in space would have an infinite size.

> **Example**
> The Space Shuttle SSME LH2/LOX engine has I_{sp} (vacuum) = 455 s, I_{sp} (1 bar) = 363 s, and $c^* = 2330\ \text{m s}^{-1}$. Therefore, during ascent through Earth's atmosphere, one gets $1.53 \leq C_f \leq 1.91$ and thus $1.23 \leq C_n \leq 1.53$. So, a SSME nozzle provides at sea level 23% and in space 53% more engine thrust.

4.2.6
Nozzle Shape

One essential result of our engine design considerations is that, according to Eq. (4.1.27), thrust only depends on the expansion ratio ε via the parameter η_e (see Fig. 4.7). It might be surprising to see that the design of the nozzle is only determined by the areal ratio of its end faces, and not on its exact shape in between. In fact, from a thermodynamic point of view, the shape of the nozzle casing is irrelevant as long as the gas expands steadily and adiabatically. So any smooth contour that precludes shock formation will do. This purports that the nozzle should not expand too rapidly behind the throat, implying that the angle of the nozzle casing with respect to the nozzle axis should not be too large. From an engineering point of view a well-designed shape can reduce the mechanical strain of the nozzle along its axis. We do not want to go into details about these two implications for the nozzle shape, but mention just two important cases.

The familiar bell-shaped nozzle (see Fig. 4.8) has an ideal shape because the gases quickly expand conically behind the throat. The longer the gases run along the nozzle, the less divergence occurs because of the bell shape, and at the exit the gases are expelled almost parallel to the nozzle axis.

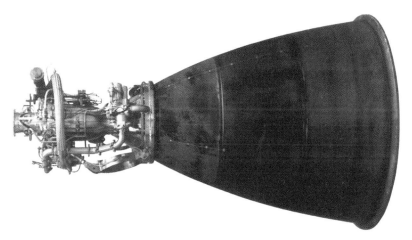

Figure 4.8 Aestus upper-stage engine from Ariane 5.

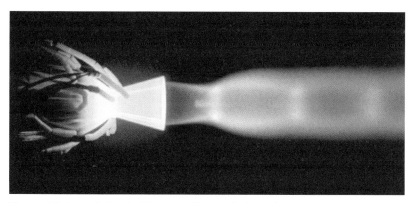

Figure 4.9 A small chemical thruster with a conical nozzle.

If mechanical strain is not an issue, for instance for engines with little thrust, simple conical nozzles are a practical solution (see Fig. 4.9). However, they lead to some thrust losses due to jet straying. Taking into account that the exit area is a ball segment (see remark following Eq. (4.1.9)), the corresponding divergence loss factor can be calculated as follows (exercise, Problem 4.3):

$$\eta_{div} = \frac{1+\cos\alpha}{2} \quad \text{momentum thrust loss factor} \tag{4.2.13}$$

where α is half the aperture angle of the cone. With typical values of $\alpha = 12°\text{--}18°$, the loss factor is about $\eta_{div} = 0.989\text{--}0.975$. So the losses amount to 1.1%–2.5%. Observe that divergence losses only affect momentum thrust, not pressure thrust. So for any straying jets Eqs. (4.1.27) and (1.2.8) should be modified to

$$F_* = \eta_{div} p_0 A_t C_\infty \sqrt{\eta_e} + (p_e - p_\infty) A_e \tag{4.2.14}$$

$$v_* = \eta_{div} v_e + (p_e - p_\infty)\frac{A_e}{\dot{m}_p} \tag{4.2.15}$$

Problems

Problem 4.1 Gas Velocity-Pressure Relation in a Nozzle

Prove explicitely with Eqs. (4.1.7) and (4.1.3) and by applying Eq. (4.1.10) that the following holds for the gas velocity–pressure dependence in a nozzle:

$$dv = -\frac{A}{\dot{m}_p} dp$$

Problem 4.2 Thrust-Pressure Relation of an Engine

Show that, for a given thermal propulsion engine with nothing changed, Eqs. (4.1.31) and (4.1.32) remain unchanged.

Problem 4.3 Thrust Loss Factor of a Conical Nozzle

(a) Assume that a conical nozzle has an aperture angle of 2α. Show that, for the thrust loss in the exit plane of the nozzle, the momentum thrust loss factor due to conical straying is

$$\eta_{div} = \frac{2\cos\alpha}{1+\cos\alpha}$$

(b) According to the remark following Eq. (4.1.9), however, the effective exhaust area is not the exit plane of the nozzle but a sphere segment, the center of which is the imaginary point where the gas flow lines converge. Show that, if integrating over this surface by using polar coordinates, the momentum thrust loss factor actually is Eq. (4.2.13)

$$\eta_{div} = \frac{1+\cos\alpha}{2}$$

(c) Convince yourself that the difference between these two results is very small for small cone angles, i.e. their series expansion

$$1 - \frac{\alpha^2}{4} + O\left(\alpha^4\right) \quad @ \ \alpha \to 0$$

differ only in $O(\alpha^4)$ such that

$$\frac{1+\cos\alpha}{2} \geq \sqrt{\cos\alpha} \geq \frac{2\cos\alpha}{1+\cos\alpha}$$

5
Electric Propulsion

5.1
Overview

Electric propulsion engines differ from thermal engines in that, amongst other things, the propellant does not serve as an energy source to heat and accelerate the propellant mass in the combustion chamber. Rather acceleration is achieved by accelerating ions in an electric field, the energy of which needs to be provided externally by an electric current source. This is both an advantage and a disadvantage at the same time. The advantage is that, theoretically, any amount of energy can be applied to the propellant mass, which would in principle permit unlimited exhaust speeds, hence unlimited specific impulse, and therefore unlimited efficiency of the engine. The disadvantage is that the structural mass of the rocket stage increases due to the additional mass of the electric generator, which directly trades with payload mass. Massive generators are required especially for high-I_{sp} engines, so their additional mass may outweigh propellant savings. Therefore, comparisons between different propulsion systems always need to consider the total propulsion system mass: propulsion system, consumed propellant, plus energy supply system.

Another disadvantage of electric propulsion is that ions repel each other, permitting only very low particle densities in the engine chamber, which in turn leads to mass flow densities many orders of magnitude lower than those of chemical engines. This results in very small thrusts. For this reason electric propulsions will not replace launch thrusters in the long run, as thrust is a key variable for launch. This is apart from the problem that their exit pressure is much lower than ambient pressure, which by itself rules out their employment for launch. On the other hand, once outer space has been reached, and especially with interplanetary flights with long flight times, continuous operation with a highly effective ion engine often pays off in comparison with a two-impulse transfer with low-efficiency chemical propulsions. This is shown in Tab. 5.1 with the example of a Mars mission.

Astronautics. Ulrich Walter
Copyright © 2008 WILEY-VCH Verlag GmbH & Co. KGaA, Weinheim
ISBN: 978-3-527-40685-2

Table 5.1 Comparison of all-chemical and ion–solar engines for the example of a Mars mission.

Parameter	Spacecraft description	
	All-chemical voyager (190 days transit)	Ion–solar voyager (250 days transit)
Injected weight (not including launch vehicle)	3540 kg	4350 kg
Power level	all chemical	23 kW
Approach velocity	4.3 km s^{-1}	1.8 km s^{-1}
Weight at approach	2400 kg	2330 kg
Weight in orbit (excluding retro inert weight)	840 kg	1630 kg
Orbit spacecraft fraction	0.35	0.70
Lander weight	1040 kg	1040 kg
Scientific payload	210 kg	810 kg
Percent scientific payload weight at approach	8.9%	34.9%

5.2
Ion Engine

Let's have a closer look at ion propulsion as it becomes more and more of practical importance. It is based on the acceleration of cold plasma in a high electric field (see Fig. 5.1). The inflowing propellant atoms are ionized in a relatively voluminous reaction chamber by hitting them with circulating elec-

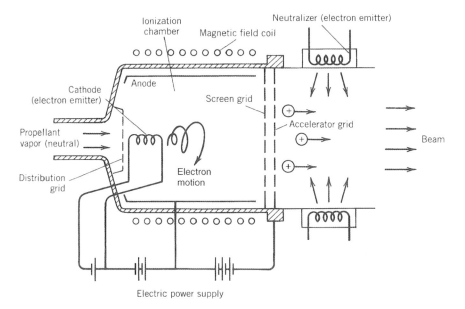

Figure 5.1 Schematic of an ion engine. The electrostatic zone is just between the screen grids.

trons, which knock outer-shell electrons out of the atoms. "Cold" here means that, during the ionization process, no internal states of the atom are excited. This is achieved by using noble gas atoms, which occur only as single atoms and hence quite naturally lack any rotational or vibrational modes to be excited. Cold, singly charged ions then enter a very narrow electrostatic zone where they are accelerated in a high electric field. After emerging from this zone, they are neutralized with the electrons separated earlier in the reaction chamber.

What is the thrust of such an ion engine? Because the mass flow density is extremely small, the exit pressure is comparable to vacuum pressure. So we only have to consider momentum thrust, for which we have already derived a general expression in Eq. (1.2.10)

$$F_* = F_e = \dot{m}_p v_e = \rho_e A_e v_e^2 \qquad (5.2.1)$$

where ρ_e is the mass density at the exit of the engine with cross section A_e, and $\dot{m}_p > 0$ the propellant mass flow rate.

5.2.1
Ion Acceleration

The crucial part of the engine for thrust generation is the acceleration zone, which determines the required parameters $\dot{m}_p, \rho_e, v_e$. In order to derive them, we need to understand the charge distribution in the electrostatic zone in detail (see Fig. 5.2). We assume the engine axis as the x axis along which the ions are moving. They enter the zone through bores in the anode plate at $x = 0$. Then they are accelerated by an electric field $V(x)$ (the form of which still has

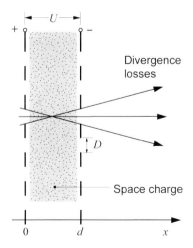

Figure 5.2 Geometrical and electric relations in the electrostatic zone.

to be derived) between anode and cathode with separation d. The acceleration voltage across the plates is $U := V(d)$. Finally, the accelerated ions exit the zone through bores in the cathode plate at $x = d$.

The charges flowing through this zone with velocity v generate a space charge with density q, which reduces the electric field $V(x)$ so that the inflow of the charges is slightly reduced. The balance between charge and electric potential is physically described by the Poisson equation (cf. Eq. (7.1.2))

$$\frac{d^2 V}{dx^2} = \frac{q(x)}{\varepsilon_0} \tag{5.2.2}$$

with ε_0 being the absolute permittivity. Because of charge conservation along the whole acceleration distance $0 \leq x \leq d$, the resulting charge flow density $j = qv$ must be constant

$$j = qv = const \quad \textbf{charge continuity equation} \tag{5.2.3}$$

This is the equivalent equation to the mass continuity equation (1.2.9) for mass conservation. The particle velocity v at any place within the acceleration zone is determined by the balance between kinetic energy and potential electric energy, i.e.

$$\frac{1}{2} m_{ion} v^2 = eV \tag{5.2.4}$$

with m_{ion} the mass of an ion, and e the charge of the single charged ion, which is the elementary charge. At the end of the acceleration distance, the exhaust velocity is determined from Eq. (5.2.4) and $U := V(d)$ as

$$v_e = \sqrt{\frac{2eU}{m_{ion}}} \tag{5.2.5}$$

With Eq. (5.2.4), the charge density at any position in the zone can be determined through Eq. (5.2.3) as

$$q = j \sqrt{\frac{m_{ion}}{2eV}} \tag{5.2.6}$$

Inserting this expression into the master equation (5.2.2) for the electric potential V we obtain

$$\sqrt{V} \frac{d^2 V}{dx^2} = \frac{j}{\varepsilon_0} \sqrt{\frac{m_{ion}}{2e}} = const$$

This is a differential equation of second degree for $V(x)$. As the equation is purely multiplicative, we can make a general power-law ansatz

$$V = V_0 x^n$$

This results in

$$\sqrt{V}\frac{d^2V}{dx^2} = n(n-1) V_0^{\frac{3}{2}} x^{\frac{n}{2}+n-2} = \frac{j}{\varepsilon_0}\sqrt{\frac{m_{ion}}{2e}}$$

For the second term to be independent of x, the exponent of x has to vanish, resulting in

$$n = \frac{4}{3} \quad \text{and} \quad V_0^{3/2} = \frac{9}{4}\frac{j}{\varepsilon_0}\sqrt{\frac{m_{ion}}{2e}}$$

From this it follows that

$$V^{3/2} = \frac{9}{4}\frac{j}{\varepsilon_0}\sqrt{\frac{m_{ion}}{2e}} x^2 \qquad (5.2.7)$$

This is the wanted electric potential distribution $V(x) \propto x^{4/3}$.

Remark: *In case we would not have any charges in the electrostatic zone, Eq. (5.2.2) would read $d^2V/dx^2 = 0$ and hence the familiar $V(x) \propto x$ in vacuum would result. The weak modification of the linear electric field distribution to a $x^{4/3}$ behavior is obviously caused by the space charge of the transiting ions.*

5.2.2
Thrust of an Ion Engine

At location $x = d$ the acceleration potential is $V = U$ and we derive from Eq. (5.2.7) for the flow density

$$j_e = \frac{4\varepsilon_0}{9}\sqrt{\frac{2e}{m_{ion}}}\frac{U^{3/2}}{d^2} \quad \textbf{Child–Langmuir law} \qquad (5.2.8)$$

The charge density at the exit we are looking for is calculated with Eq. (5.2.6) for $x = d$ and $V = U$ as

$$q_e = \frac{4}{9}\frac{\varepsilon_0 U}{d^2} \qquad (5.2.9)$$

To calculate the mass flow rate $\dot{m}_p = \rho_e A_e v_e$, we need a relationship between the mass density ρ and charge density q of singly charged ions. This can be derived if we consider that both follow quite generally from the particle density n: $\rho = m_{ion} n$ and $q = en$. Therefore the following relation holds:

$$\rho = \frac{m_{ion}}{e} q \qquad (5.2.10)$$

in particular at the exit. With this and Eqs. (5.2.9) and (5.2.5), we find for the mass flow rate

$$\dot{m}_p = \frac{4}{9}\sqrt{\frac{2m_{ion}}{e}\frac{\varepsilon_0 A_e}{d^2}} U^{3/2} \tag{5.2.11}$$

We are now able to calculate the thrust. Before doing so, it should be mentioned that because of Eq. (5.2.10), we may express the thrust in Eq. (5.2.1) with Eqs. (5.2.5) and (5.2.10) in a simpler way as follows:

$$F_* = F_e = 2q_e A_e U \tag{5.2.12}$$

If we insert Eq. (5.2.9) here, we immediately get the wanted result:

$$F_* = \frac{8}{9}\varepsilon_0 \left(\frac{U}{d}\right)^2 A_e = \frac{8}{9}\varepsilon_0 A_e E^2 = \frac{16}{9} A_e \varepsilon_E \tag{5.2.13}$$

Here $E = U/d$ is the electric field, and $\varepsilon_E = \varepsilon_0 E^2/2$ is the energy density of the free electric field. The exit surface is obviously the sum of all n bores with the diameter D in the cathode plate, i.e.

$$A_e = \frac{n}{4}\pi D^2$$

Thrust thus increases quadratically with the applied acceleration voltage and inversely proportionally to the square of the distance between the electrodes. So the perfect ion engine has the highest possible voltage with the smallest possible acceleration distance. A practical limit is reached when electrical flash-overs happen at about 100 kV cm^{-1}. So with exit areas of typically 300 cm^2 for today's thrusters, their thrust is limited to 10–100 mN.

Why is this thrust smaller by several orders of magnitude than the thrust of chemical propulsions of comparable size, although the exhaust velocity of ion thrusters is $v_e = $ 30–40 km s^{-1} and thus about ten times larger? Owing to $F_e = \dot{m}_p v_e$ this must be due to an extremely low mass flow rate, which in turn is due to a low gas density, which is due to the strong charge repulsion in the acceleration zone, as already mentioned. If the propellant gas density is extremely low, the gas exit pressure is far below ambient pressure. Ambient gas would flow into the ion chamber bringing the ionization to a stall. Therefore ion thrusters are only applicable if the environmental gas pressure is much lower than the gas exit pressure – they work best in vacuum. Even there the differential pressure at the exit is so small, that it makes the pressure thrust negligible compared to momentum thrust.

5.2.3
Internal Efficiency

Note that ion thrust is independent of ion mass. So, the kind of propellant does not matter for thrust. However it does matter for engine efficiency because the energy applied goes into kinetic energy of the accelerated ions and into internal excitation energy $E_{ion,int}$ of the ions. The engine efficiency, which is the ratio of the exhaust jet energy to the total applied energy, therefore reads for all electric propulsions (apart from resistojets) according to Eq. (2.7.2) as follows:

$$\eta_{int} = \frac{\frac{1}{2}m_{ion}v_*^2}{E_0} = \frac{\frac{1}{2}m_{ion}v_e^2}{\frac{1}{2}m_{ion}v_e^2 + E_{ion,int}} \qquad (5.2.14)$$

From this we derive that the engine efficiency increases with increasing ion mass. Since the efficiency is constant in time and because power physically is the time derivative of energy, we can rewrite Eq. (5.2.14) as

$$\eta_{int} = \frac{P_e}{P_{tot}}$$

We recall from Eq. (1.3.1) that we have quite generally

$$\frac{F_e}{P_e} = \frac{2}{v_e}$$

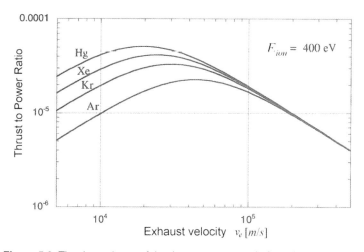

Figure 5.3 The dependence of the thrust-to-power ratio from the exhaust velocity.

From this and with Eq. (5.2.14) we find for the thrust-to-total-power ratio, which describes the thrust received from the total electrical power consumed,

$$\frac{F_e}{P_{tot}} = \frac{F_e}{P_e}\eta_{int} = \frac{m_{ion}v_e}{\frac{1}{2}m_{ion}v_e^2 + E_{ion,int}} \quad (5.2.15)$$

This important relation is depicted in Figure 5.3. It has a maximum at $v_{e,max} = \sqrt{2E_{ion,int}/m_{ion}}$ with thrust maximum at $F_{e,max}/P_e = 1/v_{e,max} = \sqrt{m_{ion}/2E_{ion,int}}$. The engine efficiency therefore increases with increasing ion mass and decreasing exhaust velocity. For this reason, ion thrusters are fueled with noble gases with large ion masses, in particular xenon. This yields more thrust at the same input power.

5.3
Electric Propulsion Optimization

Electric propulsion is special in that the exhaust velocity and thus the specific impulse depends on the acceleration voltage and hence is variable. Due to the rocket equation (2.2.4) the propellant demand decreases exponentially with I_{sp} (see Fig. 5.4). Would it then be feasible to get a steadily increasing payload ratio with an increasing acceleration voltage? This, unfortunately, is not the case, because with an increasing voltage also the mass of the power

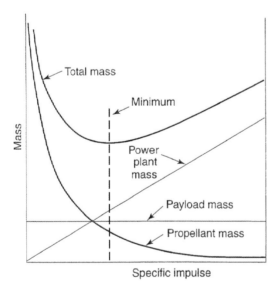

Figure 5.4 The mass of a S/C with electrical engines has a minimum, as with increasing specific impulse the required propellant exponentially decreases, but the mass of the electric generator linearly increases.

supply system increases which trades directly with payload mass. So, with a given I_{sp}, there is an optimum where the total engine plus propellant mass of a spacecraft becomes minimal (see Fig. 5.4). We now want to explore the S/C system layout at this optimum.

The following contributions add to the total mass of a S/C with an electrical propulsion system

$$m_0 = m_p + m_s + m_L + m_g = m_p + m_f \qquad (5.3.1)$$

with m_s the structural mass, m_L the payload mass, and m_g the mass of the power plant. The latter can be, for instance, a radio-isotope thermoelectric generator (RTG) or solar cells. If P_g is the electric power provided, then the so-called

$$\alpha := \frac{P_g}{m_g} \quad \textbf{specific power} \qquad (5.3.2)$$

describes the mass-specific power output of the plant. Current plants are of order 100–200 W kg^{-1}. The supplied power is converted into exhaust jet energy $\frac{1}{2}\dot{m}_p v_e^2$ (see Eq. (1.3.2)), at a given efficiency of the thruster η_t, i.e.

$$\frac{1}{2}\dot{m}_p v_e^2 = \frac{m_p}{2t_p}v_e^2 = \eta_t \alpha m_g \qquad (5.3.3)$$

Here we have assumed a continuous mass flow along the total combustion time t_p. If we define the so-called

$$v_c := \sqrt{2\alpha t_p \eta_t} \quad \textbf{characteristic velocity} \qquad (5.3.4)$$

we get the following relation between propellant mass and power plant mass

$$m_p = \frac{m_g}{\gamma^2} \qquad (5.3.5)$$

with

$$\gamma := \frac{v_e}{v_c}$$

From Eqs. (5.3.1) and (2.8.1) and with relation (5.3.5) we find for the payload ratio μ

$$\mu = \frac{m_s + m_L}{m_0} = 1 - \frac{m_p}{m_0} - \frac{m_g}{m_0} = 1 - \frac{m_p}{m_0}\left(1 + \gamma^2\right)$$

Because of Eq. (2.2.4)

$$e^{-\Delta v/v_e} = \frac{m_f}{m_0} = 1 - \frac{m_p}{m_0} \qquad (5.3.6)$$

we get

$$\mu = e^{-\lambda/\gamma}\left(1+\gamma^2\right) - \gamma^2 \tag{5.3.7}$$

with

$$\lambda := \frac{\Delta v}{v_c}$$

We now want to find the dependence of the propulsion demand (here λ) provided by the electrical engine on the variable exhaust velocity and hence on γ at a given payload ratio μ. For this purpose we solve Eq. (5.3.7) with regard to λ and finally get

$$\lambda = \gamma \ln\left(\frac{1+\gamma^2}{\mu+\gamma^2}\right) \tag{5.3.8}$$

The curves $\lambda = \lambda(\gamma)$ with parameter μ are shown in Fig. 5.5. It is now our goal to find the maximum Δv provided by the engine at a given μ, where v_e is our variable. To find this maximum, we have to differentiate Eq. (5.3.8) and find its root. This leads to the following conditional equation for γ:

$$\ln\frac{1+\gamma^2}{\mu+\gamma^2} = \frac{2\gamma^2(1-\mu)}{(\mu+\gamma^2)(1+\gamma^2)} \tag{5.3.9}$$

Solving for γ gives the optimized $v_e/v_c = \gamma$ and via Eq. (5.3.8) the maximized $\Delta v/v_c = \lambda$ as a function of μ as shown in Fig. 5.6. Having found the optimum parameters, the following calculation scheme can be given to optimize an electrical propulsion system.

Calculation scheme

1. Determine from Eq. (5.3.9) or from Fig. 5.6 for a given μ the optimum $\gamma = v_e/v_c$ and from Eq. (5.3.8) the corresponding maximized $\lambda = \Delta v/v_c$

2. Determine the propellant mass m_p from Eq. (5.3.6) and the power plant mass m_g from Eq. (5.3.5)

3. Determine at a given Δv the optimum v_e, or vice versa, through $v_e = \gamma \cdot \Delta v/\lambda$

4. Determine $v_c = v_e/\gamma$

5. For a given α and η_t, determine the optimum burn time from $t_p = v_c^2/(2\alpha\eta_t)$.

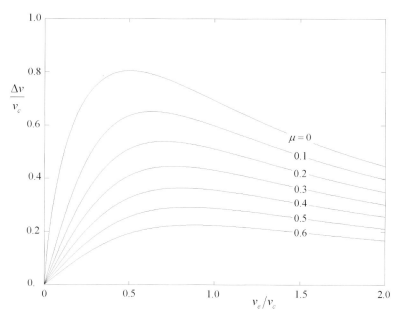

Figure 5.5 The available normalized propulsion demand of an electric propulsion as a function of the normalized exhaust velocity.

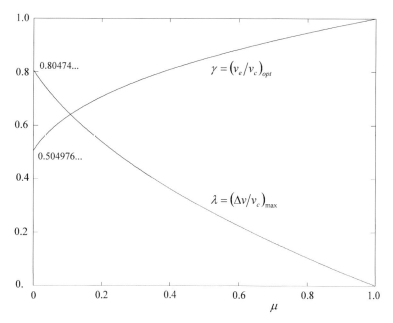

Figure 5.6 The normalized optimium exhaust velocity and the maximum available normalized propulsion demand of an electrical engine as a function of the payload μ.

As the payload mass is typically only a few percent, approximate solutions can be provided for the limiting case $\mu \to 0$ (exercise, Problem 5.1, difficult)

$$\begin{aligned}
\gamma &= 0.5050 \cdot (1 + 4.145\mu) \\
\lambda &= 0.8047 \cdot (1 - 2.461\mu) \\
m_p &= 0.7968 m_0 \, (1 + 2.685\mu) \\
m_g &= 0.2032 m_0 \, (1 + 10.975\mu) \\
v_e &= 0.6275 \cdot (1 + 6.606\mu) \, \Delta v \\
t_p &= 0.7721 \cdot (1 + 4.922\mu) \, \Delta v^2 / \alpha \eta_t
\end{aligned} \qquad @ \, \mu < 0.04 \qquad (5.3.10)$$

Note: *These are approximations in first order of μ. Therefore Eqs. (5.3.10) apply only as long as $\mu \ll 1$. It can be shown that for $\mu < 0.04$ the error $\delta\gamma/\gamma$ (and with it also that of the other quantities) remain below 5%.*

Problems

Problem 5.1 Electrical Engine Optimization
Prove that the linearized solution of Eq. (5.3.9) for $\mu \to 0$ are Eqs. (5.3.10) by first showing that

$$\gamma = \gamma_0 \cdot \left[1 + \varepsilon\mu + O\left(\varepsilon^2\right)\right]$$

$\gamma_0 = 0.504\,976 \ldots$ for $\mu = 0$ by Newton's method, and

$$\varepsilon = \frac{(1 + \gamma_0^2)^2}{2\gamma_0^2 \, (1 - \gamma_0^2)} = 4.145\,344\ldots$$

6
Ascent Flight

Now that we know the technical and physical properties of a rocket and the general equation of motion, which governs its flight, we are ready for a mission to the planets in our solar system. Before we investigate the rocket's motion in interplanetary space, it first has to ascent in the Earth's gravitational field through the atmosphere. As will be shown later, ascent and reentry are subject to identical physical laws treated by the science called *flight mechanics*. The differences between the two is that reentry is powerless, and the initial conditions of both mission phases are drastically different. This is why the problems we have to deal with are much different, and we therefore devote a chapter of its own (Chapter 10) to reentry after we consider orbital motion (Chapter 7), orbit transitions (Chapter 8), and come back from an interplanetary journey (Chapter 9). For ascent, as well as for reentry, the properties of the Earth's atmosphere are crucial. This is why we will first examine here the atmosphere's condition (Section 6.1) and the general laws of motion through the atmosphere (Section 6.2). Only after that shall we go (Sections 6.3 and 6.4) into the specifics of how to optimize an ascent into space.

6.1
Earth's Atmosphere

From a space flight point of view, the atmosphere plays an important role during ascent and reentry because of the aerodynamics at lower altitudes. It also impacts low Earth orbits because of the residual atmospheric drag at high altitudes. To determine these influences quantitatively, we have to derive expressions that describe the density distribution in the atmosphere as a function of altitude.

6.1.1
Density Master Equation

It is well known that atmospheric pressure, starting from sea level, decreases with increasing altitude. To describe its quantitative dependence in mathe-

Astronautics. Ulrich Walter
Copyright © 2008 WILEY-VCH Verlag GmbH & Co. KGaA, Weinheim
ISBN: 978-3-527-40685-2

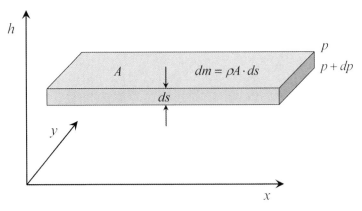

Figure 6.1 Characteristics of an infinitesimally thin atmospheric layer.

matical terms, we first imagine that the atmosphere is a stack of infinitesimally thin layers each with thickness ds (see Fig. 6.1). Without loss of generality, we assume each layer to have a finite surface area A of arbitrary size. The volume of the layer is then $A\,ds$ and the mass of air within it is $dm = \rho A\,ds$, where ρ is the atmospheric density. The additional infinitesimal pressure it generates on the ones below is the weight force per square unit area

$$dp = d\left(\frac{mg}{A}\right) = \frac{g}{A}dm = \frac{g}{A}d(\rho As) = \rho g \cdot ds$$

Here s measures the height, which is along the gravitational force direction, with increasing pressure. But atmospheric pressure is usually given as a function of altitude h relative to sea level, i.e. in the opposite direction. So $ds = -dh$. Therefore we find for the pressure change of the atmospheric pressure

$$\frac{dp}{dh} = -\rho g(h) \qquad (6.1.1)$$

where we have taken into account that with the gravitational acceleration g decreases with altitude according to

$$g(h) = g_0\frac{R_\oplus^2}{r^2} = g_0\frac{R_\oplus^2}{(R_\oplus + h)^2}$$

with $g_0 = 9.798\,29\ \mathrm{m\,s^{-2}}$ and $R_\oplus = 6378$ km the Earth's radius. We now need to know how a gas behaves under external pressure. Earth's atmosphere can be described in a very good approximation by the ideal gas law (cf. Eq. (4.1.1))

$$p = \rho R_s T \qquad \text{ideal gas law} \qquad (6.1.2)$$

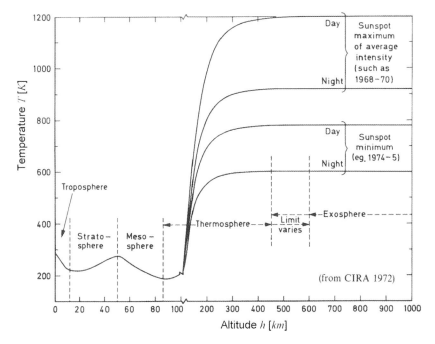

Figure 6.2 The temperature distribution in Earth's atmosphere.

and where T is the layer's temperature.

$$R_s = \frac{R}{M_{air}} = 286.91 \ \frac{J}{K \cdot kg} \quad \text{specific gas constant of standard atmosphere}$$

Strictly speaking, R_s holds only for the standard atmosphere, i.e. for the standard molecular composition. As we will see, only particular parts of the atmosphere fulfill this requirement. We now differentiate Eq. (6.1.2) with regard to the altitude h. Because ρ as well as T depend on h, this yields with Eq. (6.1.1)

$$\frac{dp}{dh} = \frac{d\rho}{dh} R_s T + \rho R_s \frac{dT}{dh} = -\rho g(h)$$

From this follows that

$$\frac{d\rho}{\rho} + \frac{dT}{T} = -\frac{g(h)dh}{R_s T} \tag{6.1.3}$$

This equation is the master equation to calculate the density function $\rho(h)$ for a given $T(h)$. By applying Eq. (6.1.2) $p(h)$ may then be derived. So all we need to know is the temperature profile $T(h)$. This is depicted in Fig. 6.2. Obviously the atmosphere can be divided in two quite different zones:

1. The so-called **homosphere** extends to an altitude of roughly 120 km and shows only modest variations in temperature. It includes the troposphere, stratosphere, and mesosphere, which behave meteorologically differently due to their different temperature gradients (see Fig. 6.2). The homosphere is constituted by a homogeneous mixture of the standard atmospheric components, and therefore $R_s = const$.

2. The so-called **heterosphere** above about 120 km displays much higher levels of and tremendous variations in temperature, which are even time-dependent. Its temperature deviates significantly from that in the homosphere. The heterosphere includes the thermosphere and exosphere (see Fig. 6.2)). As its name indicates, the molecular constituents are heterogeneous with height and become partly ionized. This results in $R_s \neq const$.

The detailed $T(h)$ profile in the homosphere as well as in the heterosphere is not analytical, so an exact solution to Eq. (6.1.3) can be obtained only by numerical integration. Because this is too intricate for practical applications, we have to look for ways to find approximate solutions.

6.1.2
Homosphere (Barometric Formula)

For general purposes the temperature within the homosphere can be considered roughly as constant with a mean value of $T = T_0 = 250$ K$= const$. For a constant temperature $dT = 0$, reducing Eq. (6.1.3) to

$$\frac{d\rho}{\rho} = -\frac{R_\oplus^2}{(R_\oplus + h)^2} \frac{dh}{H}$$

Here we have introduced the so-called scale height as $H := T_0 R_s / g_0$. The solution is found by direct integration of both sides yielding

$$\rho = \rho_0 \exp\left[-\frac{R_\oplus h}{(R_\oplus + h) H}\right] \quad @ \ 0 \leq h \leq 120 \text{ km} \tag{6.1.4}$$

where h now is considered relative to sea level and ρ_0 is the density at sea level. Compared to our assumption $T = const$ we can also safely assume $h \ll R_\oplus$ which leads to the well-known barometric formula

$$\boxed{\rho = \rho_0 \exp\left(-\frac{h}{H}\right)} \quad @ \ 0 \leq h \leq 120 \text{ km} \quad \textbf{barometric formula} \tag{6.1.5}$$

If one fits this formula to the actual atmospheric data in the range $0 \leq h \leq 120$ km, one obtains the following optimum values for ρ_0, H with errors

$\Delta\rho/\rho < 50\%$: $\rho_0 = 1.752 \text{ kg m}^{-3}$ and

$$\boxed{H = 6.7 \pm 0.1 \text{ km}} \quad @ \; 0 \leq h \leq 120 \text{ km} \quad \textbf{scale height} \tag{6.1.6}$$

Equations (6.1.5) and (6.1.6) are the most convenient and hence the most common form to describe the density distribution in the homosphere. They will be used throughout this book for ascent and reentry of a S/C.

6.1.3
Heterosphere

Spacecrafts orbit Earth at altitudes $h > 100$ km where the atmospheric drag slowly brakes their speed and thus spins them down into lower and lower orbits. To find out their orbit lifetimes (see Section 12.6.5) the detailed density profile at those altitudes needs to be known. Since in the heterosphere temperature varies strongly with height and time and R_s is not constant, a barometric formula like Eq. (6.1.5) does not hold. More appropriate atmospheric models need to be provided which by nature are, however, considerably more complex. Today's quasi-standard is the MSIS-86 model (identical to CIRA-86) and its newer version MSISE-90, but also the Jacchia 1977 (J77) model and its older variants and the Harris–Priester model from 1962 (see e.g. Montenbruck and Gill (2000)) are still frequently used. They are all based on piecewise analytical expressions for different altitudes, whose coefficients have been adapted to measured values. So there are no closed analytical expressions and the densities have to be derived numerically. Figure 6.3 depicts the mean atmospheric density profiles as derived from the MSISE-90 model above 100 km.

As an illustrative example we present the relatively simple Harris–Priester model. It is based on the properties of the upper atmosphere derived from a solution of the heat conduction equation. It takes into account the daily, but not the yearly, temperature variations in the atmosphere. The upper atmosphere expands because of daily insolation and runs about two hours behind, which corresponds to 30° degrees of longitude towards the east. The density distributions of the corresponding density peaks (maxima, M) and valleys (minima, m) are described by the functions $\rho_M(h)$ and $\rho_m(h)$ by means of piecewise exponential interpolation between interpolation altitudes h_i:

$$\begin{aligned} \rho_m(h) &= \rho_m(h_i) \exp\left(\frac{h_i - h}{H_m}\right) \\ \rho_M(h) &= \rho_M(h_i) \exp\left(\frac{h_i - h}{H_M}\right) \end{aligned} \quad @ \; h_i < h < h_{i+1} \tag{6.1.7}$$

where h is the altitude above the Earth's reference ellipsoid (see Section 12.2.1).

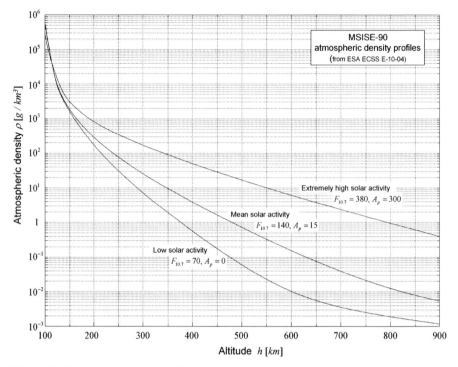

Figure 6.3 Mean atmospheric density in the heterosphere as derived from the MSIE-90 model.

The scale heights H_m and H_M are given as

$$H_m(h) = \frac{h_i - h_{i+1}}{\ln \rho_m(h_{i+1})/\ln \rho_m(h_i)}$$

$$H_M(h) = \frac{h_i - h_{i+1}}{\ln \rho_M(h_{i+1})/\ln \rho_M(h_i)}$$

The daily density variation due to insolation is modeled as a cosine variation

$$\rho(h) = \rho_m(h) + [\rho_M(h) - \rho_m(h)] \cdot \cos^n\left(\frac{\Psi}{2}\right)$$

where Ψ is the angle between the position vector of the orbiting S/C and the vector to the density peak. Density variations in geographical latitude are represented by a declinational dependence of Ψ and the exponent n: for small declination $n = 2$, and for polar orbits $n = 6$. Table 6.1 shows the density coefficients $\rho_M(h)$ and $\rho_m(h)$ at different interpolation points. Note that above 300 km the atmospheric density differs by a factor of more than two between day-time $\rho_M(h)$ and night-time $\rho_m(h)$ values. This is caused by the large temperature differences at these altitudes between day and night (see Fig. 6.2)

Table 6.1 Atmospheric density coefficients of the Harris–Priester model valid for a mean solar activity.

h [km]	ρ_m [g km^{-3}]	ρ_M [g km^{-3}]	h [km]	ρ_m [g km^{-3}]	ρ_M [g km^{-3}]
100	497400.0	497400.0	420	1.558	5.684
120	24900.0	244900.0	440	1.091	4.355
130	8377.0	8710.0	460	0.7701	3.362
140	3899.0	4059.0	480	0.5474	2.612
150	2122.0	2215.0	500	0.3916	2.042
160	1263.0	1344.0	520	0.2819	1.605
170	800.8	875.8	540	0.2042	1.267
180	528.3	601.0	560	0.1488	1.005
190	361.7	429.7	580	0.1092	0.7997
200	255.7	316.2	600	0.080 70	0.6390
210	183.9	239.6	620	0.060 12	0.5123
220	134.1	185.3	640	0.045 19	0.4121
230	99.49	145.5	660	0.034 30	0.3325
240	74.88	115.7	980	0.026 32	0.2691
250	57.09	93.08	700	0.020 43	0.2185
260	44.03	75.55	720	0.016 07	0.1779
270	34.30	61.82	740	0.012 81	0.1452
280	26.97	50.95	760	0.010 36	0.1190
290	21.39	42.26	780	0.008 496	0.097 76
300	17.08	35.26	800	0.007 069	0.080 59
320	10.99	25.11	840	0.004 680	0.057 41
340	7.214	18.19	880	0.003 200	0.042 10
360	4.824	13.37	920	0.002 210	0.031 30
380	3.274	9.955	960	0.001 560	0.023 60
400	2.249	7.492	1000	0.001 150	0.018 10

6.1.4
Piecewise-Exponential Atmospheric Model

In later chapters we will study the ascent and reentry of spacecrafts through the atmosphere and the orbit life time of satellites in low Earth orbits. For these studies the barometric formula for the homosphere is too inaccurate, while for the heterosphere the common atmospheric models are too complex to handle analytically. By examining the functional dependence of the atmospheric density in Fig. 6.3 in logarithmic representation, one recognizes that the density can be expressed quite well by piecewise straight lines corresponding to piecewise exponential functions of the form

$$\rho(h) = \rho_i \exp\left(-\frac{h - h_i}{H_i}\right) \quad @ \ h_i < h < h_{i+1} \tag{6.1.8}$$

where h is the altitude above sea level and h_i are the base altitudes above sea level for a given altitude interval, ρ_i the corresponding nominal base density, and H_i the scale height holding for the entire interval. They are given for the different altitude intervals in Table 6.2. Equation (6.1.8) is the density model, which we will use in the following for our general analytical studies related to the atmosphere.

Table 6.2 Altitude intervals and corresponding atmospheric coefficients for the piecewise exponential model based on the CIRA-72 atmospheric model. Adopted from Vallado (2001).

Altitude h (km)	Base Altitude h_i (km)	Nominal Density ρ_i (kg m^{-3})	Scale Height H_i (km)	Altitude h (km)	Base Altitude h_i (km)	Nominal Density ρ_i (kg m^{-3})	Scale Height H_i (km)
0–25	0	1.225	7.249	150–180	150	2.070×10^{-9}	22.523
25–30	25	3.899×10^{-2}	6.349	180–200	180	5.464×10^{-10}	29.740
30–40	30	1.774×10^{-2}	6.682	200–250	200	2.789×10^{-10}	37.105
40–50	40	3.972×10^{-3}	7.554	250–300	250	7.248×10^{-11}	45.546
50–60	50	1.057×10^{-3}	8.382	300–350	300	2.418×10^{-11}	53.628
60–70	60	3.206×10^{-4}	7.714	350–400	350	9.158×10^{-12}	53.298
70–80	70	8.770×10^{-5}	6.549	400–450	400	3.725×10^{-12}	58.515
80–90	80	1.905×10^{-5}	5.799	450–500	450	1.585×10^{-12}	60.828
90–100	90	3.396×10^{-6}	5.382	500–600	500	6.967×10^{-13}	63.822
100–110	100	5.297×10^{-7}	5.877	600–700	600	1.454×10^{-13}	71.835
110–120	110	9.661×10^{-8}	7.263	700–800	700	3.614×10^{-14}	88.667
120–130	120	2.438×10^{-8}	9.473	800–900	800	1.170×10^{-14}	124.64
130–140	130	8.484×10^{-9}	12.636	900–1000	900	5.245×10^{-15}	181.05
140–150	140	3.845×10^{-9}	16.149	1000–	1000	3.019×10^{-15}	268.00

6.2
Equations of Motion

It is our goal to derive the equations of motion of a S/C flight in the atmosphere of a celestial body. We start out with the general rocket equation of motion (1.4.2)

$$m\dot{v} = F_* + F_{ext}$$

where $F_* = \dot{m} v_*$ is the thrust of the rocket, and F_{ext} comprises all external forces, in particular the drag and lift forces, which are distinctive for this situation. Figure 6.4 shows the flight path of a rocket in the atmosphere with flight direction v and all relevant forces at a given point in flight. Given these forces, the equation of motion can be explicitly written as

$$m\frac{d\ddot{r}}{dt} = F_*(t) + mg(r) + D(v,r) + L(v,r) \qquad (6.2.1)$$

For a given S/C the height- and velocity-dependent drag D, lift L, and the time-dependent thrust F_* are known, and they can be used to solve Eq. (6.2.1) numerically. The solution is the wanted $r(t)$ and $v(t) = dr/dt$. For real missions, this is indeed the only possibility to determine the solution with adequate accuracy.

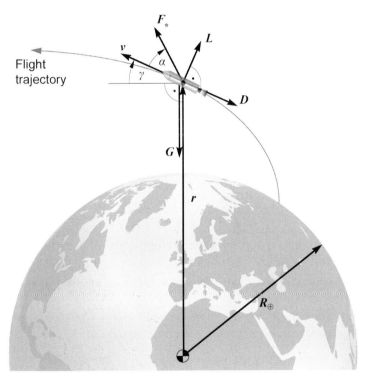

Figure 6.4 The flight path of an S/C through the atmosphere near the Earth with the effective forces thrust F_*, lift L, drag D, and gravitational force $G = mg$.

This would bring us to the end of this chapter, were it not for the need to gain a physical understanding of the processes and typical flight stages. For that we first introduce appropriate coordinate systems, which is always an essential step when exploring physical processes. Figure 6.5 describes an inertial Earth-based coordinate system (u_x, u_y) and the non-inertial S/C coordinate system (u_t, u_n) (tangential component points into the direction of motion), on which our investigations are based.

Note: *The general physical equations and in particular our equation of motion (6.2.1) hold only in an inertial, i.e. unaccelerated, and especially non-rotating, coordinate system. A rocket that ascends on a curved trajectory therefore is a non-inertial system. This is why we have to start out with the inertial Earth-based coordinate system.*

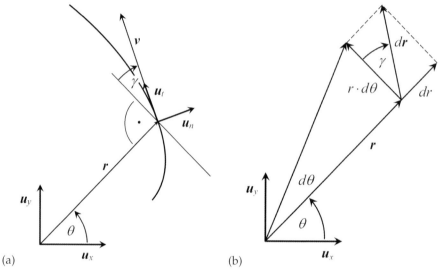

Figure 6.5 (a) Inertial Earth coordinate system (u_x, u_y) and non-inertial S/C system (u_t, u_n). (b) The components of the radial vector in the inertial Earth coordinate system.

We presume that all thrust and lift forces lie in the (u_t, u_n) plane. In this case the S/C moves only in this plane and our treatment is reduced to the two-dimensional case. By choosing the inertial Earth coordinate system, we neglect the Earth's rotation, which leads to three errors to be considered in practice:

1. The transition to a rotating Earth coordinate system changes the coordinates of the S/C trajectory relative to an observer on ground.

2. The atmosphere moves with the Earth's surface, which leads to cross-components of drag and lift. They are, however, negligible compared to wind forces. Even those we do not take into account here.

3. At launch Earth's rotation causes a tangential velocity, which adds to the S/C velocity for a launch in the eastern direction. So placing a launch-pad somewhere near the equator and launching the vehicle in an eastern direction saves propellant or alternatively enables a bigger payload. For example, the new Soyuz-2 rocket will deliver 8.5 tons of payload into LEO from Baikonur at 45.9°N, but about 9.1 tons from Kourou at 5.1°N!

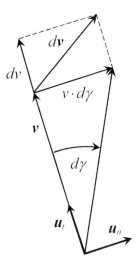

Figure 6.6 The decomposition of the velocity change dv in the S/C system.

Equation (6.2.1) is valid in the Earth inertial coordinate system. Because lift and drag are defined in the (u_t, u_n) reference frame we have to transform to that rotating S/C system. Let ω be the angular velocity of the two coordinate systems relative to each other, i.e. $\omega = \dot\theta$. For the azimuthal change of the radial vector, it follows from Fig. 6.5b $r\,d\theta = dr\cos\gamma$. This yields

$$\dot\theta = \omega = \frac{v}{r}\cos\gamma$$

where γ is the so-called **path flight angle**. Once ω is known, the S/C velocity transforms between the inertial and the S/C coordinate system as (see e.g. Kaplan (1976, p. 15))

$$\left.\frac{dv}{dt}\right|_i = \left.\frac{dv}{dt}\right|_{S/C} + \omega \times v$$

where $\omega \times v$ is the centrifugal acceleration acting on the S/C. This transformation brings us to the equation of motion in the S/C coordinate system

$$m\,(\dot v + \omega \times v) = F_* + mg + D + L \qquad (6.2.2)$$

We now want to decompose all vectors into the two axial components of the S/C system

$$\omega \times v = -\omega v \cdot u_n = -\frac{v^2}{r}\cos\gamma \cdot u_n \qquad (6.2.3)$$

According to Fig. 6.6 the decomposition of the velocity vector is

$$dv = dv \cdot u_t + v \cdot d\gamma \cdot u_n$$

6 Ascent Flight

From this it follows that

$$\dot{\vec{v}} = \dot{v} \cdot u_t + v\dot{\gamma} \cdot u_n \qquad (6.2.4)$$

and finally the following decomposition is valid for the external forces

$$\begin{aligned} F_* &= F_* \cos\alpha \cdot u_t + F_* \sin\alpha \cdot u_n \\ g &= -g\sin\gamma \cdot u_t - g\cos\gamma \cdot u_n \\ D &= -D \cdot u_t \\ L &= L \cdot u_n \end{aligned} \qquad (6.2.5)$$

where $\alpha = \angle(v, v_*)$ is the **thrust angle** (a.k.a. **angle of attack**, see Fig. 6.4), i.e. the angle between the thrust vector and the tangent to the trajectory (velocity vector). When inserting Eqs. (6.2.3), (6.2.4), and (6.2.5) into the equation of motion (6.2.2) for each component, the sum of all terms must vanish due to Newton's third law (see Eq. (7.1.11). In this way we derive two scalar equations of motion for a path through the atmosphere:

$$\dot{v} = \frac{F_* \cos\alpha - D}{m} - g\sin\gamma \qquad (6.2.6)$$

$$v\dot{\gamma} = \frac{F_* \sin\alpha + L}{m} - \left(g - \frac{v^2}{r}\right)\cos\gamma \qquad (6.2.7)$$

In addition, and according to Fig. 6.5, the following holds in the Earth coordinate system:

$$\dot{r} = \dot{h} = v\sin\gamma \qquad (6.2.8)$$

with

$$L = \frac{1}{2}\rho(r)v^2 C_L(\alpha, Re, v) A_\perp \qquad (6.2.9)$$

$$D = \frac{1}{2}\rho(r)v^2 C_D(\alpha^2, Re, v) A_\perp \qquad (6.2.10)$$

$$g(r) = g_0 \frac{R^2}{r^2} \qquad (6.2.11)$$

$$\rho(r) = \rho_0 \exp\left(-\frac{r - R}{H}\right) \qquad (6.2.12)$$

Here R is the radius of the celestial body in question, and $\rho_0 = 1.752\,\mathrm{kg\,m^{-3}}$ and $H = 6.7\,\mathrm{km}$ for the Earth. Equation (6.2.9) and (6.2.10) for drag and lift are derived from Eq. (12.6.1). These forces caused by the impacting air depend on the effective impact (wetted) area $A_\perp$, on the orbit altitude because of the atmospheric density, and via the drag and lift coefficients also on the thrust angle α. In addition there exists a dependence on the Reynolds number Re, and, to a minor extent, on the hypersonic velocity v via C_D and C_L, which are however only important during reentry.

It is important to mention that these equations implicitly require that the thrust axis coincides with the aerodynamic axis of the S/C, relative to which the thrust angle is defined, and that the center of mass lies on this axis. If this is not the case, e.g. for the Space Shuttle, then the equations of motion are far more complex, and can no longer be treated analytically. In that case one is left to solve the full equation of motion with six degrees of freedom numerically.

Only with the above form of the equation of motion is it now possible to understand flight mechanics, the science of ascent and reentry. We will, however, not solve the equations for ascent numerically, as this would hardly increase our understanding of the matter. We will only qualitatively discuss the basic types of launch profiles.

6.3
Ascent Phases

The flight mechanics of continuously powered (so-called non-ballistic) ascent flights deals with the question of how to steer a rocket from a launch-pad optimally to a predetermined target orbit. This is a very general task, which will lead, as we will see, to quite complex ascent strategies.

If you approach the problem quite naively from an orbit-mechanical point of view (see Section 8.1), you might consider the path of the rocket as a transfer orbit between two Keplerian orbits where the launch-pad is a point on the initial Keplerian orbit and the target orbit is the final orbit. But of course this is not correct, as the rocket is not in a Keplerian orbit at lift-off. Nevertheless, this orbit-mechanical approach is still quite sensible. This is because, if the rocket were in a Keplerian orbit at lift-off, and if the initial temporary aerodynamic drag did not occur, the solution of the problem would instantly be obvious: according to orbit mechanics, the optimum transfer orbit regarding propulsion demand is a Hohmann orbit. A Hohmann orbit is a two-impulse transfer. The first impulse carries the S/C into an elliptical transfer orbit, while the second impulse at apogee of the transfer orbit kicks it into the target orbit. Now, if we interpret the powered ascent phase as an "extended impulse maneuver" from zero velocity at lift-off to the entry of the powerless transfer orbit, we have found an important partial optimization of the ascent trajectory problem. So, we divide our optimum ascent into three phases (see Fig. 6.7):

1. **Thrust phase** – beginning with lift-off, the thrust phase is the path through the atmosphere until thrust is terminated, which typically lasts just a few minutes.

2. **Coasting phase** – this follows the thrust phase, and is a powerless, weightless flight without aerodynamic drag on an elliptical transfer orbit to the target orbit.

3. **Kick-burn phase** – this occurs into the target orbit.

Our optimization problem has now been reduced to determining the Hohmann transfer orbital elements and the optimum trajectory in the thrust phase.

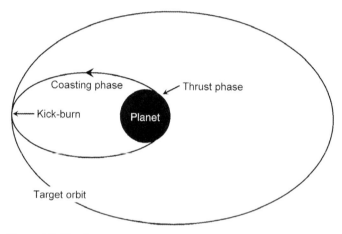

Figure 6.7 The three ascent phases: thrust phase, coasting phase (in transfer orbit), and kick-burn.

Hohmann-Transfer Orbital Elements

The orbital elements of the elliptic transfer orbit are determined on one hand by the requirement to touch the target orbit at apogee, i.e. to meet the boundary condition $r_{apo} = a_T(1 + e_T)$ (see Eq. (7.4.8b)). On the other hand, the transfer ellipse must touch the Earth's surface at the launch-pad, i.e. $r_{per} = a_T(1 - e_T)$ (see Eq. (7.4.8a)). These are two conditional equations for the two orbital elements a and e. So we are left with the final problem: What is the path during the thrust phase that smoothly transits into the coasting phase and consumes the least propellant?

> **Remark:** *Intercontinental ballistic missiles follow ballistic trajectories. They don't have their perigee at the launch-pad and therefore have an additional degree of freedom, which is used to adjust the trajectory to the target distance.*

Example

Let's take a Space Shuttle launch as an example to describe how these three phases are traversed. The launch-pad at Kennedy Space Center is at sea level, so $r_{per} = 6378$ km. The target orbit typically is at an altitude of 300 km, i.e. $r_{apo} = 6678$ km. So, according to the above equations, the transfer orbit has the orbital elements $a_T = 6528$ km and $e_T = 0.02298$. If the shuttle ascended without aerodynamic drag on this ideal transfer orbit, the ascending time until reaching the target orbit would be $t_T = 43.5$ min according to Eq. (8.1.3).

Now, let's have a look at the ascent in reality: The thrust stage lasts 8.5 min, and it takes the Shuttle seamlessly into the transfer orbit at an altitude of 120 km. If the injection into this transfer orbit was not perfect, the shuttle can adjust by a so-called OMS-1 burn. Then the shuttle is in a powerless flight for 35.5 min on the elliptical transfer orbit until perigee at an altitude of 300 km. Here a so-called OMS-2 kick-burn brings it into the circular target orbit. The total flight time is 44.0 min, just a little more than the Hohmann transfer time.

6.4 Optimum-Ascent Problem

6.4.1 Formulation of Problem

How is an optimum trajectory in the thrust phase determined? An ascent trajectory in general is determined by the equations of motion (6.2.6) to (6.2.8), plus the initial condition that the rocket rests on the launch-pad,

$$v(0) = 0$$
$$\gamma(0) = 90° \qquad \text{initial conditions} \qquad (6.4.1)$$
$$h(0) = h_0 \approx 0$$

and by the final condition that the rocket has to meet the transfer point to the transfer orbit at engine shutdown. According to Eq. (10.1.4) the final condition is expressed as

$$v(t_f) = v_f = \sqrt{\frac{\mu}{a_T(1-e_T^2)}} \sqrt{1 + 2e_T \cos\theta_T + e_T^2}$$

$$\cos\gamma(t_f) = \cos\gamma_f = \frac{1 + e_T \cos\theta_T}{\sqrt{1 + 2e_T \cos\theta_T + e_T^2}} \qquad \text{final conditions} \quad (6.4.2)$$

$$h(t_f) = h_f = r_f - R = \frac{a_T(1-e_T^2)}{1 + e_T \cos\theta_T} - R$$

Since a_T, e_T are already determined by the transfer orbit, we only have the variables $F_*(t)$, $m(t)$, $\alpha(t)$, θ_T to find an ascent trajectory with the final conditions (6.4.2). However, we also have to take into account that

$$F_* = \text{set}$$
$$m(t) = m_0 - \dot{m}_p t = \text{set} \tag{6.4.3}$$

This is because the ascent time should be as short as possible due to the gravitational loss (see Section 2.4.2 and below). This implies maximum thrust throughout ascent and therefore the thrust F_* and rocket mass m are predetermined functions of time (except temporary throttling for the Space Shuttle due to excessive drag forces). Therefore we only have the two variables $\alpha(t)$, θ_T to attain the optimum-ascent trajectory. An optimum-ascent trajectory then implies that its fuel demand is lowest. Since the fuel demand is increasing monotonically with time, and due to the gravitational loss, we find the following problem setting for an optimum ascent:

> **Formulation of the Optimum-Ascent Problem**
> Determine that functional relationship $\alpha(t)$ and that θ_T for which the ascent trajectory – given by the equations of motion (6.2.7) to (6.2.8) – satisfies the initial and final conditions (6.4.1) and (6.4.2), and for which $m_p(t_f) = \min$.

Note that the variable θ_T to be optimized can be substituted via the final condition equations (6.4.2) by either v_f, γ_f or h_f to be optimized.

The optimum-ascent problem is a typical problem of optimal control theory. Optimum control problems are inherently so complex that in general they can be solved only numerically. But are there any general design rules for $\alpha(t)$ and θ_T? To derive them we formally write $D = D(h)$, $L = L(h)$, by which we indirectly include the solution of Eq. (6.2.8), so we need only to focus on the optimization treatment of the first two equations. We start this treatment by formally integrating the equation of motion (6.2.6)

$$v = \int_0^{t_f} \frac{F_* \cos\alpha}{m} dt - \int_0^{t_f} \frac{D}{m} dt - \int_0^{t_f} g \sin\gamma \cdot dt$$

$$= F_* \int_0^{t_f} \frac{dt}{m} - F_* \int_0^{t_f} \frac{1 - \cos\alpha}{m} dt - \int_0^{t_f} \frac{D}{m} dt - \int_0^{t_f} g \cdot \sin\gamma \cdot dt$$

from which we derive with Eq. (6.4.3), with $F_* = \dot{m}_p v_*$ (see Eq. (1.1.3)), and with Eq. (10.2.7)

$$v_f = \underbrace{v_* \ln \frac{m_0}{m_f}}_{\text{rocket equation}}$$

$$-2F_* \underbrace{\int_0^{t_f} \frac{\sin^2(\alpha/2)}{m} dt}_{\text{steering losses}} - \underbrace{\frac{\kappa_D}{H} \int_0^{t_f} v^2 e^{-\frac{h}{H}} dt}_{\text{drag losses}} - \underbrace{\int_0^{t_f} g \sin\gamma \cdot dt}_{\text{gravitational losses}} \quad (6.4.4)$$

where κ_D is the dimensionless drag coefficient. We find that there are three contributions, which reduce the velocity gain: steering losses, drag losses, and gravitational losses.

To investigate the optimum flight path angle we formally integrate Eq. (6.2.7) and find with Eq. (10.2.8)

$$\gamma_f = 90° + F_* \underbrace{\int_0^{t_f} \frac{\sin\alpha}{mv} dt}_{\text{steering}} + \underbrace{\frac{\kappa_L}{H} \int_0^{t_f} v e^{-\frac{h}{H}} \cdot dt}_{\text{lift}}$$

$$- \underbrace{\int_0^{t_f} \frac{g}{v} \cos\gamma \cdot dt}_{\text{gravitation}} + \underbrace{\int_0^{t_f} \frac{v}{r} \cos\gamma \cdot dt}_{\text{centrifugal term}} \quad (6.4.5)$$

where κ_L is the dimensionless lift coefficient. There are four effects that contribute to a changing flight path angle: steering, lift, gravitation, and centrifugal force. Note that these terms don't imply losses. They just cause changes in flight path angle. A symmetric rocket body does not cause lift. In contrast a winged body such as the Space Shuttle does generate lift, which can be harnessed to reduce the flight path angle more quickly by turning the Shuttle upside down ($\kappa_L \to -\kappa_L$), which actually is done. So lift can be looked at as a kind of additional steering option, which we will neglect in the following.

With Eqs. (6.4.4) and (6.4.5) at hand we are able to ponder about an optimum-ascent trajectory. Taking all the contributions adequately into account is quite an engineering feat, and we are not able to discuss it extensively in this book. However, we want to analyze at least the essential aspects of optimization.

During the thrust phase the spacecraft changes its state vector from a vertical launch direction with zero initial velocity to a nearly horizontal flight direction and maximum velocity at engine shutdown at the transition point. We first want to investigate the losses due to velocity direction changes. Two direction changes have to be taken into consideration: the turn into the desired target orbit inclination, and the turn from the vertical launch direction into the incline of the transfer orbit. Let's have a look at the delta-v to change the orbit inclination by an angle ϕ. According to Eq. (8.4.2) this is given by $\Delta v = 2v \cdot \sin(\phi/2)$. This states that the effort for an impulse maneuver Δv is low when the changes are performed at a low speed, i.e. right after lift-off. So turning into the right orbit inclination is usually the first steering maneuver during ascent. This maneuver is irrelevant with rockets. Because of their axial symmetry, the inclination can be directly approached after lift-off without turning the rocket along its longitudinal axis. The Space Shuttle, on the other hand, because of the launch-pad orientation at Kennedy Space Center, which is a remnant of the Apollo era, first has to roll by 120° around its longitudinal axis to match its body symmetry plane (x-z plane) with the orbital plane of the International Space Station. This is the famous 120° roll maneuver.

What about the turn into horizontal flight? In line with the inclination turn, Eq. (6.4.5) claims that to reduce the flight path angle only small thrust angles $\alpha < 0$ would be needed at low v, i.e. right after lift-off. In addition Eq. (6.4.4) states that in this case the steering losses would be minimum, because α is small. A small γ early on would also reduce the gravitational loss term $g \sin \gamma$. But you don't want to turn too early, because drag is very high at low altitudes (see exponential contribution in drag term in Eq. (6.4.4)). It seems that there is a wide range of possible paths into space due to these contradictory requirements. Yet this is not the case because there is gravity turn.

6.4.2
Gravity-Turn Maneuver

Why wasting propellant to steer the rocket into horizontal flight when gravity does it for you?

According to Eq. (6.4.5) the flight path angle would reduce just by itself without any steering. It's like throwing a stone forwards and upwards. Gravity bends its path until it flies horizontally at its apex. Now, accelerate the stone in flight, which will shift the apex into space, and the increasing centrifugal force will prevent it from falling back. Thus you have the gravity-turn maneuver of an ascending rocket. Since the required thrust angle is $\alpha = 0$, there are no steering losses for the velocity gain (second term in Eq. (6.4.4)). Mathematically the gravity-turn maneuver can be described by setting $\alpha = 0$ in Eq. (6.2.7) and neglecting lift. We then get for the flight path angle rate

$$\dot{\gamma} = -\left(\frac{g}{v} - \frac{v}{r}\right) \cos \gamma \qquad (6.4.6)$$

We see that the initial rate and hence the gravity turn is big at low speeds (but zero for vertical ascent). With rising altitudes, velocity increases, so gravity turn diminishes while centrifugal forces become stronger until $g/v - v/r = 0 \to v \approx \sqrt{gR} = 7.92 \text{ km s}^{-1}$. When this happens $\dot{\gamma} \approx 0$, and if the gravity turn was initiated just right, also $\gamma \approx 0°$, i.e. the trajectory is nearly circular at the transit point. This can somehow be gleaned from the image of an ascending rocket at night in Fig. 6.8. The gravity-turn ascent is equivalent to the opposite case of a ballistic entry, as discussed in the Section 10.4.

Figure 6.8 Gravity-turn maneuver of an ascending Delta II rocket with Messenger spacecraft on August 3, 2004.

You might think that gravity turn is the philosopher's stone for ascent. This is not the case, because a gravity-turn-only ascent just eliminates steering losses. But are drag and gravity losses also minimal for such an ascent? This is generally not the case. So, although an optimized ascent is close to a gravity-turn ascent, it needs some additional ingredients. Moreover, with vertical lift-off a gravity turn does not happen all by itself. It needs to be kicked off.

6.4.3
Pitch Maneuver

For static reasons, the S/C is in a vertical position at lift-off. So just after lift-off the flight path angle is $\gamma = 90°$, $\alpha = 0$ and $L = D = 0$. From Eq. (6.4.6), we get $\dot{\gamma} = 0$: the S/C will ascent vertically. In order to subject it to a gravity-turn maneuver, we need a so-called initial kick angle (a.k.a. pitch angle, i.e. the angle between flight direction and the vertical), which may be small, but not zero. This pitch angle is brought about by the so-called pitch maneuver or pitch program, and it amounts to approximately 3°–5° (see Fig. 6.9). Only after receiving the kick angle the pitch will increase further due to the gravitational force according to Eq. (6.4.6) until it acquires about 20°–30° at an altitude of 10–15 km. Note that for small celestial bodies without any atmosphere a timely pitch maneuver plus a gravity-turn maneuver together make up an optimum ascent. So the ascent of the Apollo landing module from the Moon was virtually an ideal pitch and gravity-turn maneuver.

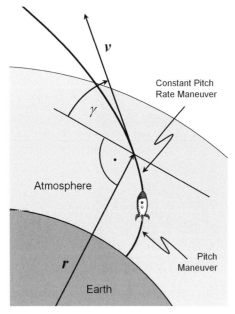

Figure 6.9 Pitch maneuver and constant-pitch-rate maneuver just after lift-off.

6.4.4
Constant-Pitch-Rate Maneuver

How are drag!losses minimized? According to Eq. (6.2.10) drag is small, if at low altitudes, despite a high atmospheric density, the velocity is very low. This is always the case after lift-off. But flight velocity increases rapidly, so the maximum aerodynamic pressure, so-called q_{max} (a.k.a. max-q), is achieved at medium altitudes, and it may become quite big. At increasing altitudes, aerodynamic pressure decreases due to the exponential decline of the atmospheric density with altitude. Apart from substantial drag losses, aerodynamic pressure also puts on high dynamic loads. This is why at max-q the Space Shuttle temporarily throttles down its three liquid propulsion engines to about 60%

thrust. So from the sole perspective of drag losses, the ascent should be as slow as possible and with the shortest path through the atmosphere.

To account for this requirement the so-called constant-pitch-rate (CPR) maneuver is frequently used rather than the gravity-turn maneuver. The **pitch angle** is defined as $\theta := \alpha + \gamma$. A CPR would therefore imply $\dot\theta = \dot\alpha + \dot\gamma = const < 0$. We want to know how the vehicle needs to be steered, i.e. what $\alpha(t)$ looks like just after lift-off, for a constant pitch rate. This problem is closely related to the problem, what the steering after lift-off is for a constant flight path angle rate (CFPAR), $\dot\gamma = const < 0$. We first investigate the latter problem before we come to the CPR problem.

Constant Flight Path Angle Rate

After lift-off we are at low speed, so we can neglect drag, lift, and centrifugal forces. In this case the flight path angle would decrease according to $\gamma = 90° + \dot\gamma t$ with $\dot\gamma < 0$. Because thrust angles are always small we get from Eq. (6.2.7)

$$v\dot\gamma = \frac{F_*\alpha}{m} - g\cos\gamma \tag{6.4.7}$$

In order to find an expression for v we consider Eq. (6.2.6). We find for $D \ll F_*$

$$\dot v = \frac{F_*}{m} - g\sin\gamma$$

We integrate both sides with respect to time. With $m(t) = m_0 + \dot m t$ from Eq. (6.4.3) and with $v(t=0) = v_0$ and the condition $\dot\gamma = const$, we find

$$v - v_0 = \int_0^t \frac{F_*}{m_0 + \dot m t'} dt' - g\int_{90°}^{\gamma} \frac{\sin\gamma'}{\dot\gamma} \cdot d\gamma'$$

$$= \frac{F_*}{m_0}\int_0^t \frac{dt'}{1 + \dot m t'/m_0} - \frac{g}{\dot\gamma}\int_{90°}^{\gamma} \sin\gamma' \cdot d\gamma'$$

$$= \frac{F_*}{\dot m}\ln\left(1 + \frac{\dot m}{m_0}t\right) + \frac{g}{\dot\gamma}\cos\gamma$$

Inserting this into Eq. (6.4.7) yields

$$\dot\gamma\left[\frac{F_*}{\dot m}\ln\left(1 + \frac{\dot m}{m_0}t\right) + v_0\right] + g\cos\gamma = \frac{\alpha F_*}{m} - g\cos\gamma$$

As $\gamma = 90° + \dot\gamma t$, $\cos\gamma = -\sin\dot\gamma t$, so we finally obtain for the thrust angle to steer a constant FPA rate

$$\alpha = \dot\gamma\left[\frac{m}{\dot m}\ln\left(1 + \frac{\dot m}{m_0}t\right) + \frac{mv_0}{F_*}\right] - 2\frac{mg}{F_*}\sin\dot\gamma t \quad \textbf{CFPAR steering law} \tag{6.4.8}$$

This law applies whenever $D, L \ll F_* = const$, $\dot\gamma = const$, and v small. After lift-off, when $\dot\gamma t \ll 1$ and $\dot m t \ll m_0$, that is $m \approx m_0$, we derive the approximate result

$$\alpha(t) = \dot\gamma \left[t\left(1 - 2\frac{m_0 g}{F_*}\right) + \frac{m_0 v_0}{F_*} \right] \qquad (6.4.9)$$

So the thrust angle increases linearly with time.

Note: *In the special case $F_* = 2m_0 g$, that is, if the thrust is twice as strong as the weight of the vehicle, the thrust angle to steer becomes constant, and zero for $v_0 = 0$. This means that, if $F_* = 2m_0 g$, then any $\dot\gamma$, which was set initially, will be maintained throughout further flight without any steering, as long as $\dot m t \ll 2m_0$ and $\dot\gamma^2 t^2 \ll 6$.*

Constant Pitch Rate

We now turn back to the CPR problem. Quite generally, from Eq. (6.2.7) it follows that $\dot\alpha + \dot\gamma(\alpha) = \dot\theta = const$. This is a differential equation for $\alpha(t)$, which however is too complex to solve analytically. But we can simplify this problem by making the special choice $\dot\gamma = const$, i.e. applying the CFPAR steering law. From $\dot\alpha = \dot\theta - \dot\gamma = const$ it follows that

$$\alpha(t) = \left(\dot\theta - \dot\gamma\right) t + \alpha_0$$

Solving Eq. (6.4.9) for $\dot\gamma$ and inserting it yields

$$\alpha \frac{2 - 2m_0 g/F_*}{1 - 2m_0 g/F_*} = \dot\theta t + \frac{m_0 v_0 \dot\gamma / F_*}{1 - 2m_0 g/F_*} + \alpha_0$$

from which we finally obtain

$$\alpha(t) = \frac{F_*/2 - m_0 g}{F_* - m_0 g} \left(\dot\theta t + \alpha_0\right) + \frac{1}{2} \frac{m_0 v_0 \dot\gamma}{F_* - m_0 g} \qquad @\ \dot\gamma = const \quad \textbf{CPR steering law} \quad (6.4.10)$$

Here, as well, we find that, if $F_* = 2m_0 g$ then $\alpha(t) = const$. No steering is needed and for $v_0 = 0$ there are even no steering losses.

6.4.5
Optimum-Ascent Trajectory

In summary, the following qualitative picture of an optimum ascent can be given. Immediately after vertical lift-off the S/C is rolled if required (Space

Shuttle) into the target orbit inclination. It is then subjected to a pitch and constant-pitch-rate maneuver, which results in a low propulsion demand at these low speeds. This brings the S/C into a relatively steep trajectory to altitudes where drag has reduced to a level that a loss-free gravity-turn maneuver bends the trajectory more and more horizontally. The cross-over from constant pitch rate with $\alpha \neq 0$ to gravity turn with $\alpha = 0$ of course is steady. Detailed investigations have shown that a good approximation to the ideal thrust phase trajectory is a piecewise constant thrust angle rate profile of the empirical form

$$\dot{\alpha} = \dot{\gamma} \cdot e^{-\kappa t} \qquad (6.4.11)$$

with form factors $\dot{\gamma}$, κ to be determined by optimization. At the end of the thrust phase the ascent trajectory passes smoothly into the elliptical transfer orbit, which finally touches and transits into the target orbit.

For such an optimized ascent trajectory the delta-v losses for an ascent into a low earth orbit are typically:

- Steering losses $\Delta v \approx 0.05$ km s^{-1}
- Drag losses $\Delta v \approx 0.4$ km s^{-1}
- Gravitational losses $\Delta v \approx 1.0$ km s^{-1}
- Earth's rotational gain $\Delta v \approx -0.464 \cdot \cos \lambda$ km s^{-1}

Earth's rotational gain is the surface speed of the launch-pad at latitude λ due to the rotation of the Earth, which directly adds to the total delta-v as a negative (for a prograde orbit) contribution. In total the delta-v demand for a typical 250 km parking orbit is

$$\Delta v_{tot} = 7.75 \mid 0.05 \mid 0.4 \mid 1.0 \quad 0.464 \cos \lambda \; \left[\text{km s}^{-1}\right]$$

$$= 9.2 - 0.464 \cos \lambda \; \left[\text{km s}^{-1}\right]$$

So, as a rule of thumb the delta-v into LEO without rotational gain is $\Delta v_{tot} \approx 9.2$ km s^{-1}.

To determine an optimum-ascent path with such optimized losses, for instance by determining the form factors $\dot{\gamma}$, κ, is a brilliant feat, in particular when also staging, variations in thrust, the aerodynamic properties of the vehicle, and winds are taken into account. In the end, a good ascent optimization is based on sophisticated software, on the knowledge of the basic ascent maneuvers, but also a lot on the skills of experienced flight mechanics engineers as well as on trial and error.

7
Orbits

7.1
Equation of Motion

After ascent, we are now in outer space. How does a spacecraft move under the influence of the gravitational forces of the Sun, planets, and moons? This is the question we will deal with in this chapter, and we are pursuing general answers to it. Let's face reality from the start: the details of motion are usually very complicated and can be determined sufficiently accurately only numerically on a computer. This is exactly how real missions are planned. But, for us, the goal is not numerical accuracy, but to understand the basic behavior of a spacecraft. To achieve this, it suffices to study some crucial cases. The easiest and by far the most important case is the motion of a spacecraft in the gravitational field of just one massive body, which we study in this chapter. Many complicated cases can be traced back to this case by minor simplifications.

Before we derive the corresponding equation of motion, solve it, and thus describe the motion of orbiting bodies, we want to gain insight into the basic principles of gravitation and show that even Newton's laws follow from these.

7.1.1
Gravitational Potential

The existence of forces seems to be so self-evident that we deem them to be the foundation of nature. But appearance can be deceptive, and Newton also succumbed to this in the late 17th century. It is not forces that are fundamental, but so-called potentials, which cause such forces. This was shown by Laplace one century later. The gravitational potential U is a property of space induced by the mass of a body and surrounding it. Like a force, you cannot see it by itself. Only if you insert a test mass into this space does the potential act on it and generate an attractive force. To better understand this mechanism, let's imagine a body with a (large) mass M at position O, which we define as the origin of our reference system (Fig. 7.1). This is why M is also called the *central body*. A vector r is the radial vector to any position outside O.

Astronautics. Ulrich Walter
Copyright © 2008 WILEY-VCH Verlag GmbH & Co. KGaA, Weinheim
ISBN: 978-3-527-40685-2

7 Orbits

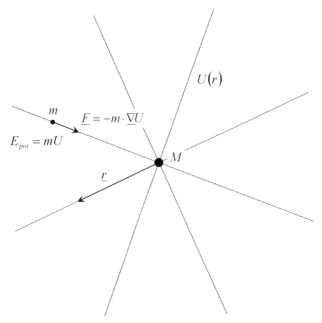

Figure 7.1 The gravitational potential of a central body M and the resulting force and energy of a test mass m in its surrounding.

The basic mutual interaction between masses and the space in which they are embedded is described by the famous field equations of Einstein's theory of general relativity:

$$G_{ik} = \frac{8\pi G}{c^4} T_{ik} \quad \textbf{Einstein field equations} \tag{7.1.1}$$

Remark: *The Einstein field equations are the components of a tensor equation, a system of 10 partial differential equations of second order in the coordinates to determine G_{ik} from the given T_{ik}. Here the cosmological constant, which more recently turned out to be very significant on a cosmological scale, has been neglected. You do not really have to understand this equation and the meaning of its terms. We start our considerations with Einstein's equations to show that the origin of Newton's gravitational field is the theory of general relativity.*

In Eq. (7.1.1) G_{ik} is the so-called Einstein tensor, which describes the basic geometric structure of space, its curvature; T_{ik} is the so-called stress–energy tensor, which describes the energy and the inertial moment distribution of matter or fields in space, and corresponds to the classic energy and mass density ρ; G is the *gravitational constant*; and c is the velocity of light. The Einstein field equations tell us that the matter and energy of the universe, on the one hand, and the curvature of space, on the other, determine each other. To put it in a different way: masses tell space how to curve, and space tells masses how to move. In contradiction to classical Newtonian mechanics, space without masses (Newton's absolute space) cannot exist.

If the curvature of space is weak, the planetary motions are far below relativistic speeds, and the pressure in the state equation of the local matter/energy distribution is small, it is possible to show that the Einstein field equations turn into the classic potential equation – also called the Poisson equation:

$$\left(\frac{\partial}{\partial x^2} + \frac{\partial}{\partial y^2} + \frac{\partial}{\partial z^2}\right) U(r) = 4\pi G \cdot \rho(r) \quad \textbf{Poisson equation} \quad (7.1.2)$$

This is now only one differential equation of second order. It states that every given mass M, or its mass density $\rho(r) = dM/dV$, respectively, generates a gravitational potential $U(r)$, which reflects the curvature of space. The gradient of space's curvature in turn acts on another body by accelerating it. Mathematically the gradient of a potential is a force field. It is exactly this gravitational force as a gradient of a gravitational potential which we commonly interpret as the cause of the acceleration. We thus have related to the central statement of general relativity: mass-induced gravitation and the curvature of space is the same. They are just different appearances (see Fig. 7.2).

Figure 7.2 According to the theory of general relativity, mass curves space, and the curvature of space is exactly the gravitational potential of this mass.

Newton proved on mere mathematical grounds that, from a gravitational point of view, every spherically symmetric mass distribution M with a finite extension R, i.e. any celestial body, outside this body of mass behaves like a point mass M at the origin O. So $\rho(r) = M\delta(r)$, with delta function $\delta(r)$.

Remark: Here $\delta(r)$ is the so-called Dirac delta function. It is defined as

$$\delta(r) = \begin{cases} \infty & \text{for } r = 0 \\ 0 & \text{for } r \neq 0 \end{cases} \quad \text{and} \quad \int_{-\infty}^{+\infty} \delta(r) \, dr = 1$$

For a homogeneous mass density $\rho(r) = M/V$ the gravitational potential thus can be determined from Eq. (7.1.2) to be

$$U(r) = -\frac{\mu}{r} \tag{7.1.3}$$

with

$\mu := GM$ **standard gravitational parameter**

For a beginner it is an easy and illustrative exercise to plug Eq. (7.1.3) into Eq. (7.1.2) and see that it works. If a second body with mass m (small mass again with the gravitational property) is placed into this potential, it acquires potential energy

$$E_{pot} = mU(r) = -\frac{\mu m}{r} \tag{7.1.4}$$

Motivated by this relation one could consider the gravitational potential also as potential energy per mass.

Note: Here M and m characterize the gravitational property of the masses, in contrast to their inertial property, which they also bear and to which we come in a moment.

7.1.2
Gravitational Field

Because the potential energy varies from point to point in space and since a body tries to minimize its energy, the test mass m will move along the gradient of the potential energy. We interpret the gradient of the potential energy as a force, the gravitational force, which mathematically is described as

$$F(r) = -\frac{d}{dr} E_{pot} = -m \frac{d}{dr} U(r) \tag{7.1.5}$$

where

$$\frac{d}{dr} \equiv \text{grad} \equiv \nabla := \left(\frac{\partial}{\partial x}, \frac{\partial}{\partial y}, \frac{\partial}{\partial z} \right)$$

The negative sign occurs in Eq. (7.1.5) because the gravitational force F points into the direction of decreasing energy E_{pot}. As an illustrative example for the calculation of a gradient let us calculate the gradient of the radial distance

$$r = \sqrt{x^2 + y^2 + z^2}$$

$$\frac{dr}{d\mathbf{r}} = \left(\frac{\partial r}{\partial x}, \frac{\partial r}{\partial y}, \frac{\partial r}{\partial z}\right) = \left(\frac{1}{2r}2x, \frac{1}{2r}2y, \frac{1}{2r}2z\right) = \frac{\mathbf{r}}{r} = \hat{\mathbf{r}} \tag{7.1.6}$$

where $\hat{\mathbf{r}}$ is the unit vector in the direction of $\mathbf{r}$. Applying $U(r)$ from Eq. (7.1.3) to Eq. (7.1.5) and because of Eq. (7.1.6), for the gravitational force (force field) we get Newton's law of gravitation

$$\mathbf{F}(\mathbf{r}) = \mu m \frac{d}{d\mathbf{r}} \frac{1}{r} = -\frac{\mu m}{r^2} \frac{dr}{d\mathbf{r}} = -\frac{\mu m}{r^2} \frac{\mathbf{r}}{r}$$

or

$$\mathbf{F}(\mathbf{r}) = -\frac{\mu m}{r^2} \hat{\mathbf{r}} \quad \textbf{Newton's law of gravitation} \tag{7.1.7}$$

which states that the absolute value of the gravitational force declines with the square of the distance to the mass at the origin. We call $\mathbf{G}(\mathbf{r}) = \mathbf{F}(\mathbf{r})/m$ *gravitational field*, which is formally the force field per unit test mass.

In addition to Newton's law of gravitation, the motion of a body in a gravitational field is determined by the body's inertial properties. We will now show that Newton's laws, which are closely linked to the inertia of a body, are also based on very fundamental properties of our universe.

7.1.3
Conservation Laws

It is common in the literature to assume Newton's laws and Newton's equation of motion as given, and then to apply them to gravitation and to derive the conservation of angular momentum and energy. This might be correct on mathematical grounds, but it *does not mean* that the conservation laws *result* from Newton's laws. It just shows that the conservation laws also hold for motion in a gravitational field (or other so-called conservative fields. See remark after Eq. (7.2.7)). Could that mean that they would not be valid in other cases? The conservation laws actually are very first principles in nature: conservation laws are always valid. This property stems from very basic features of our universe, namely that time t and space x, y, z are homogeneous, and the direction φ in space is isotropic.

Remark: *According to Einstein's equations, space and time is homogeneous and isotropic because on a cosmic scale the masses are distributed evenly in the universe. All masses in the universe have to be considered here, because only in their entirety do they determine the gross spatial structure of the cosmos.*

The so-called **Noether's theorem** (Emmy Noether, 1918) of physics tells us that these basic features result in the following conservation laws:

- Homogeneity of time, i.e. the invariance of the physical action integral against continuous time shifts $t \to t + \delta t$, results in the conservation of energy:

$$\sum_{\substack{\text{all energy forms } j \\ \text{of all bodies } i}} E_{ji} = const \qquad \textbf{conservation of energy} \qquad (7.1.8)$$

- Homogeneity of space, i.e. the invariance of the action integral against continuous spatial shifts $r \to r + \delta r$, results in the conservation of momentum:

$$\sum_{\text{all bodies } i} p_i = const \qquad \textbf{conservation of momentum} \qquad (7.1.9)$$

- Isotropy of the direction in space, i.e. the invariance of the action integral against continuous spatial rotations $\varphi \to \varphi + \delta\varphi$, results in the conservation of angular momentum:

$$\sum_{\text{all bodies } i} L_i = const \qquad \textbf{conservation of angular momentum} \qquad (7.1.10)$$

- Gauge invariance of the action integral of electrodynamics results in the conservation of electric charge.

Remark: *You do not need to understand why symmetries correspond to conservation laws. Here is just a short summary. The variables (energy, time), (linear momentum, location), and (angular momentum, rotation angle) are so-called "canonically conjugated parameters," generally written as (p_i, q_i) for every particle i. If one takes the difference in kinetic and potential energies for all particles under consideration, which is called the Lagrange function L, then from the energy minimization principle Euler's equation follows: $d(\partial L/\partial \dot{q}_i)/dt - \partial L/\partial q_i = 0$ with $p_i \equiv \partial L/\partial \dot{q}_i$. The invariance of the universe and hence of its Lagrange function L with regard to the shifts $q_i \to q_i + \delta q_i$ implies that $d\left(\sum_i \partial L/\partial \dot{q}_i\right)/dt = 0$, which in turn implies that $\sum_i \partial L/\partial \dot{q}_i = \sum_i p_i = const$. These are the said conservation laws.*

7.1.4
Newton's Laws and Equation of Motion

We are now set to derive the equation of motion in a gravitational field. First, it is important to note that Eq. (7.1.5) quite generally describes the relation between any type of energy and the force derived from it. So when taking the gradient of the energy conservation Eq. (7.1.8) and employing Eq. (7.1.5) we get for our test mass m ($i = 1$)

$$\boxed{0 = \sum_j -\frac{dE_j}{dr} = \sum_j F_j}\qquad \text{Newton's third law} \qquad (7.1.11)$$

The running index j indicates all relevant energies. This equation states that the sum of all forces acting on a point mass vanishes. This is a generalization of Newton's third law: action equals reaction. The energies relevant to our point mass are: potential energy in the gravitational field, E_{pot}, and kinetic energy, E_{kin}; there are possibly also other energies from electric, magnetic, or chemical potentials, which we will however neglect for our further considerations. The gravitational force derived from the potential energy has already been described in Eq. (7.1.7). What is still missing though is the force derived from kinetic energy. It results from

$$\frac{dE_{kin}}{dr} = \frac{1}{2}m\frac{dv^2}{dr} = mv\frac{dv}{dr} = m\frac{dr}{dt}\frac{dv}{dr} = m\frac{dv}{dt} = m\ddot{r}$$

Note: *Here m characterizes now the inertial property of the mass.*

> The inertial force is the force field of the kinetic energy.

Because velocity is a property a body always acquaints independently of the actual external forces, we quite generally can insert this result into Eq. (7.1.11), and one gets Newton's well-known second law

$$\boxed{m\ddot{r} = \sum_j F_j}\qquad \text{Newton's second law} \qquad (7.1.12)$$

where the summation is over all external forces.

Remark: *To be precise, Newton's second law states that $F = dp/dt$. But since $p = mv$, this together with Eq. (7.1.11) is equivalent to Eq. (7.1.12).*

If the external forces vanish, Eq. (7.1.12) reduces to $\ddot{r} = 0$ with the solution

$r = v_0 t + r_0$ Newton's first law

where v_0, r_0 are the initial values of our mass m. This equation states that:

> Every body persists in a state of rest or of uniform motion in a straight line unless it is compelled to change that state by forces impressed on it.

These are the words of Newton, by which he described his first law. Assuming solely a gravitational force as described in Eq. (7.1.7) and apply Newton's second law, we therefore finally get

$$\ddot{r} = -\frac{\mu}{r^3}r = -\frac{\mu}{r^2}\hat{r} \tag{7.1.13}$$

the equation of motion for an idealized two-body problem. The radial vector r to the body as the solution of this equation describes the body's motion in the gravitational field of the mass M ($\mu = GM$) fixed at point O in space.

Note: *Observe that the body's mass m no longer appears in this equation! The motion's path is thus independent of m.*

Remark: *In order that the masses cancel out in Eq. (7.1.13), we have to assume that gravitational mass and inertial mass are identical. Newton's theory is not able to explain why the gravitational and inertial masses of a body should be identical. They could just as well be different. Only the theory of general relativity provides us with a seamless explanation: acceleration forces (inertial forces) and gravitational forces are two sides of the same coin – the curvature of space. So a body must react to acceleration and gravitation in exactly the same way: inertial force = weight force. Let's illustrate this with an example due to Einstein. If you were standing in an elevator at an unknown place in outer space, you couldn't tell whether your weight is due to external gravitation or due to an acceleration of the elevator.*

> In conclusion, we have shown that classical Newtonian physics, in particular the equation of motion in a gravitational field, is an outcome of the theory of general relativity by taking into account the homogeneity and isotropy properties of space and time in our universe.

7.1.5
Real Two-Body Problem

To assume that the central body M is fixed, and the body m moves within its potential – which implies that the body m is negligibly small with respect to the central body M, $m \ll M$ – is a constraint that can easily be eliminated. Let's have a look at two bodies with unrestricted masses m_1 and m_2, which move around each other under the influence of their mutual gravitational potential. Now that we have two bodies there is no exceptional point for the origin O of our reference system. We can place it wherever we want. Let r_1 and r_2 be the radial vectors from O to m_1 and m_2, and $r := r_2 - r_1$ the connecting vector. According to Eq. (7.1.13) the vectorial equation of motion for each of the two bodies are then as follows:

$$m_1 \ddot{r}_1 = -\frac{Gm_1m_2}{r^3}(r_1 - r_2) = +\frac{Gm_1m_2}{r^3}r$$

$$m_2 \ddot{r}_2 = -\frac{Gm_1m_2}{r^3}(r_2 - r_1) = -\frac{Gm_1m_2}{r^3}r$$

(7.1.14)

It is possible to trace back these equations to that of the idealized two-body problem. To do that, one cancels m_1 from the first equation, and m_2 from the second, and then subtracts both equations from each other. This yields

$$\ddot{r} = -\frac{\mu}{r^3}r \tag{7.1.15}$$

with

$$\mu := G(m_1 + m_2) \quad \text{and} \quad r = r_2 - r_1$$

Equation (7.1.15) is the equation of motion for the connecting vector between m_1 and m_2 with the origin at m_1. This vector passes through their center of mass (CM), so according to Eq. (7.1.15) both bodies move synchronously around their common CM. If

$$\omega = \sqrt{\frac{G(m_1 + m_2)}{r^3}} = const \tag{7.1.16}$$

we have a uniform joint circular motion, $\ddot{r} = -\omega^2 r$, of the two masses around the CM.

Motion of the Center of Mass

The vector r_{cm} to the CM by definition is the mass-weighted average of the position vectors to both bodies:

$$r_{cm} := \frac{m_1 r_1 + m_2 r_2}{m_1 + m_2} \tag{7.1.17}$$

Because of Eq. (7.1.14) $m_1 \ddot{r}_1 + m_2 \ddot{r}_2 = 0$. This implies $\ddot{r}_{cm} = 0$ and hence

$$r_{cm} = v_0 t + r_0 \tag{7.1.18}$$

with initial conditions v_0, r_0. So, with no external forces acting, the CM moves along a straight line in space. This is Newton's third law applied to the CM.

Motion with regard to the Common Center of Mass

Equation (7.1.15) has the drawback that the origin is at m_1, which is not inertial, but is subject to acceleration due to the coupled motion. This is why we want to describe the motion of both masses in an inertial system – in their CM. To do this, we move (see Fig. 7.3) the origin of our coordinate system into the CM of both masses: $r_{cm} = 0$. Now r_1 and r_2 are the relative vectors with regard to the CM. Then Eq. (7.1.17) results in

$$r_{cm} = \frac{m_1 r_1 + m_2 r_2}{m_1 + m_2} = 0$$

From this it follows that

$$m_1 r_1 = -m_2 r_2 \quad \text{and} \quad m_1 r_1 = m_2 r_2$$

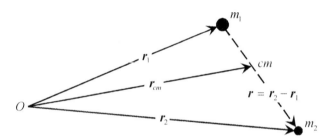

Figure 7.3 Relevant vectors in the general two-body system.

In addition, for the value of the connecting vector, $r = r_1 + r_2$ holds. With this and from Eqs. (7.1.14) we get after some simple modifications

$$\ddot{r}_1 = -\frac{\mu_1}{r_1^3} r_1, \quad \ddot{r}_2 = -\frac{\mu_2}{r_2^3} r_2$$

where

$$\mu_1 = \frac{G m_2^3}{(m_1 + m_2)^2}, \quad \mu_2 = \frac{G m_1^3}{(m_1 + m_2)^2}$$

Let m be the mass under consideration and M the other mass. Then the equation of motion for mass m in the gravitational field of mass M for following relevant considerations and hence for the rest of this book reads

$$\boxed{\ddot{r} = -\frac{\mu}{r^3} r = -\frac{\mu}{r^2} \hat{r}} \qquad \text{equation of motion} \qquad (7.1.19)$$

with

$$\mu = \frac{GM}{(1+m/M)^2}$$

This is the equation for the motion of mass m in the *common center-of-mass inertial system* of m and M.

Approximation for a Small Mass m

In order to account for small mass m moving around a large mass M (e.g. a small moon orbiting a planet) we linearly approximate Eq. (7.1.19) in m/M, which yields

$$\ddot{r} = -\frac{\mu}{r^3}r \qquad (7.1.20)$$

with

$$\mu := G(M - 2m)$$

This is the Newtonian equation of motion relevant for all planets in the solar system. The factor μ differs from the one in Eq. (7.1.15) by $3mG$, which in the case of the Moon circling the Earth amounts to a non-negligible 3.7%. For the case $m \to 0$, e.g. a spacecraft circling Earth, consider the following values:

$G = 6.67259 \, 10^{-11} \, \text{kg m}^3\text{s}^{-2}$ (gravitational constant)

$\mu_\oplus \approx GM_\oplus = g_0 R_\oplus^2$

$\quad = 3.986\,006 \times 10^{14} \, \text{m}^3\text{s}^{-2}$ (standard gravitational parameter)

$g_0 = 9.798\,28 \, \text{m s}^{-2}, \quad R_\oplus = \text{Earth's radius} = 6378.14 \, \text{km}$

The μ values for all the other planets can be found in Appendix A.

7.2 Conservation Laws in a Gravitational Field

7.2.1 Angular Momentum Conservation

Let's consider the most general case that a body m moves around the common inertial CM, which we define as the origin of our coordinate system. Its radial vector be **r** and its velocity **v**. Under these conditions its motion is determined by the equation of motion (7.1.19). Its angular momentum L, according to physics, is a conserved quantity, which reads

$$\boxed{h := \frac{L}{m} = r \times v = const} \qquad \textbf{angular momentum (invariant)} \qquad (7.2.1)$$

where h is the *mass-specific angular momentum*. In order to check whether for motion in a gravitational field the angular momentum is indeed conserved, we take the time-derivative of h: $\dot{h} = v \times v + r \times \ddot{r}$. As according to Eq. (7.1.19) $\ddot{r} \parallel r$ and the cross product of a vector with itself always vanishes, so does $\dot{h}$. Test passed.

Let's assume that the body m has initial velocity v_0 at the initial position r_0. Vectors r_0 and v_0 span a plane. Because of Eq. (7.2.1) the initial angular momentum h_0 is vertical to r_0 and v_0, and also at later times $h \cdot r = h \cdot v = 0$ holds. So, r as well as v is always vertical to h. In other words, because $h = const$, the body m always maintains its motion in the plane, which was spanned by the initial r_0, v_0.

Note: *Strictly speaking, the motion in a plane with $r, v \perp h$ is valid only for $h \neq 0$. For $h = 0$ the motion is on a line. (see Section 7.4.5).*

Therefore, at any point in time the plane spanned by r, v is the same. As shown in Section 14.4.1 any pair (r, v) also determines the shape of an orbit. Because (r, v) fully determines the state of an orbit it is called **state vector**. We conclude:

> The motion of a body m always takes place in a constant plane through the center of mass common with M, perpendicular to the angular momentum h, spanned by r and v.

As conservation of angular momentum is a very general property, independent of the details of gravitational force or its potential, it is even true for spaces with dimension other than three. We will come back to this peculiarity in Section 7.5.

7.2.2
Motion in a Plane

Before we go into more physical details, we need to know more about the mathematics of a motion in a plane. We assume that the body m moves on an arbitrary curved path. As its motion is restricted to a plane, we adopt a co-rotating two-dimensional co-rotating polar coordinate system (r, θ) as our reference frame with the instantaneous radial unit vector $u_r = \hat{r}$ and u_θ perpendicular to it. With respect to this instantaneous frame, the position vector r

will quite generally change both its length, dr, and its direction, $d\theta$. According to Fig. 7.4, the differential change vector $d\mathbf{r}$ can be decomposed as follows:

$$d\mathbf{r} = \mathbf{u}_r \cdot dr + \mathbf{u}_\theta r \cdot d\theta$$

So, the following holds for the velocity vector $\mathbf{v}$:

$$\mathbf{v} = \frac{d\mathbf{r}}{dt} = \dot{r} \cdot \mathbf{u}_r + r\dot{\theta} \cdot \mathbf{u}_\theta =: \mathbf{v}_r + \mathbf{v}_\theta$$

With the definition of the angular velocity

$$\omega := \dot{\theta} \qquad \text{angular velocity} \tag{7.2.2}$$

one finally gets

$$\mathbf{v} = (v_r, v_\theta) = (\dot{r}, \omega r) \tag{7.2.3}$$

from which (c.f. Eq. (7.2.5)) it follows that

$$v^2 = \dot{r}^2 + \omega^2 r^2 = \dot{r}^2 + \frac{h^2}{r^2} \tag{7.2.4}$$

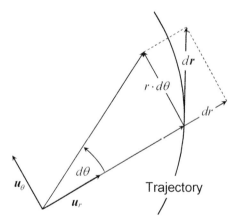

Figure 7.4 Decomposition of the differential position vector $d\mathbf{r}$ in the co-rotating coordinate system.

Note: $\dot{r}$ is only the radial part of the velocity vector $\dot{\mathbf{r}}$, and <u>not</u> its value, $\dot{r} \neq |\dot{\mathbf{r}}|$. To avoid confusion, we will therefore always write $\mathbf{v}$ rather than $\dot{\mathbf{r}}$.

7.2.3
Kepler's Second Law

With the decomposition of the velocity into a radial and a tangential component, $\mathbf{v} = \mathbf{v}_r + \mathbf{v}_\theta$, and since $\mathbf{r}$ is parallel to $\mathbf{v}_r$, the angular momentum can be written as $h = |\mathbf{r} \times (\mathbf{v}_r + \mathbf{v}_\theta)| = r \cdot v_\theta$, and because of Eq. (7.2.3) we have

$$h = \omega\, r^2 = \dot\theta\, r^2 = \text{const} \tag{7.2.5}$$

Note: *It is the angular momentum h and its conservation in time, which makes a body to orbit (ω!) steadily around a central mass and prevents the masses in our universe to instantly collapse (cf. Section 7.4.5 for trajectories with $h = 0$).*

Equation (7.2.5) is of notably significance. It states that the further a body on its orbit departs from the origin, the less its instantaneous angular velocity becomes, and vice versa. This has an immediate and important implication: If we calculate the infinitesimal area dA, which the position vector $\mathbf{r}$ sweeps by advancing through $d\theta$, it is determined according to Fig. 7.4 and Eqs. (7.2.5) and (7.2.2) by

$$dA = \frac{1}{2} r\,(r \cdot d\theta) = \frac{1}{2} r^2 \cdot d\theta = \frac{1}{2} r^2 \omega \cdot dt = \frac{h}{2} dt \tag{7.2.6}$$

That is, $h = 2\dot A = \text{const}$. So if we consider a constant time interval Δt, we find

$$\Delta A = \frac{h}{2} \int dt = \frac{1}{2} h \cdot \Delta t = \text{const} \qquad \textbf{Kepler's second law} \tag{7.2.7}$$

> The angular momentum can be interpreted as a constant areal velocity of the body: The area, which the position vector sweeps in equal time intervals, is constant.

Note that Kepler's second law is valid for any possible motion of a body in a conservative force field relative to any origin of the coordinate system and not only for bound orbits (ellipses, circles), as Kepler postulated. This is because up to here we have considered solely the law of momentum conservation and not any of Newton's laws.

Remark: *A conservative force is a force, which conserves the mechanical (= potential plus kinetic) energy of a body subject to it, such as a gravitational force. Dissipative forces or eddy fields, which arise from time-dependent magnetic fields, are non-conservative.*

7.2.4
Energy Conservation

For the motion of a body in a gravitational field, the general law of energy-conservation (Eq. (7.1.8)) reduces to

$$\varepsilon_{kin} + \varepsilon_{pot} =: \varepsilon = \text{const} \tag{7.2.8}$$

where $\varepsilon_{pot} := E_{pot}/m = U(r)$ is the *specific potential energy*, $\varepsilon_{kin} = E_{kin}/m = \frac{1}{2}v^2$ is the *specific kinetic energy*, and $\varepsilon := E_{tot}/m$ is the so-called *specific orbital energy* (a.k.a. *specific mechanical energy*). This results in the following important equations:

$$\frac{1}{2}v^2 = \frac{\mu}{r} + \varepsilon \qquad (7.2.9)$$

$$\boxed{v^2 = \mu\left(\frac{2}{r} - \frac{1}{a}\right)} \qquad \text{vis-viva equation} \qquad (7.2.10)$$

$$\frac{1}{2}\dot{r}^2 = \frac{\mu}{r} - \frac{h^2}{2r^2} - \frac{\mu}{2a} \qquad (7.2.11)$$

with $a := -\mu/2\varepsilon$. As we will see that later (see Eq. (7.3.14)), a is geometrically the semi-major axis of the orbit (that is $a > 0$ for ellipses and $a < 0$ for hyperbolas). The last of the three equations results from inserting Eq. (7.2.4) into Eq. (7.2.10).

Remark: *The German mathematician Leibniz (1646–1716) coined the name "vis viva" (Latin, meaning "living force") for the kinetic energy mv^2 (in those days) to distinguish it from Newton's gravitational force, called just "vis", and later also called "vis morte" (meaning "dead force") to distinguish it from "vis viva." You can image that this caused a serious fight between Newtonians and Lebnizians on whether their forces were dead or alive. Actually, all this only shows that at that time people did not really understand the difference between force, linear momentum, and energy, and their inter-relation: $F = dp/dt = -dE/dr$*

7.2.5
Rotational Potential

On pure mathematical grounds, the expression $-h^2/2r^2$ in Eq. (7.2.11), like its preceding μ/r, could be considered a potential: a radial angular momentum potential (rotational potential)

$$U_{rot}(r) := \frac{h^2}{2r^2} = \frac{1}{2}\omega^2 r^2 \qquad (7.2.12)$$

The latter follows from Eq. (7.2.5). Also physically this makes sense. Because in accordance with Eqs. (7.1.5), (7.2.5), and (7.1.6), the corresponding centrifugal force would be

$$F_{rot} = -m\frac{d}{dr}\left(\frac{h^2}{2r^2}\right) = m\frac{h^2}{r^3}\frac{dr}{dr} = m\frac{h^2}{r^3}\frac{r}{r} = m\omega^2 r \qquad (7.2.13)$$

This is the well-known formula for the centrifugal force in physics. It pushes the orbiting body towards the outside (positive sign). For example, for a circular orbit $r = const$, or $\dot{r} = 0$, it follows from Eq. (7.2.11) that $\mu/r = \mu/a = -h^2/2r^2$ meaning that the centrifugal force compensates the gravitational force at any point of the orbit.

Generalizing Eq. (7.1.4), the corresponding rotational energy would be

$$E_{rot} = m U_{rot}(r) = \frac{1}{2} m \omega^2 r^2 \qquad (7.2.14)$$

which is also a quite familiar expression.

7.3
Motion in a Gravitational Field

7.3.1
Orbit Equation

So far, by applying the equation of motion and general conservation laws, we have been able to determine general features of the motion without knowing the explicit solution. To obtain further details of the orbit, we have to solve the equation of motion. We can do that directly by applying some tricks. In preparation for these tricks, we differentiate $r^2 = \mathbf{r} \cdot \mathbf{r}$ which results in $2r\dot{r} = vr + rv = 2rv$ and therefore $\mathbf{r} \cdot \mathbf{v} = r \cdot \dot{r}$, of which we will make use in a moment.

After these preparatory steps, we will apply the first trick. Take the cross product of $\mathbf{h}$ with Eq. (7.1.19). This yields

$$\mathbf{h} \times \ddot{\mathbf{r}} = -\frac{\mu}{r^3}(\mathbf{h} \times \mathbf{r}) = -\frac{\mu}{r^3}[(\mathbf{r} \times \mathbf{v}) \times \mathbf{r}] = -\frac{\mu}{r^3}[\mathbf{v}(\mathbf{r} \cdot \mathbf{r}) - \mathbf{r}(\mathbf{r} \cdot \mathbf{v})]$$

$$= -\frac{\mu}{r^3}\left(r^2 \cdot \mathbf{v} - r\dot{r} \cdot \mathbf{r}\right) = -\mu \left(\frac{1}{r}\mathbf{v} - \frac{\dot{r}}{r^2}\mathbf{r}\right) \qquad (7.3.1)$$

$$= -\mu \frac{d}{dt}\left(\frac{\mathbf{r}}{r}\right) = -\mu \frac{d\hat{\mathbf{r}}}{dt}$$

with $\hat{\mathbf{r}} = \mathbf{r}/r$ the unit vector in the $\mathbf{r}$ direction. Because $\dot{\mathbf{h}} = 0$ this equation can be integrated directly to give

$$\mathbf{h} \times \mathbf{v} = -\mu \hat{\mathbf{r}} - \mathbf{A} = -\mu (\hat{\mathbf{r}} + \mathbf{e}) \qquad (7.3.2)$$

with $\mathbf{A} = \mu \mathbf{e}$ the integration constant, which is determined by the initial conditions. Apart from $\mathbf{h}$ and ε, $\mathbf{e}$ (or $\mathbf{A}$ respectively) is also an *invariant* of the system: $\mathbf{e}$ is called the *eccentricity vector*, and $\mathbf{A}$ is called the *Laplace–Runge–Lenz vector* (a.k.a. *Runge–Lenz vector* or *Laplace vector*). From Eq. (7.3.2) we

7.3 Motion in a Gravitational Field

get

$$e = \frac{1}{\mu}(v \times h) - \hat{r}$$

eccentricity vector (7.3.3)

$$= \left(\frac{1}{r} - \frac{1}{a}\right)r - \frac{1}{\mu}(rv)v = \text{const}$$

where the latter follows from Eqs. (7.2.1) and (7.2.10). Because $v \times h$ and $\hat{r}$ lie in the plane of motion, so must e (and also A), that is $h \cdot e = 0$.

We now use another trick to directly derive the equation for the orbit trajectory. We multiply Eq. (7.3.2) with r and get

$$-\mu(r + e \cdot r) = r \cdot (h \times v) = -h(r \times v) = -h^2 \tag{7.3.4}$$

from which we derive

$$\frac{h^2}{\mu} = r(1 + e \cdot \cos\theta)$$

with $\hat{e}\hat{r} = \cos\theta$. This equation takes us finally to

$$\boxed{r = \frac{p}{1 + e \cdot \cos\theta}} \quad @\ h \neq 0 \qquad \text{orbit equation} \tag{7.3.5}$$

with

$$p := h^2/\mu =: a\left(1 - e^2\right) \qquad \text{semi-latus rectum} \tag{7.3.6}$$

$$\boxed{h^2 = \mu a\left(1 - e^2\right)} \tag{7.3.7}$$

A geometrical analysis of Eq. (7.3.5) shows that it describes four conic sections: circle ($e = 0$), ellipse ($0 < e < 1$), parabola ($e = 1$) and hyperbola ($e > 1$). These will be discussed in detail in Section 7.4. The geometric interpretation of their characteristic parameters a, e, b, p is depicted in Figure 7.5. As we will see later, the parameter a defined in Eq. (7.3.6) is the semi-major axis.

Note 1: *The orbit equation is only valid as long as $h \neq 0$. If the body m takes on $h = 0$, then according to Eq. (7.2.5) $\omega = 0$ and therefore $\theta = \text{const}$. Thus the body directly falls towards the central body M (see also note above and Section 7.4.5 for details).*

7 Orbits

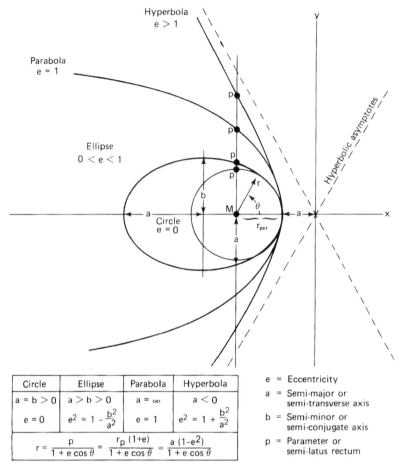

Figure 7.5 Geometrical representation of the parameters a, e, b, p for the four conic sections.

Note 2: *Having achieved geometrical interpretations of a and e, we see that the equations (7.2.10)*

$$v^2 = \mu\left(\frac{2}{r} - \frac{1}{a}\right) \quad \text{vis-viva equation}$$

and Eqs. (7.3.5) with (7.3.7)

$$h = \omega r^2 = \sqrt{\mu a (1 - e^2)}$$

are very important, simple, and useful equations, which link orbit state properties, r, v, ω, to geometric properties of the orbit, a, e.

The eccentricity vector **e** points from the focal point (central body) to its smallest approach distance, the so-called periapsis. Its absolute value is the orbit eccentricity e, which describes how elongated the orbit is. Because $\hat{e}\hat{r} = \cos\theta$, θ measures the angle between the radial vector to the periapsis and the radial vector to the current position of the body. We hence can determine the exact position on the orbit with θ. This important parameter θ is called the **true anomaly** (a.k.a. *orbit angle*). In summary we figured out that:

> The eccentricity vector **e** is the base vector relative to which the position of the orbit, the true anomaly θ, is measured: $\hat{e}\hat{r} = \cos\theta$. It points from the orbit's focal point (central body) in the direction of the periapsis (see Fig. 7.7 later). Its absolute value describes the elongation of the orbit.

How did we succeed in solving the apparently difficult vectorial equation of motion so swiftly? We made use of our previous knowledge that the momentum is a constant of motion and of its relation to **r** and **v**, namely $\mathbf{h} = \mathbf{r} \times \mathbf{v}$. Therefore we just had to integrate only once to find, besides **h**, the second integral of motion **e**.

7.3.2
Orbit Velocity, Flight Path Angle

Position vector **r** plus velocity vector **v** make up the all-important state vector $(\mathbf{r}, \mathbf{v})$, which unequivocally defines the orbit, as we will see later. We want to determine it in the co-rotational reference system (see Fig. 7.6) as a function of the orbital parameters a, e, θ.

Orbit Velocity

While in this reference system the radial vector is just $\mathbf{r} = (r, 0, 0)$, the velocity vector is given according to Eq. (7.2.3) as

$$\mathbf{v} = (\dot{r}, r\dot{\theta}, 0)$$

To calculate its components, we differentiate the orbit equation (7.3.5) in the form $h^2/\mu r = 1 + e\cos\theta$, and with $h = \dot{\theta}r^2$ (see Eq. (7.2.5)) we get

$$-\frac{h^2}{\mu r^2}\dot{r} = -e\sin\theta \cdot \dot{\theta} = -e\sin\theta \frac{h}{r^2}$$

From this follows that

$$\dot{r} = \frac{e\mu}{h}\sin\theta \tag{7.3.8}$$

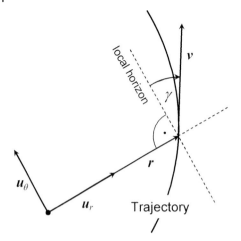

Figure 7.6 Flight path angle γ in the co-rotational reference system.

Because of Eq. (7.2.5) we find

$$r\dot\theta = r\frac{h}{r^2} = \frac{h}{r} = \frac{\mu}{h}(1+e\cos\theta)$$

If we plug this into $v = (\dot r, r\dot\theta, 0)$ we finally obtain

$$v = \frac{\mu}{h}(e\sin\theta,\ 1+e\cos\theta,\ 0) \tag{7.3.9}$$

By squaring its components we also derive the absolute value of the velocity vector to be

$$v = \frac{\mu}{h}\sqrt{1+2e\cos\theta + e^2} \tag{7.3.10}$$

Flight Path Angle

We now define the much used **flight path angle** γ. This is the angle that the velocity vector v makes with the *local horizon*, the vertical on the radial vector. According to Fig. 7.6

$$v = v(\sin\gamma, \cos\gamma, 0) \tag{7.3.11}$$

Comparing this with Eq. (7.3.9) we find

$$\cos\gamma = \frac{1+e\cos\theta}{\sqrt{1+2e\cos\theta + e^2}}$$
$$\sin\gamma = \frac{e\sin\theta}{\sqrt{1+2e\cos\theta + e^2}} \tag{7.3.12}$$

7.3.3
Orbital Energy and Angular Momentum

We intuitively know that the orbital energy must somehow depend on its shape and/or size. To derive the relationship we square Eq. (7.3.2) on both sides

$$\mu^2 (e + \hat{r})^2 = \mu^2 \left(e^2 + 2e\hat{r} + 1\right) = (h \times v)^2 = h^2 \cdot v^2$$

The latter holds because of $h \perp v$. From Eq. (7.3.4) it follows that $e\hat{r} = h^2/\mu r - 1$. The vis-viva equation (7.2.9) states $v^2 = 2\mu/r + 2\varepsilon$. This applied to the above equation leads to

$$2h^2\varepsilon = \mu^2 \left(e^2 - 1\right) \qquad (7.3.13)$$

From this we derive the following:

> The specific orbital energy ε is negative for $e < 1$, zero for $e = 1$, or positive for $e > 1$.

With Eq. (7.3.7), Eq. (7.3.13) can be transformed into the following simple expression for the specific orbital energy:

$$\boxed{\varepsilon = -\frac{\mu}{2a}} \qquad (7.3.14)$$

Equation (7.3.14) proves that the parameter a (semi-major axis) defined in Eq. (7.3.7) is identical with the one in the vis-viva equations (7.2.10) and (7.2.11), and thus the geometrical interpretation of the semi-major axis formerly announced is correct.

Equation (7.3.14) not only is very simple, but also is remarkable because like Eq. (7.3.7) and the vis-viva equation ((7.2.10)) it relates the important physical quantity "total orbital energy" to the geometrical size of the orbit, the semi-major axis. Thereby:

> The orbital energy can be directly read from the orbital size: the larger the orbit, the larger (negative sign!) the orbital energy.

Does such a direct relation also exist for the angular momentum? Absolutely: According to Eq. (7.3.6) the semi-latus rectum p is a direct measure of the angular momentum. However, p is a quite uncommon parameter to describe an orbit, except for a parabola. Rather a and e are used more often, because they are Keplerian elements (see Section 7.3.4). In fact, and according to Eq. (7.3.13), the eccentricity can be employed to gauge the angular momentum *at a given orbital energy e*: If $e = 1$ (rectilinear orbit, see Section 7.4.5) then the angular momentum vanishes, and for $e = 0$ (circular orbit) the angular momentum becomes maximum, $h = \mu\sqrt{a}$. At a given orbital energy the circular orbit hence is that bound orbit with the biggest angular momentum.

Viral Theorem

A body on its trajectory continuously changes position and velocity, and with it the potential and kinetic energy. Since the orbital energy is constant, kinetic energy is transformed into potential energy and the other way round, i.e. $\dot{\varepsilon}_{kin} \propto -\dot{\varepsilon}_{pot}$. Only on a circular orbit ε_{kin} and ε_{pot} remain constant as (see Eq. (7.4.5))

$$\varepsilon_{pot} = -2\varepsilon_{kin} = const \quad @ \text{ circular orbit}$$

However, it is possible to show that for any periodic orbit in a gravitatively bound many-body system: If $\langle \varepsilon_{kin} \rangle$ and $\langle \varepsilon_{pot} \rangle$ are the kinetic and potential energy time-averaged over an orbital period, then the following holds

$$\langle \varepsilon_{pot} \rangle = -2\langle \varepsilon_{kin} \rangle = const \quad \textbf{virial theorem} \quad (7.3.15)$$

Reference Zero

It is important to realize that the gravitational potential as a solution to the Poisson equation (7.1.2) is of the general form

$$U(r) = -\frac{\mu}{r} + U_0$$

with

$$\left(\frac{\partial}{\partial x^2} + \frac{\partial}{\partial y^2} + \frac{\partial}{\partial z^2} \right) U_0(r) = 0$$

A particular solution is $U_0 = const$. So the reference zero of the gravitational potential and, due to $\varepsilon_{pot} = E_{pot}/m = U(r)$ (see Eq. (7.1.4)), hence also of the specific potential energy, can be chosen freely. For instance, the potential energy of a body near the Earth is usually measured in terms of its altitude above the Earth's surface: $U(h) = E_{pot}/m = gh$ for $h \ll R_\oplus$. In astrophysics, though, the vis-viva equation is based on the assumption $U_0 = 0$. So for $r \to \infty$ the potential energy is defined as becoming zero, that is to say:

> In astrodynamics the potential energy is measured with regard to $r = \infty$.

For instance, in astrodynamics a body near the Earth would have the potential $U(h) = -\mu_{earth}/(r_{surface} + h) \approx -\mu_{earth}/r_{surface} + gh$. Things become more complicated if we consider embedded potentials. For an interplanetary flight we have to consider the Earth's potential at position r_{earth} in the Sun's potential. In this case the body would have $U(h) = -\mu_{sun}/r_{earth} - \mu_{earth}/r_{surface} + gh$.

7.3.4
Orbital Elements (Keplerian Elements)

The set of differential equations (7.1.19) to be solved in general is a system of six independent equations of first order, which, written for instance in a rectangular coordinate system, read

$$\dot{x} = v_x, \quad \dot{v}_x = -\mu x/r^3$$
$$\dot{y} = v_y, \quad \dot{v}_y = -\mu y/r^3$$
$$\dot{z} = v_z, \quad \dot{v}_z = -\mu z/r^3$$

Each of these equations calls for an integral constant, in total six independent constants, which are in general called "integrals of motion," or "orbital elements" in astrodynamics. But there may be different sets of orbital elements and any set is equivalent to any other. The orbital elements are defined by the state of the body at any time during the motion. Usually the initial state (r_0, v_0) is chosen, which may be considered by itself as a set of orbital elements (3 + 3 components = 6 elements). This property is the reason for the significance of the state vector (r, v). The state vector has a major drawback: r and v change continuously with time.

We have already derived other orbital elements which are constant in time: angular momentum h, eccentricity vector e, and orbital energy ε. These, however, represent seven independent elements. Their number is reduced to six because of the condition $h \cdot e = 0$. Another condition, which links h, e and ε, is Eq. (7.3.13). So in total we have five independent constant orbital elements, which determine the orientation and shape of the orbit. One can show that this is the minimum number of elements for that. In order to determine the position of body on the orbit, we need an additional element, which we have also derived already: θ. This is the only element that is time-dependent. We now would have another set of six orbital elements.

In principle, there are many different sets of orbital elements. Usually, the so-called six **Keplerian elements** are chosen (Fig. 7.7), which bear the same nice property that five elements are constant and only one is time-dependent.

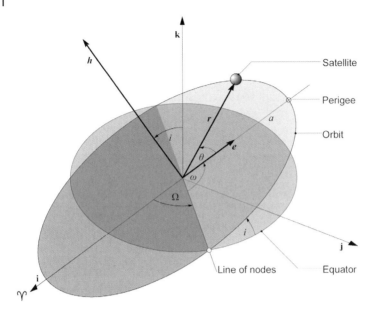

Figure 7.7 Graphical representation of the Keplerian elements of an elliptic satellite orbit in a topocentric equatorial coordinate system i, j, k, where the plane spanned by i and j with the i axis oriented towards the vernal equinox ♈ is the reference plane. The orbital plane and reference plane intersect in the so-called line of nodes. The line of nodes in turn intersects the orbit in the so-called ascending node (where the satellite moves towards the upper side with respect to h of the reference plane) and the descending node (vice versa).

Two of them describe the orientation of the orbital plane:

- Ω right ascension of ascending node (in short RAAN) = longitude of the ascending node = angle between the lines from the origin of the coordinate system to the vernal equinox and to the ascending node;
- i inclination = angle between the angular momentum vector h and the vector in z direction = angle between the corresponding planes perpendicular to these vectors.

Two elements describe the dimension and shape of the orbit:

- a length of the semi-major axis a measure of the size of the orbit.
- e eccentricity describes the elongation of the orbit.

One element determines the orientation of the orbit in the orbital plane:

- ω argument of periapsis = angle in the orbital plane between line of nodes and the periapsis measured in the direction of motion.

One element describes the current position of the body on the orbit:

- θ true anomaly = the angle between the periapsis and the current position vector r. θ can be and often is substituted by the mean anomaly M (see Eq. (7.4.15)), which according to Eq. (7.4.6), (7.4.18), and (7.4.30) is in direct correspondence to θ.

Note: Ω will loose its meaning for $i \to 0$ and ω for $e \to 0$, as then Ω or ω become undefined. Both happens for satellites in geostationary orbit.

7.3.5
Position on the Orbit

The most natural variable in our world to determine the sequence of events and measure their relative distance is time t. However, up to now time hasn't shown up as an orbital element to determine the sequence of orbit positions. Rather we found the orbit angle θ. We are therefore left to find $\theta(t)$. This is what we want to do in this section by looking for its equation of motion – or more properly "its equation of evolution" – and solve it. Maybe you didn't notice, but we have already met the equation of evolution for θ; it is Eq. (7.2.5)

$$\dot{\theta} = \frac{h}{r^2} = \frac{\mu^2}{h^3}(1 + e \cdot \cos\theta)^2 \quad \text{equation of motion for } \theta \quad (7.3.16)$$

Here the latter relation results from inserting the orbit equation (7.3.5) and with $h = \sqrt{p\mu}$. Note that, together with the orbit equation, this is a crucial equation. It is the equation of motion to determine progress for the orbit angle. Separating the variables and integrating on the left side with respect to time results in

$$\frac{\mu^2}{h^3}(t - t_0) = \int_0^\theta \frac{d\theta'}{(1 + e \cdot \cos\theta')^2} \quad (7.3.17)$$

with t_0 the time of the passage of the body through the periapsis so that $\theta(t_0) = 0$. Sometimes t_0 is called (astronomical) **epoch**. This epoch is to be distinguished from the notion "standard epoch J2000" (= January 1, 2000, 12.00h UTC (see Section 13.2)). So the general meaning of "epoch" is "a reference point in time".

Keplerian Problem

We still have to solve the integral in Eq. (7.3.17) for arbitrary e. Except for a circular and parabolic orbit this is too complicated to do by regular means. Even if we were able to solve it, the solutions would we very complicated (see Eqs. (7.4.18) and (7.4.30)). The real problem however is that these solutions do not display the time dependence explicitly, that is $\theta = \theta(t)$, but only implicitly, that is $t = t(\theta)$. One could, though, for any given point in time solve $t = t(\theta)$ for θ numerically. But this requires quite an effort. In the face of this problem, Kepler at the beginning of the 17th century proposed a method that shifts the problem analytically to a simpler one, which can be solved with less effort, though still numerically. We will also follow Kepler's elegant method in the next section. In fact, as we will see later, his method provides an easy way to solve Eq. (7.3.17) analytically.

The so-called "Keplerian problem" is historically the problem of finding the orbit position at a given time, if it was known at an earlier time. The background is the astronomical problem, even today, of finding a celestial body if it was observed at earlier times and having determined its orbital elements. Since the Keplerian elements a, e, i, Ω, ω are constant in time, Kepler's problem lies in the difficulty of determining $\theta = \theta(t)$.

In the following section we will examine the specific properties of each type of orbit and the solutions to the Keplerian problem, separately.

7.4
Keplerian Orbits

In this section we study the detailed properties of the different Keplerian orbits and display their basic results. This includes in particular the analytic solutions to $t = t(\theta)$ for the elliptical and hyperbolic orbits and the numerical algorithms to calculate $\theta = \theta(t)$.

7.4.1
Circular Orbit

For $e = 0$, Eq. (7.3.17) can be solved immediately and according to Eq. (7.3.5) and because of $h = \sqrt{\mu a}$ (Eq. (7.3.7)) we get

$$r = a = p = const \tag{7.4.1a}$$

$$\theta = \sqrt{\frac{\mu}{a^3}} \cdot (t - t_0) \tag{7.4.1b}$$

which is a circular orbit with radius a with a negative specific orbital energy according to Eq. (7.3.14). From Eq. (7.4.1b) we find $2\pi = T\sqrt{\mu/a^3}$ for a full rotation, and hence for the orbital period T (cf. Eq. (7.4.11))

$$\boxed{T = 2\pi \sqrt{\frac{a^3}{\mu}}} \quad \text{orbital period} \tag{7.4.2}$$

With Eq. (7.2.4) and $\dot{r} = 0$, we get $v = h/a = const$, i.e. a constant orbital velocity. Because $h = \sqrt{\mu a}$ and Eq. (7.4.1a) this can also be expressed as

$$\boxed{v = \sqrt{\frac{\mu}{r}} = \sqrt{\frac{\mu}{a}} = const} \tag{7.4.3}$$

The orbital velocity decreases with the root of the orbital radius. For example, the velocity of a body that circles the Earth in the theoretically lowest orbit possible, $r = R_\oplus$, is, with $\mu_\oplus = g_0 R_\oplus^2$ (see Eq. (7.1.20))

$$v_0 = \sqrt{g_0 R_\oplus} = 7.905 \text{ km s}^{-1} \quad \text{first cosmic velocity} \quad (7.4.4)$$

This is the highest possible orbital circular velocity around the Earth, as according to Eq. (7.4.3), the circular velocity decreases with an increasing orbital altitude r. In a typical low Earth orbit, such as the International Space Station, of 400 km, it is only 7.67 km s^{-1}. Though speed decreases for higher orbits you still need more energy to reach higher orbits. This is because from Eq. (7.3.14) $\varepsilon = -\mu/2a = -\mu/2r$, i.e.

$$2\varepsilon = \varepsilon_{pot} = -2\varepsilon_{kin} = const \quad (7.4.5)$$

The latter follows because $\varepsilon_{kin} = v^2/2 = \mu/2a$. In words this means:

> - At any point of a circular orbit the absolute value of the potential energy is twice that of the kinetic energy (cf. Eq. (7.3.15)).
>
> - The orbital energy of a circular orbit is negative, and its absolute value equals that of the kinetic energy.

The energy for lifting a body into a higher circular orbit therefore is determined as follows:

$$\Delta\varepsilon = \Delta\varepsilon_{kin} + \Delta\varepsilon_{pot} = -\frac{1}{2}\Delta\varepsilon_{pot} + \Delta\varepsilon_{pot} = \frac{1}{2}\Delta\varepsilon_{pot} > 0$$

> Orbit lifting reduces the kinetic energy by a given amount, but it increases the potential energy by double that amount. Therefore, although orbit lifting increases the orbital energy, the orbit velocity decreases.

On the other hand, this leads to the paradoxical situation that a S/C in a circular LEO, decelerated due to drag, in effect gains velocity, because it spirals down to lower altitudes.

7.4.2
Parabolic Orbit

For $e = 1$ we get a parabola. Its semi-major axis is $a = \infty$, according to Eq. (7.3.7), and its orbital energy $\varepsilon = 0$, according to Eq. (7.3.13). In this case one is able to directly integrate Eq. (7.3.17) with the substitution $x := \theta/2$:

$$\int_0^\theta \frac{d\theta'}{(1+\cos\theta')^2} = \int_0^\theta \frac{d\theta'}{(2\cos^2\theta'/2)^2} = \frac{1}{2}\int_0^{\theta/2} \frac{dx}{\cos^4 x}$$

$$= \frac{1}{2}\int_0^{\theta/2} \left(\frac{1}{\cos^2 x} + \frac{\tan^2 x}{\cos^2 x}\right) dx = \frac{1}{2}\left(\tan x + \frac{1}{3}\tan^3 x\right)\bigg|_0^{\theta/2}$$

$$= \frac{1}{2}\left(\tan\frac{\theta}{2} + \frac{1}{3}\tan^3\frac{\theta}{2}\right)$$

The position equation can now be provided analytically

$$2\sqrt{\frac{\mu}{p^3}}(t-t_0) = \tan\frac{\theta}{2} + \frac{1}{3}\tan^3\frac{\theta}{2} = \frac{2}{3}\left(\frac{r}{p}+1\right)\sqrt{2\frac{r}{p}-1}$$

which sometimes is called **Barker's equation**.

Note: *Because for a parabola $a = \infty$ and $e = 1$, but the semi-latus rectum $p = a(1-e^2)$ is a finite, non-zero number, the semi-latus rectum is the sole orbit element that exhaustively describes the shape of a parabolic orbit.*

By applying Cardano's method and by Descartes' rule of signs it can be shown (exercise, Problem 7.9) that there is only one real solution to this equation for θ and r, respectively, namely

$$\tan\frac{\theta}{2} = \left(\sqrt{q^2+1}+q\right)^{1/3} - \left(\sqrt{q^2+1}-q\right)^{1/3}$$

$$= 2\sinh\left(\frac{1}{3}\operatorname{arcsinh} q\right)$$

$$2\frac{r}{p} = \left(\sqrt{q^2+1}+q\right)^{2/3} + \left(\sqrt{q^2+1}-q\right)^{2/3} - 1 \qquad (7.4.6)$$

$$= 2\cosh\left(\frac{2}{3}\operatorname{arcsinh}|q|\right) - 1$$

with

$$q = 3\sqrt{\frac{\mu}{p^3}}(t-t_0)$$

where the later expressions follow from Chebyshev polynomials, which are particularly useful for small time intervals $t - t_0$. So, also for the parabolic case we are able to solve the Keplerian problem.

The parabolic orbit is a limiting orbit where the body is just able to escape the central body with $v_\infty (r = \infty) = 0$. We want to calculate the velocity, the so-called escape velocity v_{esc}, needed for a body at any position r from a central body to get to parabolic speed. The vis-viva Eq. (7.2.9) together with $\varepsilon = 0$ results in

$$\boxed{v_{esc} = \sqrt{\frac{2\mu}{r}}} \qquad \text{escape velocity} \qquad (7.4.7)$$

For example, the velocity of a body at the surface of the Earth needed to escape Earth's gravitation – the so-called second cosmic velocity – is, with $\mu_\oplus = g_0 R_\oplus^2$ (see Eq. (7.1.20))

$$v_{esc,\oplus} = \sqrt{2 g_0 R_\oplus} = 11.180 \text{ km s}^{-1} \qquad \text{second cosmic velocity}$$

7.4.3
Elliptical Orbit

For $0 < e < 1$ the orbit is an ellipse which is the most general bounded orbit in a gravitational field (**Kepler's first law**). According to Eq. (7.3.14) its orbital energy is *negative*. According to Eq. (7.3.5) and with Eq. (7.3.6), there is a minimum and a maximum distance to the central body at the focal point (Fig. 7.8):

$$r_{per} = a(1-e) \quad \text{periapsis} \qquad (7.4.8a)$$

$$r_{apo} = a(1+e) \quad \text{apoapsis} \qquad (7.4.8b)$$

The general terms periapsis and apoapsis are also called pericenter and apocenter. Depending on the central body, the specific terms are perigee/apogee (Earth), perihelion/aphelion (Sun), periselene/aposelene (Moon), etc.

From Eqs. (7.4.8a) and (7.4.8b) the semi-major axis a and the eccentricity e are derived as

$$a = \frac{r_{per} + r_{apo}}{2} \qquad (7.4.9a)$$

$$e = \frac{r_{apo} - r_{per}}{r_{apo} + r_{per}} \qquad (7.4.9b)$$

The semi-minor axis b is given by

$$b = a\sqrt{1-e^2} = h\sqrt{\frac{a}{\mu}} = \langle r \rangle_\theta \qquad (7.4.9c)$$

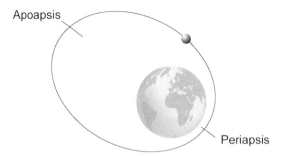

Figure 7.8 Elliptical orbit.

Geometrically, the eccentricity is the distance Δ from the center of the ellipse to its focal point normalized to a: $\Delta/a = (a - r_{per})/a = ae/a = e$. According to Appendix A the semi-minor axis b corresponds to the mean radius $\langle r \rangle_\theta$ averaged over θ.

The velocities at the periapsis and apoapsis follow from Eq. (7.3.10)

$$v_{per} = \frac{\mu}{h}(1+e) = \sqrt{\frac{\mu}{a}}\sqrt{\frac{1+e}{1-e}} \qquad (7.4.10a)$$

$$v_{apo} = \frac{\mu}{h}(1-e) = \sqrt{\frac{\mu}{a}}\sqrt{\frac{1-e}{1+e}} \qquad (7.4.10b)$$

According to Kepler's second law (Eq. (7.2.7)), the swept area ΔA during the time Δt is given by $\Delta A = \Delta t \cdot h/2$. If one integrates over a full rotation, the swept area ΔA is the area of the ellipse πab, and Δt is the orbital period T, and thus $T = 2\pi ab/h$. Because of Eq. (7.4.9c) $b = h\sqrt{a/\mu}$ we get

$$\boxed{T = 2\pi\sqrt{\frac{a^3}{\mu}}} \qquad \text{orbital period (Kepler's third law)} \qquad (7.4.11)$$

Remark 1: *Actually, Kepler stated in his third law for the relation of two orbit periods around the same central body: $T_1^2/T_2^2 = a_1^3/a_2^3$. This follows from the equation above.*

Remark 2: *It is quite remarkable that the orbital period, just as the specific orbital energy $\varepsilon = -\mu/2a$, does not depend on eccentricity, but only on the semi-major axis.*

Remark 3: *Kepler's third law can be used to determine the mass M of a celestial body: By precise determination of the orbital period and semi-major axis of a small moon or a satellite around the celestial body the standard gravitational parameter $\mu = GM$ and hence the mass M can be determined.*

Keplerian Transformation

We now seek to tackle the Keplerian problem, as already discussed in Section 7.3.5, and solve Eq. (7.3.17). For this we apply Kepler's method, which transfers the problem to a new angle parameter, the so-called **eccentric anomaly** E (see Fig. 7.9).

Remark: *You may be worried that a Roman letter rather than a Greek symbol symbolizes an angle. The answer to such "anomalies" is as always: for historic reason. Introduced by Kepler and getting used to it for centuries, nobody dares to switch to modern standards.*

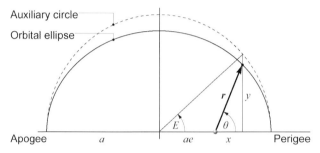

Figure 7.9 Geometric interpretation of the eccentric anomaly.

The transformation is performed geometrically by drawing a great circle around the ellipse with the radius a and projecting the position vector onto the horizontal and vertical axis. Analyzing the resulting segments x and y in Fig. 7.9 shows that

$$r \cos \theta = a \cos E - ae = a (\cos E - e) \tag{7.4.12a}$$

$$r \sin \theta = a\sqrt{1 - e^2} \sin E \tag{7.4.12b}$$

Squaring and applying the equations to each other results in

$$r = a(1 - e \cos E) \tag{7.4.13a}$$

$$\cos \theta = \frac{\cos E - e}{1 - e \cos E}, \quad \cos E = \frac{e + \cos \theta}{1 + e \cos \theta} \tag{7.4.13b}$$

$$\sin \theta = \frac{\sin E \sqrt{1 - e^2}}{1 - e \cos E}, \quad \sin E = \frac{\sin \theta \sqrt{1 - e^2}}{1 + e \cos \theta} \tag{7.4.13c}$$

$$\tan \frac{\theta}{2} = \tan \frac{E}{2} \sqrt{\frac{1 + e}{1 - e}} \tag{7.4.13d}$$

The first equation is nothing else than the orbit equation (7.3.5), only with the orbit angle substituted by the eccentric anomaly. It is exactly this linearization of the orbit equation that will make it possible to perform the final required integration, to which we will turn in a moment.

Mean Orbit Parameters

If we take the transformation literally, the body would move on a circle with a constant mean angular velocity

$$\langle \omega \rangle_t = \frac{1}{T} \int_0^T \omega \cdot dt = \frac{1}{T} \int_0^T \frac{d\theta}{dt} \cdot dt = \frac{1}{2\pi} \sqrt{\frac{\mu}{a^3}} \int_0^{2\pi} d\theta = \sqrt{\frac{\mu}{a^3}}$$

This idealized motion gives rise to two new definitions: mean motion n and mean anomaly M

$$\boxed{n := \langle \omega \rangle_t = \sqrt{\frac{\mu}{a^3}} = \frac{2\pi}{T}} \quad \text{mean motion} \quad (7.4.14)$$

$$\boxed{M := \int_{t_0}^t n \cdot dt' = n\,(t - t_0)} \quad \text{mean anomaly} \quad (7.4.15)$$

So, whereas $\omega = d\theta/dt$ is the instantaneous angular velocity, n is the time-averaged mean angular velocity. Only for circular orbits $n = \omega$. The intention behind defining M is to have a (theoretical) anomaly, which, like θ for the circle, advances monotonically in time. Just as for the true anomaly θ, the mean anomaly M is measured relative to the periapsis, and therefore $M(t_0) = 0$, where t_0 is the time of passage through periapsis. With these two definitions one attempts to convey with (n, M) the steadiness of (ω, θ) from the circle to the ellipse. But, if defined with respect to a determined epoch, in contrast to θ, M is not cyclically limited to the interval $[0,360°]$, but increases continuously with time, so it is unambiguous. In this way, like t, M is a perfect orbit position sequencer. This is why it is used in astronomy as a standard sequencing coordinate.

Note: *The mean advancing orbit angle is not E as one would intuitively have expected from the Keplerian transformation but $M = E(t) - e \sin E(t)$ (see Eq. (7.4.16) below). Therefore M does not have a general simple geometric interpretation. Only if $e \ll 1$ M can be shown to be close to the orbit angle as measured from the empty focal point.*

Remark: *The weird term "anomaly" for the angles M, E and θ goes back to the Ptolemaic astronomic system. At that time, each angle that could not be traced back to a true circular motion, appeared to be wrong or "anomalous".*

Keplerian Equation

We will now perform the final step to integrate Eq. (7.3.17). Since we switched from θ to E we also have to rewrite Eq. (7.3.17) in terms of E. That is, we need a differential equation for E as a function of time t. To get it, we differentiate $r = a(1 - e \cos E)$ with respect to time

$$\dot{r} = ae \sin E \cdot \dot{E} = \frac{e\mu}{h} \sin \theta$$

The latter is true because of Eq. (7.3.8). From this and from Eq. (7.3.7), Eq. (7.4.12b), and (7.4.13a) we get

$$\dot{E} = \frac{dE}{dt} = \frac{e\mu \sin \theta}{h \cdot ae \sin E} = \frac{1}{a}\sqrt{\frac{\mu}{a}} \frac{\sin \theta}{\sqrt{1-e^2} \sin E} = \frac{1}{r}\sqrt{\frac{\mu}{a}} = \frac{1}{1-e\cos E}\sqrt{\frac{\mu}{a^3}}$$

After separating the variables one gets

$$(1 - e \cos E) \cdot dE = n \cdot dt$$

This equation of motion is the equivalent to Eq. (7.3.16) and may be readily integrated to give

$$\boxed{M = n(t - t_0) = E(t) - e \sin E(t)} \quad \text{Keplerian equation} \quad (7.4.16)$$

Because t_0 is the time of passage through periapsis, $E_0 = E(t_0) = 0$. The Keplerian equation relates a given time t to E and according to Eq. (7.4.13b) this in turn is directly linked to θ. So, in principle, we have achieved our goal to determine the position of a body on its path as a function of time.

Note: *While the orbit equation (7.3.5) determines the shape of an elliptical orbit, the Keplerian equation determines the progress of the orbiting body on it. Therefore both are key equations for celestial mechanics.*

Remark: *You may wonder how to think up such an ingeniously simple transformation of a circle around the center of the ellipse to solve the Keplerian problem. The reason simply is this: Early on, Kepler assumed that the central body is located at the center of the ellipse. So he tried to derive the "anomalous" ellipse from this wrong assumption, which only holds for circles. So drawing circles around the ellipse's center came in quite naturally. Only later, when he realized that the central body must be located at the focal point, did he see the usefulness for a transformation of his earlier thoughts.*

Solving the Keplerian Equation

The Keplerian equation has still the drawback that it cannot be solved analytically for E. This is achieved only numerically. A common way is Newton's method. For this, one defines the function

$$f(E) = E - e \sin E - M$$

which transforms the problem to finding the root of $f(E)$. Newton's method states that you quickly get it in quadratic convergence (that means very fast if you are close to the solution) by the iteration

$$E_{i+1} = E_i - \frac{f(E_i)}{f'(E_i)} = E_i - \frac{E_i - e \sin E_i - M}{1 - e \cos E_i} \qquad (7.4.17)$$

with initial value $E_0 = M + 0.85 \cdot e \cdot \operatorname{sgn}(\sin M)$ and $0 \leq M = n(t - t_0) \leq 2\pi$. We thus have finally reached our goal:

> In order to determine the position of a body on an orbit at a given time t relative to a given t_0, we calculate E from Eq. (7.4.17) and apply it to Eqs. (7.4.13a) and (7.4.13b), whereby we get $r = r(E(t))$ and $\theta = \theta(E(t))$.

Remark: *Recently, in 1986, Conway found a more efficient algorithm than Newton's method for solving the Keplerian equation based on the cubically convergent Laguerre method (For details see Chobotov (2002, Section 4.1). Nevertheless, Newton's method is still a very good and frequently used algorithm.*

Analytical Solution

The Keplerian equation (7.4.16) provides a means to find analytical solutions to $r(t)$ and $\theta(t)$. Since from Eq. (7.4.13c) $\sin E = (\sqrt{1 - e^2} \cdot \sin \theta) / (1 + e \cos \theta)$ we immediately get from Eq. (7.4.16) the analytical solution to the Keplerian problem

$$M = n(t - t_0) = \arcsin \lambda - e\lambda \qquad (7.4.18)$$

where

$$\lambda = \frac{\sin \theta \sqrt{1 - e^2}}{1 + e \cos \theta} = \frac{r\sqrt{1 - e^2}}{ep}\sqrt{e^2 - \left(\frac{p}{r} - 1\right)^2} = \frac{\sqrt{2ar - a^2(1 - e^2) - r^2}}{ea}$$

The latter results from applying orbit equation (7.3.5). Just like the Keplerian equation, this equation suffers from the fact that the solution is only implicit. It could also be solved with the Newton's method with regard to θ or r. But to calculate the function $f(E)$ and even more so its derivative $f'(E)$ would be too much effort, even if done numerically. So when is comes down to having a fast algorithm, e.g. an orbit propagator, which every millisecond calculates the exact orbit position of a spacecraft, then Kepler plus Newton are invincible. If a high accuracy of the result is not decisive, then a graphical depiction of Eq. (7.4.18) like that given in Fig. 7.10 might be a favorable solution to the Keplerian problem. In addition it provides a good overview.

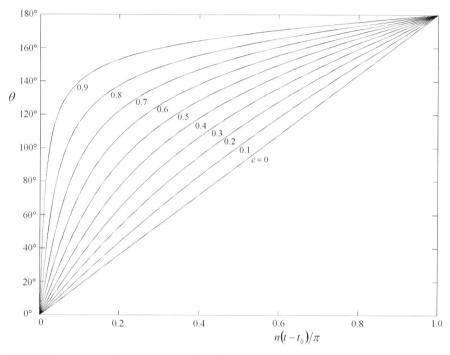

Figure 7.10 A graphical solution to the Keplerian problem.

Function Series Expansions

For small eccentricities, $e < 0.662\,743\,4\ldots$, explicit series expansions in M and hence t can be provided. Because their derivations are purely mathematical in nature and therefore do not provide any more insight into the Keplerian problem, we present some selected expansions without their derivations. These are (exercise; Problem 7.10, see e.g. Murray and Dermott (1999)):

$$\frac{r}{a} = 1 - e \cos M + e^2 \sin^2 M + \frac{3}{2} e^3 \cos M \sin^2 M + O\left(e^4\right)$$

$$\theta = M + 2e \sin M + \frac{5}{4} e^2 \sin(2M) + \frac{1}{4} e^3 \left(\frac{13}{3} \sin(3M) - \sin M\right) + O\left(e^4\right)$$

$$E = M + e \sin M + \frac{e^2}{2} \sin(2M) - \frac{e^3}{8} [\sin M - 3 \sin(3M)] + O\left(e^4\right)$$

7.4.4
Hyperbolic Orbit

For $e > 1$, the orbit is a hyperbola (Fig. 7.11) with its focal point at $(\pm ea, 0)$. Note that for a hyperbola a is defined as negatively (see Eq. (7.3.7)). Because of Eq. (7.3.14), the orbital energy is positive. According to Eqs. (7.3.5) and (7.4.10a) a hyperbola has its closest approach to the focal point at the periapsis with

$$r_{per} = \frac{h^2}{\mu(e+1)} = -a(e-1) \tag{7.4.19}$$

$$v_{per} = \frac{\mu}{h}(e+1) = \sqrt{-\frac{\mu}{a}}\sqrt{\frac{e+1}{e-1}} \tag{7.4.20}$$

A hyperbola doesn't possess an apoapsis. It reaches infinity at an asymptote, the angle of which is determined from Eq. (7.3.5) for $r \to \infty$ as

$$\cos\theta_\infty = -\frac{1}{e} \tag{7.4.21}$$

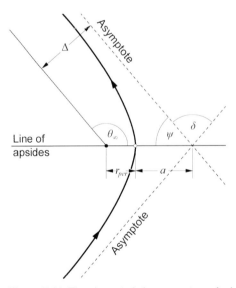

Figure 7.11 The characteristic parameters of a hyperbola.

The so-called aiming radius Δ, which is the distance between the focal point and the asymptote measured normal to the asymptote, is found from Fig. 7.11 to be

$$\Delta = -(r_{per} + a)\sin\beta = -ae\,\sin\theta_\infty$$

Here we have chosen a negative sign because Δ should be a positive value, but a is negative. With Eqs. (7.4.19) and (7.4.21) one gets

$$\Delta = -a\sqrt{e^2 - 1} = r_{per}\sqrt{\frac{e+1}{e-1}} \quad \text{aiming radius} \tag{7.4.22}$$

According to Eq. (7.3.10) the orbital velocity at infinity has the so-called hyperbolic excess velocity

$$v_\infty = \frac{\mu}{h}\sqrt{e^2 - 1} = \sqrt{-\frac{\mu}{a}} \quad \text{hyperbolic excess velocity} \tag{7.4.23}$$

For interplanetary flight the parameter

$$C_3 := v_\infty^2 = -\frac{\mu}{a} \quad \text{characteristic energy} \tag{7.4.24}$$

is often used. To attain a given hyperbolic excess velocity, for instance for an interplanetary transfer, a speed of $v > v_{esc}$ is required at the departure orbit. It is determined from the vis-viva equation (7.2.9) and Eq. (7.4.7) as

$$v^2 = \frac{2\mu}{r} + v_\infty^2 = v_{esc}^2 + v_\infty^2 \tag{7.4.25}$$

The related kick-burn to reach this departure velocity is called the *Oberth maneuver*. In fact, this equation is nothing other than the energy conservation equation, where only the kinetic energy shows up, because the potential energy vanishes at infinity.

Keplerian Equation and Solutions

One can show that the essential equations (7.4.13a)–(7.4.13d) and (7.4.16) of the ellipse can be expressed in a similar way for a hyperbola. We can shorten the derivation by accepting that there is also an eccentric anomaly F for the hyperbola, which has the following relation to the elliptical eccentric anomaly E: $E = iF$ with $i = \sqrt{-1}$. If one applies this relation to Eqs. (7.4.13a)–(7.4.13d), one directly obtains

$$r = a(1 - e\cosh F) \tag{7.4.26a}$$

$$\cos\theta = \frac{e - \cosh F}{e\cosh F - 1}, \quad \cosh F = \frac{e + \cos\theta}{1 + e\cos\theta} \tag{7.4.26b}$$

$$\sin\theta = \frac{\sqrt{e^2 - 1} \cdot \sinh F}{e\cosh F - 1}, \quad \sinh F = \frac{\sqrt{e^2 - 1} \cdot \sin\theta}{1 + e\cos\theta} \tag{7.4.26c}$$

$$\tan\frac{\theta}{2} = \tanh\frac{F}{2} \cdot \sqrt{\frac{e+1}{e-1}} \tag{7.4.26d}$$

and also

$$M = n \cdot (t - t_0) = e \sinh F(t) - F(t) \tag{7.4.27}$$

with

$$n = \sqrt{-\frac{\mu}{a^3}}, \tag{7.4.28}$$

where t_0 is the time of passage through the periapsis, and M is the mean anomaly (see Eq. (7.4.15)).

To determine the orbit position as a function of time just as with the ellipse, one has to solve Eq. (7.4.27) with Newton's method

$$F_{i+1} = F_i - \frac{e \sinh F_i - F_i - M}{1 - e \cosh F_i} \tag{7.4.29}$$

with the initial value

$$F_0 = e \sinh M - M$$

If the result is inserted into Eqs. (7.4.26a) and (7.4.26b) one obtains the parameterized orbit $r = r(F(t))$ and $\theta = \theta(F(t))$.

Analytical solution

In analogy to the elliptic case, from Eq. (7.4.27) with $\sinh F = \sqrt{e^2 - 1} \sin \theta / (1 + e \cos \theta)$ from Eq. (7.4.26c) we find an implicit solution for r and θ

$$M = n(t - t_0) = e\lambda - \operatorname{arcsinh} \lambda$$

$$= e\lambda - \ln \frac{\sqrt{e+1} + \sqrt{e-1} \cdot \tan(\theta/2)}{\sqrt{e+1} - \sqrt{e-1} \cdot \tan(\theta/2)} \tag{7.4.30}$$

with

$$\lambda = \frac{\sqrt{e^2 - 1} \cdot \sin \theta}{1 + e \cos \theta} = \frac{r\sqrt{e^2 - 1}}{ep} \sqrt{e^2 - \left(\frac{p}{r} - 1\right)^2}$$

7.4.5
Rectilinear Orbit

For a vanishing angular momentum, $h = 0$, the first trick in Section 7.3.1 is not applicable, so the general orbit equation (7.3.5) is not a solution. According to Eq. (7.2.5) this is the case whenever a transverse velocity $v_\theta = 0$ occurs (of course, as $h = r \cdot v_\theta = \text{const}$, v_θ is either always or never zero), for instance if one positions a body at an arbitrary distance from the central body with an initial velocity $v = 0$. The equation of motion (7.3.16)

$$\dot{\theta} = \frac{h}{r^2}$$

is, however, still valid. For $h = 0$ we find the solution

$$\theta = const$$

That is, the body moves on a straight line towards the central body (or away from it), until it crashes into the central body. To derive the orbit equation we resort to the vis-viva equations, which are usually a good starting point. We set $h = 0$ in Eq. (7.2.11), which results in

$$\dot{r}^2 = \frac{2\mu}{r} - \frac{\mu}{a}$$

Rectilinear Ellipse

For $r \leq 2a$ and $v < 0$ (inbound motion) this leads to the following equation of motion

$$\frac{dr}{dt} = -\sqrt{\frac{\mu}{a}}\sqrt{\frac{2a}{r} - 1} \quad \text{equation of linear motion} \tag{7.4.31}$$

Separating the variables and integration leads for $r(0) = r_0$ and $v(0) = v_0$ to the orbit equation

$$(t - t_0)\sqrt{\frac{\mu}{a^3}} = \frac{r}{a}\sqrt{\frac{2a}{r} - 1} - 2\arcsin\sqrt{\frac{r}{2a}} - \frac{r_0 v_0}{\sqrt{\mu a}} + 2\arccos\sqrt{\frac{r_0 v_0^2}{2\mu}} \tag{7.4.32}$$

Convince yourself that this indeed is the solution to Eq. (7.4.31) by differentiation.

Note: *The rectilinear orbit is a degenerate ellipse with $e = 1$. For this we have from Eq. (7.4.8a) and (7.4.8b) $r_{per} = 0$ and $r_{apo} = 2a$: The focal point coincides with the periapsis at $r = 0$ and the empty focus with the apoapsis at $r = 2a$. This is the reason for the seemingly odd condition $r \leq 2a$.*

The semi-major axis a is determined from the initial conditions via Eq. (7.4.31) as

$$a = \frac{\mu r_0}{2\mu - r_0 v_0^2} \tag{7.4.33}$$

7 Orbits

For $v_0 = 0$ we find $a = r_0/2$ and hence from Eq. (7.4.32)

$$t = \sqrt{\frac{r_0^3}{8\mu}} \left(\frac{2r}{r_0} \sqrt{\frac{r_0}{r} - 1} + 2 \arccos \sqrt{\frac{r}{r_0}} \right) \qquad (7.4.34)$$

For $r = 0$ the orbiting body crashes into the point of origin whereby the time to collision is

$$t_{col} = \frac{\pi}{\sqrt{8\mu}} r_0^{3/2} \qquad (7.4.35)$$

This result can also be derived when one applies the orbiting time of an elliptical orbit (Eq. (7.4.11)) $T = 2\pi\sqrt{a^3/\mu}$ with $a = r_0/2$ to this case. This is possible because it is independent of the orbit's eccentricity. The collision time then is half an orbit revolution.

Example
Let's assume the Earth's motion would abruptly be stopped. When would it crash into the center of the Sun (provided that the total masses of the Sun were combined in the center)? When would it crash onto the surface of the Sun?

Answer:
Mean distance from the mean Earth to Sun: $2a = r_0 = 149.6 \times 10^6$ km
Standard gravitational parameter for the Sun: $\mu_\odot = 1.327 \times 10^{11}$ km^{3}s^{-2}
According to Eq. (7.4.35) the time to the center of the Sun is

$$t_{col} = \pi \sqrt{\frac{(149.6 \times 10^6)^3}{8 \cdot 1.327 \times 10^{11}}} \text{ s} = 64.57 \text{ days}$$

The Sun has a radius of $r = 0.696 \times 10^6$ km. According to Eq. (7.4.34) the body crashes into the surface of the Sun after

$$t_{col} = \frac{64.57}{\pi} (0.1361 + 2 \cdot 1.5025) \text{ days} = 64.56 \text{ days}$$

According to Eq. (7.4.31), it would have an impact velocity of 41.7 km s^{-1}.

If the functional relationship $r = r(t)$ or $\dot{r} = \dot{r}(t)$ is in reqired, Eq. (7.4.32) is of no much use due to its implicit nature. We then have to resort to the

Keplerian equation for $e = 1$, which yields the following solution scheme:

$$M = nt$$
$$E - \sin E = M$$
$$r = a\,(1 - \cos E) \tag{7.4.36}$$
$$\dot{r} = \frac{\sqrt{a\mu}}{r} \sin E$$

Here the second equation has to be solved for E by Newton's method Eq. (7.4.17), setting $e = 1$.

Rectilinear Parabola

Comets which arrive from the edge of our solar system (from the so-called Oort's cloud) exhibit $e \approx 1$, $a \approx \infty$. This is a parabolic orbit. If in addition they exhibit $h = 0$, so Eq. (7.4.32) is no longer applicable. We then again have to resort to the vis-viva equation and carry out the limit $a \to \infty$. With $h = 0$ we then obtain

$$\dot{r} = -\sqrt{\frac{2\mu}{r}}$$

Note that as above we have chosen the negative sign here, because for a comet falling in, $\dot{r} < 0$. This differential equation can be integrated directly

$$\int_{r_i}^{r} \sqrt{r'} dr' = -\sqrt{2\mu} \int_0^t dt'$$

where t counts the time from a given initial position r_i (for instance, the point of first comet sighting) to a given later orbit position r. This yields

$$r^{3/2}(t) = r_i^{3/2} - \sqrt{\frac{9\mu}{2}} \cdot t \quad \text{orbit equation} \tag{7.4.37}$$

From this we find the collision time, the time from the initial position r_i to collision at $r = 0$, as

$$t_{col} = \sqrt{\frac{2}{9\mu} r_i^3} \tag{7.4.38}$$

7.5
Life in Other Universes?

Mathematically, and also physically, in principle it is quite possible that other universes with other spatial dimensions might exist. String theory, for example, states that at the time of the Big Bang our universe started out with nine spatial dimensions. According to current beliefs (Brandenberger and Vafa, 1989), initially all these nine dimensions were curled up on the Planck scale (10^{-35} m), that is to say, there were no macroscopic dimensions like what we have today. The so-called "strings" that make up our elementary particles were "living" on these curled-up spaces. Very shortly after the Big Bang, an antistring crashed into one of these rolled-up strings and, according to the belief, they eliminated each other and generated an uncurled space dimension: the first macroscopic dimension was born. In one dimension the probability that a string and an antistring meet is still very high. A string and antistring annihilated anew, which led to a second macroscopic dimension. The question is whether two dimensions offer enough space that the strings and antistrings no longer meet each other. Obviously not: the third dimension was born. Will another string and antistring meet again some day in our three dimensions to open up a fourth dimension? Nobody knows. But we do know that coincidences play an important role in quantum mechanics. It could have been quite possible that not only two, but even four or five macroscopic dimensions could have formed, especially when the universe was still very small. Could we live in such a universe? Life in our universe, apart from many other factors, decisively depends on whether we have stable planetary orbits around a central star. So if we want to have an answer to the question of whether life would be possible in universes with other dimensions, first of all we would have to find out whether there would be stable planetary orbits. This is exactly what we will figure out now.

7.5.1
Equation of Motion in n Dimensions

First of all, one has to consider that according to Noether's theorem (see Eq. (7.1.8) to (7.1.10)) the conservation laws, especially the law of conservation of angular momentum, are independent of the dimension of the space. They are only determined by the homogeneity and isotropy of space-time, and not by its dimensionality. The conservation laws are thus valid in *all homogeneous n-dimensional universes*. So, the law of conservation of energy is also valid:

$$E_{kin} + E_{pot} = m\varepsilon = const$$

Also the two expressions (see Eq. (7.1.4))

$$E_{kin} = \frac{1}{2}mv^2$$

$$E_{pot} = mU(r)$$

are independent of dimension. Angular momentum is defined as the cross-product of position vector and velocity vector. Vectors are one-dimensional entities. Conservation of angular momentum means that these two vectors in a given n-dimensional space open up a hyperspace. Two non-collinear vectors open up a plane. So for $n \geq 2$ the gravitationally determined motion in general is restricted to a plane spanned by the initial vectors. (If by coincidence the initial vectors were collinear, which is inevitably in the case $n = 1$, then the motion for $n \geq 1$ would be one-dimensional.) Hence, independent of n, the general gravitational motion of a body in an n-dimensional space is always in a plane, and we therefore will measure it by means of polar coordinates (r, θ). As the motion takes place in a plane, the considerations leading to Eq. (7.2.4)

$$v^2 = \dot{r}^2 + \frac{h^2}{r^2}$$

still remain correct. The only thing that changes with space dimensions is the gravitational potential. We find it by solving the corresponding Poisson equation (7.1.2). Its general solution in $n \geq 3$ dimensions is (exercise, Problem 7.1)

$$U(r) = -\frac{\mu}{r^{n-2}} \tag{7.5.1}$$

whereby μ carries the unit $[\mu] = [m^n \, s^{-2}]$. By applying all the expressions given above to the energy conservation equation, we get the vis-viva equation for n dimensions

$$\dot{r}^2 = \frac{2\mu}{r^{n-2}} - \frac{h^2}{r^2} + 2\varepsilon \quad \textbf{vis-viva equation in } n \textbf{ dimensions} \tag{7.5.2}$$

We are seeking the equation of motion, and so we differentiate Eq. (7.5.2) to get

$$\ddot{r} = -\frac{(n-2)\mu}{r^{n-1}} + \frac{h^2}{r^3} \tag{7.5.3}$$

A big advantage of this differential equation is the fact that it is not vectorial, but scalar. A disadvantage is the sum of two terms on the right side of the equation, as they both contain r. With this, we are no longer able to find a simple analytical solution for the differential equation by just separating the variables. So we have to look for other approaches. An important feature to solve differential equations in a smart way is to make a solution ansatz or a substitution, which comprises as much advance information as possible. Apparently we always have expressions of the type $1/r$. We assume that the solution is of the same type, and thus we change to the new radial variable: $\rho := 1/r$. (In the literature this substitution is known as the Burdet transformation.) The second piece of previous knowledge we have is the conservation of the angular momentum. We therefore select the fixed angular momentum as one coordinate axis z, and for the other coordinates we use the rotating system of polar coordinates (ρ, θ). All in all we now have a system of cylindrical coordinates (ρ, θ, z), which will later prove to be naturally adapted to this problem. We also know that the motion is periodic, whereas the time variable is linear. So it is a good idea to change from the time variable t to the orbit angle variable θ. Substituting $\rho = 1/r$ or $r = 1/\rho$, respectively, results with Eq. (7.2.5) in

$$\dot{r} = -\frac{1}{\rho^2}\frac{d\rho}{dt} = -\frac{1}{\rho^2}\frac{d\rho}{d\theta}\frac{d\theta}{dt} = -\frac{1}{\rho^2}\rho'\omega = -\frac{1}{\rho^2}\rho'\frac{h}{r^2} = -h\rho'$$

and

$$\ddot{r} = -h\frac{d\rho'}{dt} = -h\rho''\frac{h}{r^2} = -h^2\rho^2\rho''$$

With this we obtain for Eq. (7.5.3)

$$\rho'' + \rho = \frac{(n-2)\mu}{h^2}\rho^{n-3}$$

This equation of motion indeed looks easier. But we are still missing a relation between the second variable θ and ρ. We get the missing equation from Eq. (7.3.16)

$$\dot{\theta} = \frac{h}{r^2} = h\rho^2$$

This concludes our search for differential equations of motion in n dimensions.

$$\rho'' + \rho = \frac{(n-2)\mu}{h^2}\rho^{n-3}$$
$$\dot{\theta} = h\rho^2$$
$$\ddot{u}_i = 0 \quad i = 3,\ldots,n$$

Newton's equation of motion in n dimensions (7.5.4)

Here u_i are the remaining radial vector components, which do not lie in the motion plane.

7.5.2
The $n = 3$ Universe

In order to verify the equation of motion in n dimensions, we test for our well-known three dimensions

$$\rho'' + \rho = \frac{\mu}{h^2}$$

To solve this, we rewrite it as

$$\rho'' = -\left(\rho - \frac{\mu}{h^2}\right)$$

With the substitution $\lambda := \rho - \mu/h^2$ we get $\lambda'' = \rho''$, which results in the new simple differential equation $\lambda'' = -\lambda$. It has the solution $\lambda = \lambda_0 \cos(\theta + \varphi)$. By resubstitution we obtain

$$\rho = \frac{\mu}{h^2} + \rho_0 \cos(\theta + \varphi)$$

The two integration constants ρ_0 and φ are determined by the specific initial conditions. Resubstituting $\rho = 1/r$ results in the well-known orbit equation (see Eq. (7.3.5))

$$r = \frac{p}{1 + e \cdot \cos \theta}$$

with $p := h^2/\mu$ and $e := p\rho_0$

7.5.3
The $n = 4$ Universe

For $n = 4$ dimensions, the first equation of (7.5.4) reads

$$\rho'' = -\rho\left(1 - \frac{2\mu}{h^2}\right) = \pm k^2 \rho$$

with

$$k := \sqrt{\left|1 - \frac{2\mu}{h^2}\right|}$$

We have to distinguish the following three cases:

1. Case $1 - \frac{2\mu}{h^2} > 0$

    Here $\rho'' = -k^2\rho$ and the solution is $\rho = \rho_0 \sin(k\theta + \varphi)$ or

    $$r = \frac{r_0}{\sin(k\theta)} \tag{7.5.5}$$

where we chose $\varphi = 0°$ as an initial condition at $\theta = \pi/(2k)$. That is, for $\theta = 0$ a planet is at infinity and approaches the scene. With $r = r_0$ it attains its smallest distance to the star, and then recedes into infinity. In total, this process can be considered as a flyby at a star.

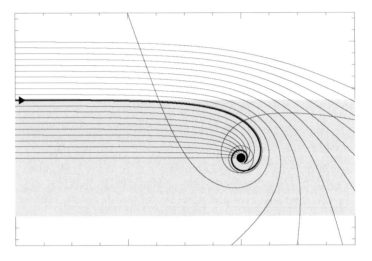

Figure 7.12 Numerical simulation of the two-body problem in a four-dimensional space. Light bodies with the same momentum, but different aiming radii, approach the central body from the left side. They either escape into infinity again, or they crash into the center, depending on whether their aiming radius is in the shaded area or not. There are no stable orbits. From Tegmark (1997).

2. Case $1 - \dfrac{2\mu}{h^2} < 0$

   Here $\rho'' = k^2 \rho$ and the solution is $\rho = \rho_0 e^{k\theta + \varphi}$ or

   $$r = r_0 e^{-k\theta} \qquad (7.5.6)$$

   where we again chose an initial $\varphi = 0°$ at $\theta = \pi/2k$. In other words, the planet spirals exponentially towards the star until it crashes into it.

   Note: *The other possible mathematical solution $r = r_0 e^{k\theta}$ is unphysical, as it would imply a repelling gravitational force.*

3. Case $1 - \dfrac{2\mu}{h^2} = 0$

   Here $\rho'' = 0$ and thus $\rho = a\theta + b$ or

   $$r = \dfrac{1}{a\theta} \qquad (7.5.7)$$

Here $b = 0$ was selected, i.e. $r(\theta = 0) = \infty$. On this borderline orbit the planet spirals towards the star, inversely proportional to θ, until it crashes onto its surface.

The orbits of three cases are depicted in Figure 7.12 taken from the literature (Tegmark, 1997), which were obtained by numerical simulation.

> **Conclusion:** In a four-dimensional universe, none of the three possible cases of orbits are stable or bounded, therefore no planetary systems can exist, and hence no life would be possible.

7.5.4
Universes with $n \geq 5$

In universes with $n \geq 5$ dimensions the inverse radial acceleration is

$$\rho'' = \frac{(n-2)\mu}{h^2} \rho^{n-3} - \rho = \rho \left[\frac{(n-2)\mu}{h^2} \rho^{n-4} - 1 \right] \quad (7.5.8)$$

Let's start our examination with a planet at $r = \infty$, i.e. $\rho = 0$, approaching the star. As long as the distance is large enough, so that

$$r^{n-4} > \frac{(n-2)\mu}{h^2}$$

or

$$\rho^{n-4} < \frac{h^2}{(n-2)\mu}$$

is valid, the expression in brackets of Eq. (7.5.8) is smaller than zero, and therefore $\rho'' < 0$. Let's assume $\rho'' = -a = const < 0$ for a short period of time. Then we get the solution

$$\rho = \rho_0 + b\theta - \frac{a}{2}\theta^2$$

Even if the inverse radial velocity b were slightly positive at the beginning, ρ will decrease further after a certain advance of the orbit angle. If ρ decreases, ρ'' will become even more negative, and we get a run-away effect with the limiting value $\rho \to 0$, implying $r \to \infty$. This means that our body is gravitationally not bound to the star. At a certain position it will approach the star to a minimum distance, depending on the incidence angle and velocity, without falling below the critical radius

$$\rho_{crit} = \frac{1}{r_{crit}} = \left[\frac{h^2}{\mu(n-2)} \right]^{\frac{1}{n-4}}$$

On its further track it recedes and disappears in the depths of the universe. In total its track will be more or less deflected. Qualitatively, this case corresponds to the first case in four dimensions.

If the body falls below the critical inverse radius due to its initial conditions, then the term in brackets in Eq. (7.5.8) becomes positive and thus $\rho'' > 0$. Let's assume $\rho'' = a = const > 0$ for a moment. Then we get the solution

$$\rho = \rho_0 + b\theta + a\theta^2$$

As the body approaches from outside, its inverse radial velocity is $b > 0$, which means that in total ρ increases even faster (i.e., r decreases more). Then the term in brackets in Eq. (7.5.8) attains even larger positive values. Therefore $\rho'' > 0$ increases further, and ρ increases even faster. So we get an opposite run-away effect with the limit value $\rho \to \infty$, implying $r \to 0$. The planet approaches the star faster and faster, until it crashes into the star.

> **Conclusion:** In universes with $n \geq 5$ dimensions, again no stable planetary systems can exist, and thus life is not possible.

How, then, would an $n \geq 4$ dimensional universe evolve? Probably after its birth and after a very short period of time, the so-called epoch of radiation, when the masses form, these masses would immediately join together to form black holes, and they would never form any gravitationally coupled stellar system or even galaxies. The black holes would merge into even bigger holes within a short period of time because of their large critical radius, and finally there would only be a huge black hole left, which would absorb all the radiation and matter of the universe. That would be the quick end of a universe, hardly had it begun to exist.

> **Remark:** *Historically, in 1917 Ehrenfest already showed qualitatively, and in 1963 Büchel showed by general energy considerations, that for $n \geq 4$ there is no possibility that stable planetary orbits can exist. They are either deflected by the central body, or they crash into it within a very short period of time.*

7.5.5
Universes with $n \leq 2$

In 1984 Deser et al. applied the theory of general relativity to $n \leq 2$ spatial dimensions, and found that the space surrounding a point mass would not have curvature (the Riemann tensor and with it Einstein's curvature tensor would vanish). This means that other particles would not experience any gravitational pull. So, in $n \leq 2$ dimensional spaces there is no gravitational attraction at all, let alone an answer to the question of stable orbits. Classic astrodynamics erroneously has a different point of view. From the Poisson equation (7.1.2) it follows that in two dimensions a gravitational potential $U(r) = -\mu \ln r$ with force $F \propto -\mu/r$ exists (exercise, Problem 7.1). Since $U(r)$ diverges for $r \to \infty$ this already shows us that this solution is quite far from reality. The inconsistency between the theory of relativity and Newton's physics can be explained by the fact that for $n \leq 2$ in the theory of general relativity a correspondence to classical physics no longer exists.

> **Conclusion:** Since in $n \leq 2$ dimensional universes a gravitational force is non-existent, also planets cannot exist, let alone life.

So, with our three spatial dimensions, we live on an island of stability, and we can only assume and hope that it is not as coincidental as the string theory currently suggests.

Problems

Problem 7.1 Solutions of the Poisson Equation

Show that the Poisson equation (7.1.2) for an n-dimensional space for $\rho(r) = M \cdot \delta(r)$ has the following solutions for $U(r)$ ($\mu = GM$):

$$U(r) = \begin{cases} -\dfrac{\mu}{r^{n-2}} & @\ n \geq 3 \\ -\mu \ln r & @\ n = 2 \\ 4\pi\mu \cdot r & @\ n = 1 \end{cases}$$

Consider in particular the case at $r = 0$.

Problem 7.2 Bounded Orbits

Show that if

$$F(r) = -\frac{\mu m}{r^{p+1}} r$$

with $p > 1$, then a body moving with negative energy cannot move indefinitely far from the origin.

Problem 7.3 Stellar Orbits

(a) *Stellar orbits in globular cluster galaxies.* Suppose we have a globular cluster galaxy, i.e. is a spherical distribution of stars, which we assume is homogeneous with constant mass density $\rho = const$. First show that the gravitational force on a star is proportional to the radial distance to its galactic center $F(r) = -Gm_*\rho r$ (G = gravitational constant, m_* = mass of the star), by taking into account Newton's finding that the masses beyond the radial sphere do not contribute to the gravitational force. Then show that orbits of stars in the globular cluster galaxy are either ellipses or circles with center (not focus) at the center of the galaxy with angular velocity $\omega = \sqrt{G\rho}$

(b) *Stellar orbits in the Milky Way.* Most galaxies such as our own Milky Way, however, are thin rotating disks. Assuming again a constant mass density $\rho = const$, and assuming no vertical movements of stars within the disk the gravitational potential of a star within the disk with radius R is

$$U(r) = -2G\rho h (R - r) [K(m) + E(m)]$$

where $K(m)$ and $E(m)$ are the complete elliptic integrals of the first and second kind, respectively, and

$$m = -\frac{4Rr}{(R-r)^2}$$

Show that for $r < R$ holds $K(m) + E(m) \approx \pi$ and therefore $\mathbf{F}(\mathbf{r}) \approx -2\pi G\rho m_* h\hat{r}$. So, the absolute value of the force per unit star mass is constant everywhere. Then show that for only small radial excursions the orbit of stars around the center of a flat galaxy are ellipses or circles with center at the center of the galaxy but now with rotational rate $\omega = \sqrt{2\pi G\rho h/\langle r \rangle}$.

Problem 7.4 Eccentricity from Eccentricity Vector

Derive from the eccentricity vector $\mathbf{e} = (\dot{\mathbf{r}} \times \mathbf{h})/\mu - \hat{\mathbf{r}}$ directly that for the absolute value of the eccentricity holds $1 - e^2 = h^2/\mu a$.

Problem 7.5 Virial Theorem of a Two-Body System

(a) Show that, in a two-body system with the origin at the central body,

$$\ddot{I} = 4E_{kin} + 2E_{pot}$$

holds, with $I = mr^2$ the moment of inertia of the orbiting body.

(b) Prove Eq. (7.4.5), $2\langle E_{kin}\rangle + \langle E_{pot}\rangle = 0$, for a bounded orbit.

Problem 7.6 Virial Theorem of an n-body system

(a) Prove that $\ddot{I} = 4E_{kin} + 2E_{pot}$ also holds for a many-body system with $I := \sum_i m_i r_i^2$.

(b) Prove the virial theorem Eq. (7.3.15) for n bodies, which are not necessarily all bounded, if $\lim_{t\to\infty} \dot{I}/t = 0$ holds.

Remark: It can be shown that $\lim_{t\to\infty} \dot{I}/t = 0$ is equivalent to $\lim_{t\to\infty} I/t^2 = 0$, which is always fulfilled in a bounded many-body system.

Problem 7.7 Sundman's Inequality

Let L be the total angular momentum and $I := \sum_i m_i r_i^2$ be the total inertial moment of a coplanar many-body system. Prove Sundman's inequality

$$E_{kin} \geq \frac{L^2}{2I}$$

which holds generally for a many-body system for the coplanar many-body case.

Hint: Look for the minimum of

$$E_{kin,\theta} = \tfrac{1}{2}\sum_i m_i v_{i,\theta}^2 \leq \tfrac{1}{2}\sum_i m_i\left(v_{i,\theta}^2 + v_{i,r}^2\right) = E_{kin}$$

with respect to the angular momentum of each body. Alternatively study

$$L = \left|\sum_i m_i \mathbf{r}_i \times \mathbf{v}_i\right|$$

and apply Cauchy's inequality

$$\left(\sum_i a_i b_i\right)^2 \leq \left(\sum_i a_i^2\right)\left(\sum_i b_i^2\right)$$

Remark: *From Sundman's inequality follows the important Sundman's theorem: A many-body system of point masses can never totally collapse ($I = 0$) unless its total angular momentum $L = 0$. Since today our solar system has $L \neq 0$ and L is conserved, it will never fully collapse. Good to know.*

Problem 7.8 Orbit Equation – Fast Track
Starting out from $r^2 = \mathbf{r} \cdot \mathbf{r}$ show that

$$\ddot{r} = \frac{h^2}{r^3} - \frac{\mu}{r^2}$$

holds. Applying the Burdet transformation (see Section 7.5.1), show that the orbit equation

$$r = \frac{h^2/\mu}{1 + e \cdot \cos\theta}$$

follows.

Problem 7.9 Solutions to Barker's Equation
Show with Cardano's method and Descartes' rule of signs that the unique real solutions to Barker's equation from Section 7.4.2 are

$$\tan\frac{\theta}{2} = \left(\sqrt{q^2+1}+q\right)^{1/3} - \left(\sqrt{q^2+1}-q\right)^{1/3}$$

and

$$2\frac{r}{p} = \left(\sqrt{q^2+1}+q\right)^{2/3} + \left(\sqrt{q^2+1}-q\right)^{2/3} - 1$$

with

$$q = 3\sqrt{\frac{\mu}{p^3}}(t-t_0)$$

Problem 7.10 Series Expansions

(a) Prove the series expansions

$$E = M + e \sin M + \frac{e^2}{2} \sin(2M) - \frac{e^3}{8}[\sin M - 3\sin(3M)] + O(e^4)$$

$$\frac{r}{a} = 1 - e \cos M + e^2 \sin^2 M + \frac{3}{2}e^3 \cos M \sin^2 M + O(e^4)$$

for an elliptical orbit by applying the *Banach fixed point theorem* to $E = M + e \sin E = f(E)$ under the constraint that f is Lipschitz continuous for $e < 0.6627434\ldots$. Then apply the result to $r/a = 1 - e \cos E$.

Remark: *This solution procedure may sound like elementary mathematics. In fact the solution algorithm, called contraction mapping, is only a generalization of Newton's method. Just the verification that it works is elementary mathematics.*

Note: *Contraction mapping is a very convenient method to solve implicit functional relations if the function is Lipschitz continuous. Practically, Lipschitz continuity is not checked beforehand, but contraction mapping is just applied and only then observed whether the series converges. In fact, when deriving Eq. (12.5.10) we made use of the contraction mapping without saying.*

(b) Prove

$$\theta = M + 2e \sin M + \frac{5}{4}e^2 \sin(2M) + O(e^3)$$

for an elliptical orbit by applying contraction mapping to the integral equation

$$\int_0^\theta \frac{dx}{(1 + e \cdot \cos x)^2} = \frac{\mu^2}{h^3}(t - t_0) = \frac{M}{(1 - e^2)^{3/2}}$$

8
Orbit Transitions

One of the most important maneuvers in space is the transfer between two orbits around the same central body. These orbits are very often circular orbits, because they provide the maximum altitude (residual atmospheric drag in low Earth orbit!) for a given achievable orbital energy, or because in geostationary orbit the satellite has to rotate exactly in accord with the Earth. Transfers between two elliptical orbits with small eccentricity, e.g. between two planetary orbits, are also interesting. Very often the question is this: How do you get from any initial Keplerian orbit to any target orbit with as little propulsion demand as possible? Because this general problem is quite complicated, we will restrict ourselves to the following conditions:

- The two orbits are elliptical and coplanar.

- The spacecraft have the same direction of rotation around the same central body, so that their angular momentum vectors are collinear, but their magnitudes are different.

- The orbital elements are such that the two orbits nowhere touch or cross each other.

Due to the last constraint, we can define an inner and an outer orbit, the radii of which we will denote by $r_\bullet$ and $r_\circ$.

8.1
Two-Impulse Transfer (Hohmann Transfer)

Throughout this section we assume in addition that the transfers are achieved with impulsive thrust maneuvers, so-called boosts or kick-burns with $F_* \gg F_{ext}$ (see Section 2.3) so that the received created Δv is determined solely by the thrust characteristics. This assumption is in general valid for today's chemical propulsion engines for orbit control.

Astronautics. Ulrich Walter
Copyright © 2008 WILEY-VCH Verlag GmbH & Co. KGaA, Weinheim
ISBN: 978-3-527-40685-2

8.1.1
General Considerations

As an example of an orbit transfer, let's consider the situation where we are in an elliptical low Earth orbit (LEO), and we want to get from any point on this orbit into an elliptical geosynchronous orbit (GSO) – secularly synchronous, i.e. synchronization with the Earth's rotation is only achieved on the average of an orbit (see Fig. 8.1). A kick-burn in LEO will first take us into an elliptical transfer orbit, the so-called GSO transfer orbit (GTO). This transfer orbit has to cross or touch the GSO at some point. Once we are at the crossing point, a second kick-burn would bring us into GSO.

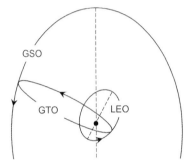

Figure 8.1 The GTO as a Hohmann transfer orbit from an elliptical LEO to an elliptical GSO (not to scale).

Now, an optimum transfer will minimize the sum of the two Δv at the two transition points. Because a kick-burn Δv at a point in space increases the kinetic energy $\varepsilon_{kin}(v + \Delta v) = \frac{1}{2}m(v + \Delta v)^2$ but leaves the potential energy unchanged, the orbital energy ε of the S/C increases by the same amount. Due to $\varepsilon = -\mu/2a$ the upshot is that the semi-major axis expands. In fact, an essential part of the transfer is to enlarge the semi-major axis of the initial LEO to that of the final GSO.

In essence, an optimum transfer will in part maximize the increase of the vehicle's orbital energy at a given amount of Δv. The question is this: How is a kick-burn performed to maximize the increase? The change of the orbital energy due to a kick-burn Δv is given by

$$\Delta \varepsilon = \varepsilon_{kin}(v + \Delta v) - \varepsilon_{kin}(v) = \frac{1}{2}(v + \Delta v)^2 - \frac{1}{2}v^2 = v \cdot \Delta v + \frac{1}{2}(\Delta v)^2 \quad (8.1.1)$$

So for a given amount of boost Δv the orbital energy is maximally increased if the boost Δv is parallel to the current velocity vector v. In other words:

> A maximum increase in orbital energy is achieved if the transfer boost is in the direction of motion, i.e. tangentially to the initial track.

Since this principle applies also for the second transition point to achieve the final orbit (another increase of a), we immediately obtain the rule for an energetically optimum transfer between to ellipses with only two boosts, the so-called Hohmann transfer:

> A **Hohmann transfer** orbit is an elliptical orbit that energetically optimizes the transfer between any two coplanar, co-rotating, non-crossing elliptical orbits. It tangentially touches these orbits at two points where the S/C transits with a kick-burn.

Such a Hohmann transfer orbit is depicted in Fig. 8.1. A Hohmann transfer of course also works the other way round, i.e. for a transfer from an outer to an inner elliptical orbit. So we have arrived at an answer to our optimization problem for a given starting point on the initial orbit. This leaves open the answer to the following question: At which point on the initial orbit should we perform the kick-burn to optimize overall the transfer between the two elliptical orbits. One could presume that due to $\Delta \varepsilon \propto v \cdot \Delta v$ a first kick-burn at the highest orbital speed, i.e. at the periapsis, would always be a good choice. But this neglects the second kick-burn at low speeds at the apoapsis of the transfer ellipse. A survey of transfers with different orbital elements shows that the ratio of the eccentricities of the inner and outer orbits is important. If the inner orbit has a higher eccentricity than the outer orbit, the optimum transfer varies. But in general we can say that:

> If the inner orbit has a lower eccentricity than the outer orbit, the transfer to or from the apoapsis of the outer orbit requires the least delta-v budget.

The Earth-Mars transfer is a nice example for this rule, because here $e_{earth} = 0.0167$ and $e_{mars} = 0.0934$. Figure 8.2 shows the delta-v budget of the transfer as a function of the orbit angle of the transition point on the Mars orbit.

The transfer optimization problem gets even more complicated if arbitrary angles between the lines of apsides or non-coplanar or crossing orbits are considered. Because these cases become too complex to lay them out in a textbook, we skip this problem here. In the following we rather focus on the transfer between circular and nearly circular orbits, which are of great practical value.

8.1.2
Transfer between Circular Orbits

Hohmann transfers are specifically interesting between two circular orbits, as Earth orbits are mostly circular for energy reasons. Here $r_\bullet = a_\bullet = const$

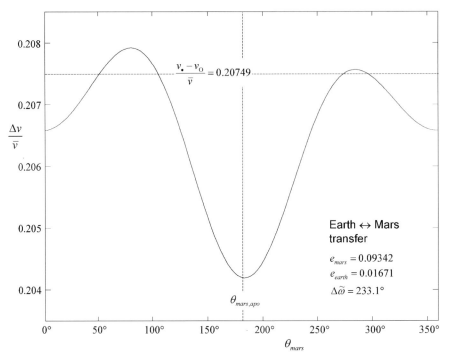

Figure 8.2 Normalized delta-v budget for Earth–Mars Hohmann transfers as a function of the orbit angle of the intersection with Mars' orbit. The angle between their lines of apsis is 233.1°. The horizontal line is the limiting value if both orbits were circular. Note the stretch of the y scale.

and $r_\bigcirc = a_\bigcirc = $ const. It is easy to find out the orbit parameters of a Hohmann transfer orbits between circular orbits. For the semi-major axis of this Hohmann transfer orbit, the following is obviously applicable:

$$a_H = \frac{1}{2}(a_\bullet + a_\bigcirc) \tag{8.1.2}$$

The transfer time is exactly half a period of the transfer ellipse, so according to Eq. (7.4.11)

$$t_H = \pi\sqrt{\frac{a_H^3}{\mu}} \tag{8.1.3}$$

The transfer ellipse with its two degrees of freedom a_H and e_H is completely and unambiguously determined by the boundary condition $r_{H,per} = a_H(1 - e_H)$ (see Eq. (7.4.8a)) and by $r_{H,per} = a_\bullet$ and $r_{H,apo} = a_\bigcirc$. As for the

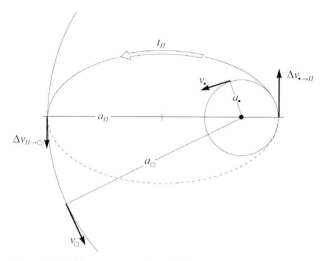

Figure 8.3 Orbit parameters for a Hohmann transfer between an inner and an outer circular orbit.

semi-minor axis $b_H = a_H\sqrt{1-e_H^2}$ applies, the other orbit parameters can be easily derived:

$$e_H = \frac{a_\circ - a_\bullet}{a_\circ + a_\bullet} \tag{8.1.4}$$

$$b_H = \sqrt{a_\bullet a_\circ} \tag{8.1.5}$$

Let's determine the Δv for a Hohmann transfer between two circular orbits. According to Eq. (2.5.1), this is calculated for two Hohmann transfer kick-burns as the sum of the individual amounts

$$\Delta v = \begin{cases} |\Delta v_{\bullet \to H}| + |\Delta v_{H \to \circ}| & @ \bullet \to \circ \\ |\Delta v_{\circ \to H}| + |\Delta v_{H \to \bullet}| & @ \circ \to \bullet \end{cases}$$

$$= (v_{H\bullet} - v_\bullet) + (v_\circ - v_{H\circ}) \quad @ \bullet \leftrightarrow \circ$$

The latter holds because $|\Delta v_{\bullet \to H}| = |\Delta v_{H \to \bullet}| = v_{H\bullet} - v_\bullet$ and $|\Delta v_{\circ \to H\circ}| = |\Delta v_{H\circ \to \circ}| = v_\circ - v_{H\circ}$. According to Eq. (8.1.4) $(1+e_H)/(1-e_H) = a_\circ/a_\bullet$ and therefore from Eqs. (7.4.10a) and (7.4.10b) it follows that $v_{H\bullet} = v_{per} = v_\bullet \sqrt{a_\circ/a_H}$ and $v_{H\circ} = v_{apo} = v_\circ \sqrt{a_\bullet/a_H}$. Inserting this into the above equation yields the two contributions

$$\Delta v = |\Delta v_{\bullet \leftrightarrow H}| + |\Delta v_{H \leftrightarrow \circ}| = v_\bullet \left(\sqrt{\frac{a_\circ}{a_H}} - 1\right) + v_\circ \left(1 - \sqrt{\frac{a_\bullet}{a_H}}\right) \tag{8.1.6}$$

which are represented individually in Fig. 8.4.

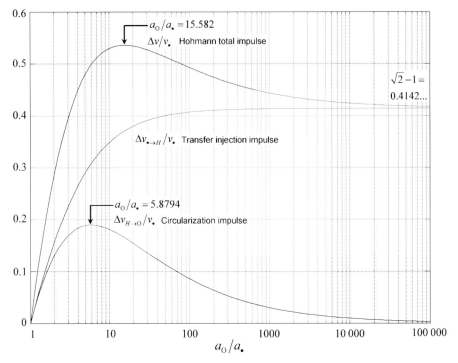

Figure 8.4 Propulsion demand for Hohmann transfers from an inner to an outer orbit.

The first (transfer injection) boost increases continuously with increasing distance between the orbits. This is quite easy to understand, as the semi-major axis a_H of the transfer orbit is determined by this distance, and the transfer orbital energy also increases non-linearly with $\varepsilon = -\mu/2a$ according to Eq. (7.3.14). The second (circularization) boost has a maximum at $a_\bigcirc/a_\bullet = 5.879\,362\ldots$ (exercise, Problem 8.3a). The reason is as follows. If the two orbits are in close vicinity to each other, the transfer orbit crosses the target orbit almost tangentially, and the second boost is then very small. For increasing distances, the propulsion demand for the second boost increases. If the target orbit is almost at infinity, the target velocity is almost zero, and the effort $\Delta v_\bigcirc$ is again very small. The second boost has to achieve a maximum somewhere in between. Because of this maximum of the second boost, the total Hohmann transfer also has a maximum $\Delta v_{\max} \approx 0.536\,258 \cdot v_\bullet$ at $a_\bigcirc/a_\bullet = 15.581\,72\ldots$ (exercise, Problem 8.3b).

If the terms in Eq. (8.1.6) are arranged skillfully and extended by $\sqrt{\mu} = v_\bullet\sqrt{a_\bullet} = v_\bigcirc\sqrt{a_\bigcirc}$, we get

$$\boxed{\Delta v = (v_\bullet - v_\bigcirc)\left(\frac{\sqrt{a_\bullet} + \sqrt{a_\bigcirc}}{\sqrt{a_H}} - 1\right) < v_\bullet - v_\bigcirc \quad @\ \bullet \leftrightarrow \bigcirc} \quad (8.1.7)$$

Equation (8.1.7) is valid for both transfer directions, i.e. $\bullet \to \circ$ and $\circ \to \bullet$. Note that propellant mass losses of the S/C because of the boosts have not been considered here (cf. Section 2.3).

Adjacent Circular Orbits

For a Hohmann transfer between adjacent circular orbits, $a_\circ \approx a_\bullet$, we can approximate (exercise, Problem 8.1)

$$\frac{\sqrt{a_\bullet} + \sqrt{a_\circ}}{\sqrt{a_H}} - 1 \approx 1 - \frac{1}{16}\left(\frac{a_\circ - a_\bullet}{a_\bullet}\right)^2 \to 1 \tag{8.1.8}$$

and hence

$$\boxed{\Delta v \approx v_\bullet - v_\circ} \quad @\ a_\circ \approx a_\bullet \tag{8.1.9}$$

> **Example**
> For $a_\circ/a_\bullet \leq 2.5$, i.e. for orbits with altitudes up to 10 000 km above Earth's surface, the error due to Eq. (8.1.9) is smaller than 5.1%.

If the orbit distance Δr is very small, $a_\circ = a_\bullet + \Delta r$, we can use the average orbit radius $a = r$ and Eq. (8.1.9) to approximate the impulse demand as

$$\Delta v \approx v_\bullet - v_\circ = \sqrt{\mu}\left(\frac{1}{\sqrt{a_\bullet}} - \frac{1}{\sqrt{a_\circ}}\right) = \frac{v}{2r}\Delta r$$

As one might expect, this result coincides with the value for an infinitesimal limit, which one derives if one differentiates the velocity of the circular orbit $v = \sqrt{\mu/r}$

$$|dv| = \left|-\frac{1}{2}\frac{\sqrt{\mu}}{r^{3/2}}\right| dr = \frac{v}{2r} dr \tag{8.1.10}$$

This result will be important for continuous thrust transfers with small thrusts in Section 8.2.

Distant Circular Orbits

If the circular orbits are distant from each other such that $a_\circ \gg a_\bullet$ we can approximate

$$\frac{\sqrt{a_\bullet} + \sqrt{a_\circ}}{\sqrt{a_H}} - 1 \approx \sqrt{2}\left(1 + \sqrt{\frac{a_\bullet}{a_\circ}}\right) - 1 \tag{8.1.11}$$

We therefore obtain in the limit of infinite distances

$$\Delta v = \left(\sqrt{2} - 1\right)(v_\bullet - v_\circ) \quad @\ a_\circ/a_\bullet \to \infty \tag{8.1.12}$$

Between these two limiting cases the expression $\Delta v/(v_\bullet - v_\circ)$ is strictly monotonically decreasing (see Fig. 8.5), and that is why in Eq. (8.1.7) the inequality strictly holds.

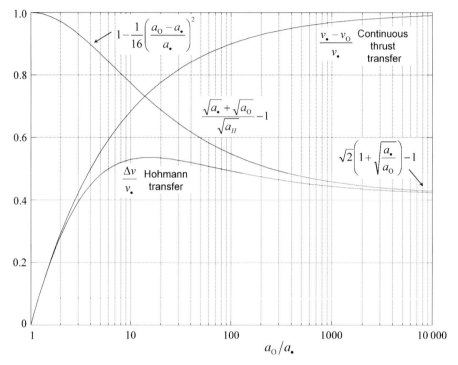

Figure 8.5 The terms of Eq. (8.1.7) and their limiting values. The factor $(v_\bullet - v_\circ)/v_\bullet$ is the propulsion demand of a continuous thrust transfer (see Section 8.2).

8.1.3
Transfer between Nearly Circular Orbits

We now consider Hohmann transfers between two coplanar, co-rotating, non-crossing ellipses, with small eccentricities $e_\bullet, e_\circ \ll 1$. The detailed analytical approximations are quite convoluted due to $e_\bullet e_\circ$ cross-terms, but the general upshot is that the dependence of delta-v as a function of $e_\bullet, e_\circ$ is very weak. This can be observed for instance from Fig. 8.6 where, for Hohmann transfers between two apsides points on two orbits with $a_\circ/a_\bullet = 1.52365$ (Earth – Mars), the normalized delta-v is depicted as a function of $e_\bullet = e_\circ$ numerically.

For $e \leq 0.1$ the variations are smaller than 2% of their corresponding value for circular orbits $(v_\bullet - v_\circ)/\bar{v} = 0.20749$. As we can see from Fig. 8.2 this is in accordance with the Earth–Mars transfer, $e_{mars} = 0.09342$ and $e_{earth} = 0.01671$, even if the transfers would be performed from any position of the

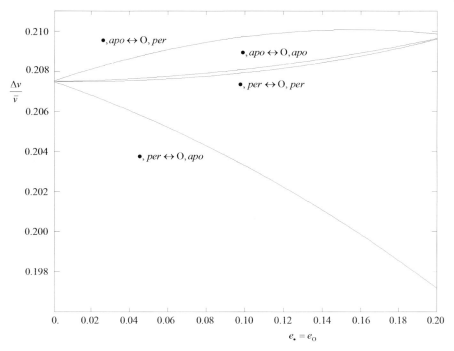

Figure 8.6 The delta-v budget as a function of the common eccentricities. The dependences are given for orbits with $a_○/a_● = 1.523\,65$ (Earth–Mars). Note the stretched Δv scale.

orbit. So for all practical purposes we get the same result as for a transition between two circular orbits:

$$\Delta v \approx (v_● - v_○)\left(\frac{\sqrt{a_○} + \sqrt{a_●}}{\sqrt{(a_○ + a_●)/2}} - 1\right) \quad @ \ e_●, e_○ \leq 0.1 \quad (8.1.13)$$

Nevertheless, the most favorable Hohmann transfer between two nearly circular orbits is always *from the periapsis of the inner orbit to the apoapsis of the outer orbit*. This is in agreement with our general rule from Section 8.1.1.

8.1.4
Sensitivity Analysis

From an energy point of view Hohmann transfers are the most favorable two-impulse transfer orbits. But they also have some disadvantages. They can be very sensitive to small inaccuracies of the transfer injection impulse. Let's have a closer look at this dependence for transfers between circular orbits.

Let r_{per} and r_{apo} be the periapsis and apoapsis radii of the transfer orbit. (For convenience we drop the index H to indicate the Hohmann transfer orbit.) The periapsis is determined by the initial orbit radius. From the vis-viva equation (7.2.10) we get for the velocity on the Hohmann orbit at the periapsis

$$v_{per}^2 = \frac{2\mu}{r_{per}} - \frac{\mu}{a} = \frac{\mu_\oplus}{a} \frac{1+e}{1-e}$$

If the transfer injection burn has a small error dv_{per} at the fixed periapsis $r_{per} = const$ we get

$$2v_{per} dv_{per} = \frac{\mu}{a^2} da$$

from which follows

$$\frac{da}{a} = 2\frac{av^2}{\mu} \frac{dv_{per}}{v_{per}} = 2\frac{1+e}{1-e} \frac{dv_{per}}{v_{per}} \qquad (8.1.14)$$

So, a thrust inaccuracy generates a certain variation of the semi-major axis. Now we want to know how this affects the position of the apoapsis. We start with Eq. (7.4.8b)

$$r_{apo} = a(1+e)$$

Its change in position is determined by differentiation

$$\frac{dr_{apo}}{r_{apo}} = \frac{da}{a} + \frac{de}{1+e} \qquad (8.1.15)$$

The two parameters a and e are not independent of each other, but are linked by the constancy of the periapsis of the initial orbit

$$const = r_{per} = a(1-e)$$

After differentiating this equation, we see how their changes depend on each other:

$$de = (1-e) \frac{da}{a} \qquad (8.1.16)$$

So together with Eq. (8.1.15), we get

$$\frac{dr_{apo}}{r_{apo}} = \frac{2}{1+e} \frac{da}{a}$$

and finally with Eq. (8.1.14), we get

$$\frac{dr_{apo}}{r_{apo}} = \frac{4}{1-e} \frac{dv}{v} = \frac{4a}{r_{per}} \frac{dv_{per}}{v_{per}} = 2\left(1 + \frac{r_{apo}}{r_{per}}\right) \frac{dv_{per}}{v_{per}} \qquad (8.1.17)$$

That is, for a given injection burn error dv_{per} the relative target point accuracy decreases with increasing transfer distances.

Example 1
Let's examine the Hohmann transfer from an initial LEO orbit ($h = 400$ km) to GEO. Because of $r_{GEO}/r_{LEO} = 6.232$ we get

$$\frac{dr_{apo}}{r_{apo}} = 14.46 \frac{dv_{per}}{v_{per}}$$

A relatively small burn error of just 0.5% would lead to an inaccuracy in the target distance of 7.2%. That is 3000 km deviation from the GEO orbit!

Example 2
Let's have a look at the Hohmann transfer from an initial LEO orbit ($h = 400$ km) to the Moon. Because of $r_{moon}/r_{earth} = 56.654$ we get

$$\frac{dr_{apo}}{r_{apo}} = 115.3 \frac{dv_{per}}{v_{per}}$$

That means that the same small burn error of just 0.5% would lead to an inaccuracy in the target distance of 58%. That would just bring us to nirvana!

Summary
Bare Hohmann transfers may be the most favorable transfer orbits from an energy point of view, but they are not employed in practice because of the long transfer times and the large possible target errors.

Already small deviations from an ideal Hohmann transfer and/or inflight corrections however strongly reduce these drawbacks as discussed in Section 9.3.2 for near-Hohmann transfers to Mars.

8.2 Continuous Thrust Transfer

If we have electrical propulsion engines, such as ion thrusters, thrust is low and continuous, so impulse transfers are impossible. An ion engine would rather have to fire permanently in the direction of motion (recall: $\Delta\varepsilon \propto v \cdot \Delta v$) to optimally but slowly spiral the satellite into higher and higher orbits. We now want to calculate the transfer orbit and the delta-v for continuous thrust orbit transfers. To do so, we have to integrate the differentially small velocity increase as given by Eq. (8.1.11)

$$\Delta v = \frac{\sqrt{\mu}}{2} \int_{r_\bullet}^{r_O} \frac{dr}{r^{3/2}} = -\sqrt{\frac{\mu}{r}}\bigg|_{r_\bullet}^{r_O} = v_\bullet - v_O \qquad (8.2.1)$$

If we compare the result with Eq. (8.1.7), we see from Fig. 8.5 that the Hohmann transfer is always more favorable than a continuous thrust transfer. But with ion propulsions and their very tiny thrusts there is no alternative to that.

In order to calculate the transfer time between an initial circular orbit with orbit radius $r_\bullet$ to the instantaneous circular orbit with radius r, we need the explicit orbital path $r = r(t)$. As the circular condition $v = \sqrt{\mu/r}$ is valid for each point of the orbital curve, it is sufficient to find $v = v(t)$. We find it with the help of the rocket equation $F_* = m \cdot \dot{v} = \dot{m}v_*$. To be able to apply it, we have to consider the mass reduction $\dot{m}$ due to the propellant consumption. We assume that the vehicle with mass m is accelerated by a constant thrust F with constant mass flow rate $\dot{m}_p = -\dot{m} = \text{const}$, so $m = m_0 + \dot{m}t$. With this relation we integrate the corresponding equation $dv = \dot{m}v_*/m \cdot dt$

$$\int_{v_\bullet}^{v} dv' = \int_0^t \frac{\dot{m}v_*}{m} dt'$$

to get

$$v - v_\bullet = v_* \int_0^t \frac{dt'}{m_0/\dot{m} + t'} = v_* \ln\left(\frac{m_0/\dot{m} + t}{m_0/\dot{m}}\right) = v_* \ln\left(1 - \frac{\dot{m}_p}{m_0}t\right) \qquad (8.2.2)$$

Because $v = \sqrt{\mu/r}$ this results in the following spiral orbit trajectory (Fig. 8.7)

$$r(t) = \mu \left[v_\bullet + v_* \ln\left(1 - \frac{\dot{m}_p}{m_0}t\right)\right]^{-2} \quad @\ \dot{m}_p = \text{const} \quad \begin{array}{l}\text{continuous}\\ \text{thrust}\\ \text{trajectory}\end{array} \qquad (8.2.3)$$

By solving for t we get the following from Eq. (8.2.2) for the transfer time

$$t_{ct} = \frac{m_0}{\dot{m}_p}\left[1 - \exp\left(\frac{v - v_\bullet}{v_*}\right)\right]$$

$$= \frac{m_0 v_*}{F_*}\left\{1 - \exp\left[\frac{\sqrt{\mu}}{v_*}\left(\frac{1}{\sqrt{r}} - \frac{1}{\sqrt{r_\bullet}}\right)\right]\right\} \qquad (8.2.4)$$

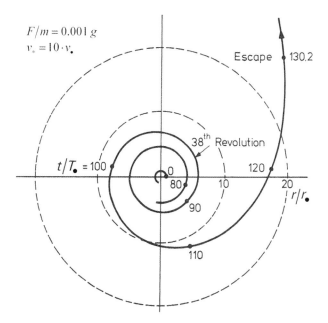

Figure 8.7 Continuous thrust trajectory as a result of a continuous tangential thrust. Index • refers to the initial orbit. Because the spiral is very narrow at the beginning, only the last two revolutions are depicted.

Example
A satellite is released from the Space Shuttle payload bay at an altitude of 300 km, and it is supposed to spiral with thrust $F_*/m_0 = 100$ µg $= 10^{-3}$ ms^{-2} and $v_* = 10\,000$ m s^{-1} to GEO. Because $v_\bullet = 7.72$ km s^{-1} and $v_{GEO} = v_\bigcirc = 3.07$ km s^{-1}, the transfer time is calculated to be $t_{ct} = 43$ days.

8.3
Three-Impulse Transfer

Bi-elliptic Transfer

The maximum of the total delta-v for Hohmann transfers (see Fig. 8.4) at $a_\bigcirc/a_\bullet \approx 15.6$ shows up because $\Delta v_{H \to \bigcirc}$ achieves its maximum at $a_\bigcirc/a_\bullet \approx 5.88$ for turning into the target orbit. This second delta-v contribution, however, vanishes for $r \to \infty$. This gives rise to the assumption that it might be possible to save propulsion demand with a total of three impulses by first transferring from the initial orbit to infinity (or at least far out), and then turning back again to the target orbit (see Fig. 8.8).

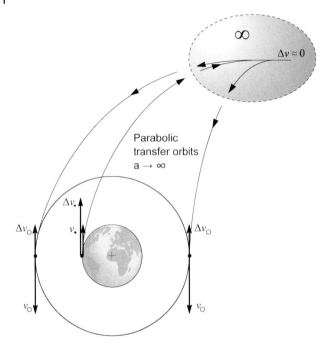

Figure 8.8 Schematic of a three-impulse transfer.

We want to determine the propulsion demand for such a bi-elliptic transfer. According to Eq. (8.1.7) and with Eq. (8.1.11) the delta-v budget to a remote intermediate circular orbit with orbit radius $a_\times$ and orbit velocity $v_\times = \sqrt{\mu/a_\times}$ can be calculated to be

$$|\Delta v_{\bullet \to \times}| \approx \left[\sqrt{2}\left(1+\sqrt{\frac{a_\bullet}{a_\times}}\right)-1\right](v_\bullet - v_\times)$$

The delta-v budget for a kick-burn that brings us back to the target orbit in turn amounts to

$$|\Delta v_{\times \to \circ}| \approx \left[\sqrt{2}\left(1+\sqrt{\frac{a_\circ}{a_\times}}\right)-1\right](v_\circ - v_\times)$$

So in total we find after a brief but substantial calculation

$$\Delta v = |\Delta v_{\bullet \to \times}| + |\Delta v_{\times \to \circ}|$$
$$= \left(\sqrt{2}-1\right)(v_\bullet + v_\circ) + 2v_\times \qquad @\ a_\times \gg a_\circ \qquad (8.3.1)$$

At infinity $v_\times = 0$ and the total delta-v budget becomes minimal

$$\Delta v = \left(\sqrt{2}-1\right)(v_\bullet + v_\circ) \qquad @\ a_\times/a_\circ \to \infty \qquad (8.3.2)$$

Equation (8.3.2) is illustrated in Fig. 8.9. For $a_\bigcirc/a_\bullet > 11.938\,765\ldots$ the propulsion demand is indeed more favorable with bi-elliptic transfers than with Hohmann transfers (exercise, , Problem 8.4). An additional benefit of a bi-elliptic transfer is that at infinity, where $v = 0$, one can also change the orbital plane, the flight direction, or even the direction of rotation of the orbital curve virtually without any propulsion demand. The serious drawback is that it takes forever to get to infinity. Even if one actually does not go to infinity, but only sufficiently far away, these transfers still take far too long. And currently there are hardly any practical applications for orbits beyond an orbit altitude of $r_\bigcirc = 11.94 \cdot R_\oplus = 76\,150$ km.

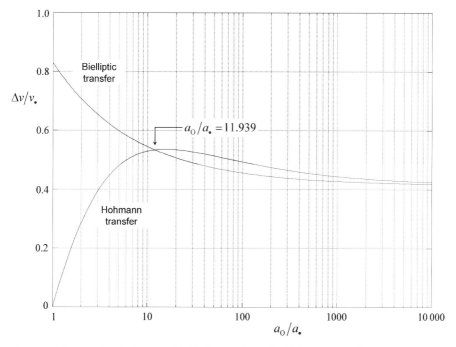

Figure 8.9 Comparison between a bi-elliptic transfer and a Hohmann transfer.

n-Impulse Transfers

Very generally it can be shown that for a transition between *any* two elliptical orbits the total delta-v budget for a three-impulse transfer (not necessarily bi-elliptic) might be smaller, but does not have to be, than with the two-impulse Hohmann transfer. The above bi-elliptic transfer for $a_\bigcirc/a_\bullet > 11.94$ is an example for this. In addition, it can also be shown that the total delta-v budget minimized by Hohmann or a three-impulse transfer cannot be further minimized by maneuvers with more than three impulses. So the Hohmann transfer or the minimum three-impulse transfer represents the absolute minimum for the propulsion demand.

8.4
One-Impulse Maneuvers

8.4.1
General Considerations

Until now we have examined only tangential kick-burns, because energetically they are most favorable either in a row of two (Hohmann) or three (bi-elliptic) impulse maneuvers. But these will not work if any specific orbital element needs to be changed, for instance the orbit inclination. For this we need very generally a one-impulse maneuver. Such a maneuver (Fig. 8.10) will transfer a vehicle with velocity v_1 at any position on its orbit by a kick-burn Δv into any other direction with velocity v_2 according to

$$\Delta v = v_2 - v_1$$

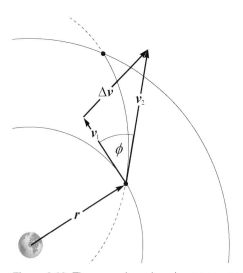

Figure 8.10 The general one-impulse maneuver.

To determine the corresponding propulsion demand, one has to square this equation and take the root on both sides, resulting in

$$\Delta v = \sqrt{v_2^2 + v_1^2 - 2v_2 v_1 \cos\phi} \tag{8.4.1}$$

where ϕ is the angle enclosed between v_1 and v_2. This is the most general equation to calculate a single-burn delta-v. From this equation we conclude some notable special cases:

1. If the kick-burn is along the direction of motion, then $\phi = 0$ and, of course,

$$\Delta v = |v_2 - v_1|$$

2. If the kick-burn only changes the flight direction ϕ, but not the instantaneous orbit velocity, i.e. $v_1 = v_2 =: v$ (see Fig. 8.11), then

$$\Delta v = v\sqrt{2(1-\cos\phi)} = 2v \cdot \sin\frac{\phi}{2} \qquad (8.4.2)$$

3. If a large velocity change is pending, for instance for a Hohmann transfer, one should (if necessary) use the trans-Hohmann injection kick-burn to simultaneously change the orbital plane, as because of $v_2 \gg v_1 \cos\phi$ it is provided at only a very small additional effort.

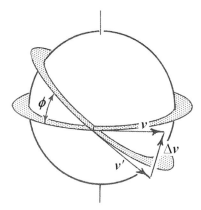

Figure 8.11 For a circular orbit the inclination is changed optimally at the nodal point.

From the second case we derive the rule:

> For a given change of flight direction ϕ, the propulsion demand Δv is smallest when the orbit velocity at kick-burn is smallest. This is why it is best to do single kick-burn maneuvers in the *apoapsis*, where this is the case.

Note: *Don't confuse this rule "delta-v smallest at lowest v" with the seemingly contradictory one in Section 8.1.1 "highest energy gain with tangential kickburn at largest v". While for two-impulse Hohmann transfers we were seeking the smallest delta-v at a required increase of the orbital energy (to increase the semi-major axis to that of the transfer orbit), we are here seeking for the smallest one-impulse delta-v at a given change of velocity direction.*

For this reason it might be preferable for a maneuver with a big propulsion demand, for instance an orbital plane rotation, to take the vehicle first into a higher orbit, change the plane there at apoapsis at low speed and low budget, and then take it back to its original orbit. It is this principle that the bi-elliptic transfer makes use of.

As we will see in a moment, not all orbital elements can be changed optimally, or not at all, at the apoapsis. Therefore the above rule is not generally applicable.

8.4.2 Orbit Correction Maneuvers

With this we come to the key question of general orbit maintenance: Which kick-burn optimally – that is, with least delta-v budget – changes a given orbital element? To find the answers to this, we first have to investigate kick-burns systematically. Most generally there are three types of kick-burn to change the orbital trajectory (see Fig. 8.12):

1. $\delta v_{\|}$ kick-burn in the direction of the orbital motion (along-track);

2. $\delta v_{\perp O}$ kick-burn perpendicular to the direction of motion, but within the orbital plane, outbound;

3. $\delta v_{\perp\perp}$ kick-burn perpendicular to the orbital direction and perpendicular to the orbital plane, in the direction of the angular momentum vector.

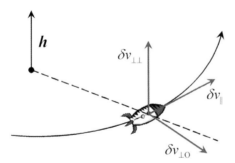

Figure 8.12 Decomposition of a kick-burn into the along-track and two cross-track directions within and outside the orbital plane.

Generally, the following applies to these kick-burns:

- Kick-burns perpendicular to the direction of orbital motion can only change the direction of motion and not its speed, and hence not the orbital energy ε. So because $\varepsilon = -\mu/2a$, $\delta v_{\perp O}$ and $\delta v_{\perp\perp}$ cannot influence the semi-major axis.

- Because $\delta v_{\|}$ and $\delta v_{\perp O}$ are in the orbital plane, they cannot change those orbital elements that determine the orientation of the orbital plane, that is Ω and i. On the other hand $\delta v_{\perp\perp}$ only changes the orientation of the orbital plane and therefore cannot change a and e.

So from these general considerations alone we are able to exclude the impact of the kick-burns on some of the orbital elements, which are indicated by empty white boxes in Table 8.1.

However, it is of great practical significance to know, how kick-burns change the orbital elements in detail. These relations can be derived (exercise, Problem 8.5a) from the Gaussian Variational Equations (12.1.2) with transformations (see Fig. 8.13)

$$a_r = \delta v_r = \cos\gamma \cdot \delta v_{\perp\odot} + \sin\gamma \cdot \delta v_{\|}$$

$$a_\theta = \delta v_\theta = -\sin\gamma \cdot \delta v_{\perp\odot} + \cos\gamma \cdot \delta v_{\|}$$

and with $\sin\gamma$, $\cos\gamma$ from Eq. (7.3.12) to be

$$\frac{\delta a}{a} = \frac{2v}{(1-e^2)v_h} \cdot \frac{\delta v_{\|}}{v_h} \qquad +0\cdot\delta v_{\perp\odot} \qquad +0\cdot\delta v_{\perp\perp} \quad (8.4.3a)$$

$$\delta e = 2(e+\cos\theta) \cdot \frac{\delta v_{\|}}{v} + \frac{(1-e^2)\sin\theta}{1+e\cos\theta} \cdot \frac{\delta v_{\perp\odot}}{v} \qquad +0\cdot\delta v_{\perp\perp} \quad (8.4.3b)$$

$$\delta i = 0\cdot\delta v_{\|} \qquad +0\cdot\delta v_{\perp\odot} \qquad +\frac{\cos(\theta+\omega)}{1+e\cos\theta} \cdot \frac{\delta v_{\perp\perp}}{v_h} \quad (8.4.3c)$$

$$\delta\Omega = 0\cdot\delta v_{\|} \qquad +0\cdot\delta v_{\perp\odot} \qquad +\frac{\sin(\theta+\omega)}{(1+e\cos\theta)\sin i} \cdot \frac{\delta v_{\perp\perp}}{v_h} \quad (8.4.3d)$$

$$\delta\omega = \frac{2\sin\theta}{e} \cdot \frac{\delta v_{\|}}{v} - \left(1 + \frac{1}{e}\frac{e+\cos\theta}{1+e\cos\theta}\right) \cdot \frac{\delta v_{\perp\odot}}{v} - \frac{\sin(\theta+\omega)\cot i}{1+e\cos\theta} \cdot \frac{\delta v_{\perp\perp}}{v_h} \quad (8.4.3e)$$

where

$$v = v_h \sqrt{1 + 2e\cos\theta + e^2}$$

and

$$v_h = \frac{\mu}{h} = \sqrt{\frac{\mu}{a(1-e^2)}}$$

which corresponds to the orbital velocity at the orbit angle $\cos\theta = -e/2$.

Note: *In a circular orbit θ and ω are undefined and therefore the impact of kickburn on most orbital elements cannot be determined from Eq. (8.4.3). In this case refer to problem 8.6.*

For practical purposes there are six special orbit positions for optimal firing, which are given in Table 8.1 in the rightmost column. The matrix of the table shows how the three different normalized kick-burns (entries in the middle

8 Orbit Transitions

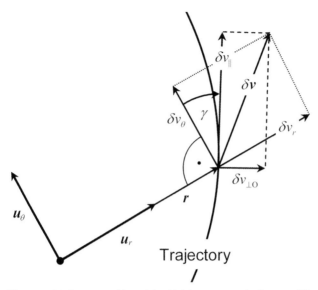

Figure 8.13 Decomposition of the kick-burn vector in the two different reference systems.

Table 8.1 The effects (matrix elements) of the three different kinds of kick-burns on the orbital elements at the special orbit positions (most right column). White boxes and relations not shown in this table denotes: There are no effects at any orbit position. Dark grey boxes give the dependences: a dash indicates that for this orbit position the effect vanishes, open circles denote complex dependences without practical use, the terms displayed are the factors, which multiplied with the kick-burn of that column gives the change of orbital element of that row. The upper/lower signs correspond to the upper/lower orbit positions in the orbit position column.

	$\delta v_\|/v_h$	$\delta v_{\perp O}/v_h$	$\delta v_{\perp\perp}/v_h$	
$\delta a/a$	$2/(1\mp e)$			Peri/Apoapsis
δe	± 2	—		$\theta = 0°$
$\delta\omega$	—	$\mp 1/e$	○	$\theta = 180°$
δi			$\pm c_\pm$	Nodes
$\delta\Omega$			—	$\omega + \theta = 0°$
$\delta\omega$	○	○	—	$\omega + \theta = 180°$
δi			—	Orthogonal to nodes
$\delta\Omega$			$\pm s_\pm / \sin i$	$\omega + \theta = +90°$
$\delta\omega$	○	○	$\mp s_\pm \cot i$	$\omega + \theta = -90°$

three columns) at the given special positions affect the different orbital elements (row headings). A dash denotes that the effect just vanishes at this position; a circle denotes that the term describing the effect is more complex and of no practical use; and the given terms are the factors, which multiplied by the normalized kick-burn from the column headings, delivers the change of orbital element given in the row heading. The important point is that at the six positions the orbital elements are affected selectively as well as optimally, where "optimally" means that for a given change of orbital elements the utilized propulsion mass is minimal. The coefficients showing up in the terms are

$$c_\pm(\omega) = \frac{1}{1 \pm e\cos\omega} \tag{8.4.4}$$

$$s_\pm(\omega) = \frac{1}{1 \pm e\sin\omega} \tag{8.4.5}$$

Note that possible changes of the orbital period are derived via Eq. (7.4.11) to be

$$\frac{\delta T}{T} = \frac{3}{2}\frac{\delta a}{a} \tag{8.4.6}$$

So, only tangential kick-burns effect the orbital period. But tangential burns at the periapsis or apoapsis always change the semi-major axis jointly with the eccentricity according to Table 8.1. The reason is as follows. As for a kick-burn in the peri/apoapsis, this orbital point is also the peri/apoapsis for the initial ellipse and the target ellipse, the following is valid:

$$const = r_{apo/per} = a(1 \pm e)$$

Differentiating this equation gives $0 = \delta a(1 \pm e) \pm a \cdot \delta e$ and hence

$$\frac{\delta a}{a} = -\frac{\delta e}{e \pm 1} \quad @ \; \delta v_\parallel \text{ kick-burn at apo/periapsis} \tag{8.4.7}$$

Selective Change of Eccentricity

There exists an orbit position not given in Table 8.1, at which the eccentricity can be changed selectively, that is without effecting other orbital elements, in particular a. This is from Eq. (8.4.3e) position $(e + \cos\theta)/(1 + e\cos\theta) = -e$ or

$$\cos\theta = -\frac{2e}{1+e^2} \tag{8.4.8}$$

The change at this position amounts to

$$\delta e = \pm\sqrt{1+e^2}\frac{\delta v_{\perp\odot}}{v_h} \quad \text{where} \quad \begin{array}{l} + : \quad 90° \leq \theta < 180° \\ - : \quad 180° < \theta \leq 270° \end{array} \tag{8.4.9}$$

Selective Change of Semi-major Axis

This allows one also to change the semi-major axis selectively. First a and e are changed with a $\delta v_\parallel$ kick-burn at the periapsis. Thereafter δe is canceled selectively at position

$$\cos\theta = -\frac{2(e+\delta e)}{1+(e+\delta e)^2} \qquad (8.4.10)$$

by the kick-burn

$$\delta v_{\perp\bigcirc} = \mp \frac{2}{\sqrt{1+(e+\delta e)^2}} \cdot \delta v_{\parallel} \quad \text{where} \quad \begin{array}{l} -: \quad 90° \leq \theta < 180° \\ +: \quad 180° < \theta \leq 270° \end{array} \qquad (8.4.11)$$

Selective Change of RAAN

This trick can also be used to change the right ascension of the ascending node (RAAN) Ω selectively: With a $\delta v_{\perp\perp}$ kick-burn at a position $\pm 90°$ from the ascending node, Ω and ω are changed. Thereafter $d\omega$ is canceled selectively at the new *periapsis* by the kick-burn

$$\delta v_{\perp\bigcirc} = e \cdot \cot i \cdot \delta v_{\perp\perp} \cdot \begin{Bmatrix} +s_+ (\omega + \delta\omega) & @ \ \omega + \theta = +90° \\ -s_- (\omega + \delta\omega) & @ \ \omega + \theta = -90° \end{Bmatrix} \qquad (8.4.12)$$

or alternatively also at the *apoapsis*, where the kick-burn is fired with the same force as in Eq. (8.4.12) but in the reverse direction.

Selective Change of Periapsis and Apoapsis Radius

Finally we mention that the periapsis and apoapsis radii r_{per}, r_{apo} can be changed selectively and optimally by kick-burns $\delta v_{\parallel, apo}$ and $\delta v_{\parallel, per}$

$$\frac{\delta r_{per}}{r_{per}} = 4\frac{\delta v_{\parallel, apo}}{v_h}, \quad \frac{\delta r_{apo}}{r_{apo}} = 4\frac{\delta v_{\parallel, per}}{v_h} \qquad (8.4.13)$$

at that apsis *opposite* to the one to be changed.

Problems

Problem 8.1 Adjacent Circular Orbit Approximation
Prove Eq. (8.1.8)

$$\frac{\sqrt{a_\bullet} + \sqrt{a_\circ}}{\sqrt{a_H}} - 1 \approx 1 - \frac{1}{16}\left(\frac{a_\circ - a_\bullet}{a_\bullet}\right)^2 \quad @\ a_\circ \to a_\bullet$$

Problem 8.2 Transfer Between Aligned Ellipses
Consider a Hohmann transfer between two coplanar and coaxial ellipses. Show that the propulsion demand for the transition between the periapsis of the inner ellipse to the apoapsis of the outer ellipse is

$$\Delta v = \sqrt{\mu}\left(\frac{1}{\sqrt{r_{\bullet,per}}} - \frac{1}{\sqrt{r_{\circ,apo}}}\right)$$

$$\left(\frac{\sqrt{r_{\circ,apo}} + \sqrt{r_{\bullet,per}}}{\sqrt{a_T}} - \frac{\sqrt{r_{\circ,apo}}\sqrt{1+e_\bullet} - \sqrt{r_{\bullet,per}}\sqrt{1-e_\circ}}{\sqrt{r_{\circ,apo}} - \sqrt{r_{\bullet,per}}}\right)$$

$$\leq \sqrt{\mu}\left(\frac{1}{\sqrt{r_{\bullet,per}}} - \frac{1}{\sqrt{r_{\circ,apo}}}\right)$$

Remark: *For circular orbits this expression passes over into Eq. (8.1.7).*

Problem 8.3 Hohmann Transfer Maxima
(a) Prove that the maximum of the circularization impulse $\Delta v_{HO\to O}$ (see Fig. 8.4) is the root of the equation $x^3 - 5x^2 - 5x - 1 = 0$. Find the root $x = r_\circ/r_\bullet = 5.879\,362\ldots$ by Newton's method.
(b) By the same token, prove that the propulsion demand of a Hohmann transfer achieves a maximum at $x = a_\circ/a_\bullet = 15.581\,72\ldots$ which is the root of $x^3 - 15x^2 - 9x - 1 = 0$.

Problem 8.4 Hohmann Versus Bi-elliptic Transfer
Prove that for $x = r_\circ/r_\bullet > 11.938\,765\,4724\ldots$ the bi-elliptic transfer has a lower propulsion demand than the Hohmann transfer. Show that $x = r_\circ/r_\bullet$ is the root of the equation $x - 1 = \sqrt{1+x}(\sqrt{x} - \sqrt{2} + 1)$.

Problem 8.5 Variations of Orbital Elements by Kick-Burns
(a) Prove Eq. (8.4.3) as described in the text.
(b) Prove Table 8.1 by applying Eq. (8.4.3) at the given positions.

Problem 8.6 Kick-Burns in a Circular Orbit
Show with Eq. (7.3.3) that kick-burns in a circular orbit cause the following changes of orbital elements

$$\begin{pmatrix} \delta a/a \\ \delta e \\ \delta i \\ \delta \Omega \end{pmatrix} = \sqrt{\frac{a}{\mu}} \begin{pmatrix} 2 & 0 & 0 \\ 2 & 1 & 0 \\ 0 & 0 & 1 \\ 0 & 0 & 0 \end{pmatrix} \begin{pmatrix} \delta v_{\|} \\ \delta v_{\perp \circ} \\ \delta v_{\perp \perp} \end{pmatrix}$$

where at firing the position in the new orbit is

$$\theta = \begin{cases} 0° & @ \ \delta v_{\|} \\ 90° & @ \ \delta v_{\perp \circ} \\ - & @ \ \delta v_{\perp \perp} \end{cases}$$

Note that this true anomaly and the position where the kick-burn was performed determine the induced argument of periapsis ω.

Problem 8.7 Orbital Phasing
Suppose two satellites are flying a close formation on the same orbit at relative distance s and orbital period T. Show that, if s needs to be corrected, a kick-burn $\delta v_{\|}$ at the periapsis will cause a position shift of

$$\delta s_{per} = 3T \frac{1+e}{1-e} \delta v_{\|}$$

at the periapsis after one orbit, while a kick-burn $\delta v_{\|}$ at the apoapsis will cause a position shift of

$$\delta s_{apo} = 3T \frac{1-e}{1+e} \delta v_{\|}$$

at the apoapsis after one orbit.

9
Interplanetary Flight

In the last chapter we had a look at the transfer between two Keplerian orbits, and we saw how we can purposefully alter an orbit with selective impulse maneuvers. So we know how to head for targets with as little effort as possible, and we are generally prepared to embark on flights to other planets in our solar system. This entails two problems:

1. Shortly after launch we will mainly move in the gravitational field of the Earth. However, the further away we get from Earth, the smaller its force becomes, until during the transit flight we reach the influence of the Sun. Moreover, when we approach the planet of destination, we will enter the domain of its gravitational field, and there we will move on totally different orbits compared to the orbits during transit. How can we describe our orbits under these changing gravitational influences?

2. The second problem is that we no longer simply have to get from one planetary orbit to another. We now also want to meet the planet on the target orbit. This phasing problem is a further difficulty that has to be considered.

In fact, the first problem – to determine the orbit in the spheres of influence of different celestial bodies – is so serious that we cannot solve it exactly by analytical means. So, in practice, all interplanetary flights are determined only by complex numerical simulations. This also enables one to include even more complex maneuvers into the total flight trajectory, so-called gravity-assist and weak-stability-boundary maneuvers, which we will get to know at the end of this chapter. But as the important goal here is the basic understanding of orbit mechanics, we are seeking a method to describe the essentials of the processes, albeit not precisely. This is indeed possible. The method is called "patched conics".

Astronautics. Ulrich Walter
Copyright © 2008 WILEY-VCH Verlag GmbH & Co. KGaA, Weinheim
ISBN: 978-3-527-40685-2

9.1 Patched-Conics Method

Patched conics is a sequential domain-by-domain method. It simply separates the problem of a transition between two gravitational fields into two independent spatial domains: one domain in which one gravitational field dominates the other, and thus the other is neglected, and a second domain in which the situation is the other way round. The transition between them is, of course, not abrupt in reality; rather, it is very gradual. If you take the whole flight trajectory, however, the transition phase is rather short, and this is why the patched-conics method works so well in practice. It can be shown that, albeit the precision of the orbit trajectories derived by this method is only mediocre, the derived delta-v budget precision is very good, so that it is possible to carry out a fairly good mission planning with patched conics, which is later merely refined by numerical methods. Now let's have a closer look at the patched-conics method.

The baseline of the patched-conics method is that in any space domain the trajectory of a vehicle is determined by only one gravitational field, namely that which dominates. According to patched conics, if we start in low Earth orbit (LEO), we exclusively move in the gravitational field of the Earth, and we neglect the gravitational field of the Sun, which is about 1600 times weaker (see Fig. 9.1). The further we go into interplanetary space, the more the gravitational influence ratio is shifted towards the Sun, until we reach a point where the two gravitational forces, and thus also the two accelerations affecting our spacecraft, have the same size. That's where we transit from the so-called sphere of influence (SOI) of the Earth to the SOI of the Sun. The practical simplification is that the orbit calculation takes into account only one gravitational field and then the other. It is, of course, important to connect steadily and differentiably, that is to patch, the orbital conic segment in one SOI with that in the other. That's no problem, as long as you know where the transition point is. Where do you find the edge r_{SOI} of an SOI?

9.1.1 Sphere of Influence

To calculate the edge of the SOI, let's first consider at a test mass m at rest near the planet (index p), which orbits the Sun (index sun). It is exposed purely to their gravitational forces, which are given by

$$F_p(r) = \frac{Gm_p}{r^2} \quad (9.1.1)$$

and

$$F_{sun}(R) = \frac{Gm_{sun}}{R^2} \quad (9.1.2)$$

where we denote distances with regard to the planet by r and those with regard to the Sun by R. In fact, the test mass is not at rest but, with the planet, orbits the Sun. So, in addition to the gravitational forces, we have to consider centrifugal forces. At the center of mass of the planet, its centrifugal force just cancels out the gravitational force of the Sun. A planetary orbit is just in balance with these two forces. Any departure r of the test mass m from the planet's orbit in the direction to the Sun leads to a reduction of the centrifugal force and at the same time to an increase of the gravitational force of the Sun, which for small distances effectively comes to a factor $\sqrt{r/R}$ for F_{sun}, and thus Eq. (9.1.2) must be rewritten as

$$F_{sun}(R) = \frac{G m_{sun}}{R^2} \sqrt{\frac{r}{R}}$$

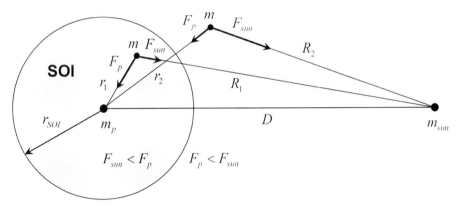

Figure 9.1 The gravitational force F of a planet (mass m_p) and of the Sun (mass m_{sun}) affecting a test mass m at two points near the SOI boundary.

We now consider two points near the edge of the SOI (see Figure 9.1). At point 1 the gravitational force of the planet dominates that of the sun, $F_{sun} < F_p$, and at point 2 it's just the other way round, $F_p < F_{sun}$. Right on the edge of the SOI the equilibrium equation

$$\frac{F_{sun}(R_1)}{F_p(r_1)} = \frac{F_p(r_2)}{F_{sun}(R_2)}$$

must hold, and $R_1 = R_2 =: R$, $r_1 = r_2 =: r_{SOI}$. If one inserts the above expressions into this equilibrium equation one obtains

$$r_{SOI} = R \left(\frac{m_p}{m_{sun}} \right)^{2/5}$$

We first consider the point on the SOI edge, which lies on the line connecting the planet and the Sun at mutual distance D. Then $R = D - r_{SOI}$ and we find

$$r_{SOI} = \frac{D}{(m_p/m_{sun})^{-2/5} + 1} \tag{9.1.3}$$

If $m_p = m_{sun}$ then for symmetry reasons $r_{SOI} = \frac{1}{2}D$ must hold, which is also derived from this equation. Since usually $m_p \ll m_{sun}$, we arrive at the approximate result

$$\boxed{r_{SOI} = D\left(\frac{m_p}{m_{sun}}\right)^{2/5}} \quad @ \; m_p \ll m_{sun} \tag{9.1.4}$$

So far, our argumentation was based on forces on the connecting line between planet and Sun. Because the test mass experiences slight variation of the Sun's gravitation and a centrifugal force when being at off-line positions on the SOI boundary, the SOI of a planet actually is a rotational ellipsoid with the connecting line as its symmetry axis. However, this induced anisotropy is so small that for all practical cases the SOI in good approximation is a sphere with radius r_{SOI}.

The result Eq. (9.1.4) was already derived by the French mathematician Lagrange around 1800, and it is still the best practical estimate of the SOI of a small celestial body orbiting another. If one applies the well-known planetary parameters of our solar system, one gets the SOI radii as given in Table 9.1.

Table 9.1 The radii of the planet's spheres of influence.

Planet	r_{SOI} (10^6 km)
Mercury	0.113
Venus	0.618
Earth	0.932
Mars	0.579
Jupiter	48.35
Saturn	54.71
Uranus	51.91
Neptune	87.01
Pluto	3.36

9.1.2
Patched Conics

Let's suppose that we make an interplanetary flight. As long as the S/C is within the SOI of the departure planet, the impact of the Sun can be neglected, and we get a hyperbolic departure orbit. When passing the SOI boundary the planetocentric hyperbolic orbit changes into a heliocentric transit trajectory. Because the SOI of a planet is much smaller than its distance to the Sun, the transit trajectory is a Hohmann ellipse between the two planetary orbits. Finally, the S/C approaches the target planet on a planetocentric hyperbola in its SOI. For the patched-conics process each of these three Keplerian orbits has to be tuned so that they pass smoothly into each other with regard to location and orbital velocity (Fig. 9.2).

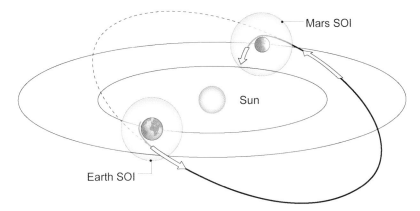

Figure 9.2 Patched-conics trajectory for a transit between Earth and Mars.

Moon's SOI

To derive the SOI radius of the Moon in the gravitational field of the Earth, one has to consider that $m_{earth}/m_{moon} = 81.300$. So $(m_{moon}/m_{earth})^{-2/5} = 5.808$ is no longer much bigger than unity. The Moon's SOI radius therefore has to be calculated from Eq. (9.1.3) to be $r_{SOI} = 56\,500$ km.

Note: *In the literature one finds $r_{SOI} = 66\,200$ km, which is derived from Eq. (9.1.4). However, Eq. (9.1.3) is more suited in this case.*

Though we can provide an SOI radius, the patched-conics method (a Hohmann ellipse around the Earth patched with a hyperbola around the Moon) is not appropriate for lunar trajectories. The reasons are as follows:

1. The lunar SOI is no longer negligibly small compared to the Earth–Moon distance. Therefore the elliptical orbit no longer is a Hohmann orbit to the Moon.

2. The Earth and the Moon move around a common center of mass, which is 4670 km away from the geocenter, and thus the position of the Earth shifts in the course of time.

3. The impact of the Sun cannot be neglected in the transfer zone.

These additional complications can be described by an extended patched-conics method with a so-called Michielsen diagram. For details please refer to Kaplan (1976, Section 3.5). Rapprochement (recurrent) orbits as well as "free return orbits" for Moon missions will be explained in detail in Sections 9.6 and 9.7 of this book.

9.2
Departure Orbits

To elaborate the patched-conics method for interplanetary flights let's take a flight from Earth to Mars as a running example. Without much loss of generality we assume that both planetary orbits are in the same plane and are circular. As with every interplanetary flight, the departure of the S/C is from a circular LEO parking orbit with radius r_0 (see Fig. 9.3). Thus its velocity is

$$v_0 = \sqrt{\frac{\mu_\oplus}{r_0}} \qquad (9.2.1)$$

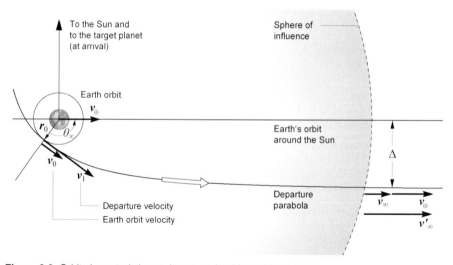

Figure 9.3 Orbit characteristics at departure (parking orbit and departure hyperbola).

To get the S/C from here to the heliocentric transfer orbit, an impulse maneuver $\Delta v = v_1 - v_0$ is required at the right position θ_∞ of the LEO, such that the departure velocity v_1 causes an excess velocity v_∞ at the edge of the SOI, as the initial velocity for the predetermined heliocentric transfer orbit. We now want to calculate the required Δv and θ_∞, supposing v_∞ is known, which we will determine later.

From Eq. (7.4.25) the departure velocity for a hyperbola can be calculated as follows:

$$v_1 = \sqrt{\frac{2\mu_\oplus}{r_0} + v_\infty^2} = \sqrt{2v_0^2 + v_\infty^2} \qquad (9.2.2)$$

So for the injection burn, we get

$$\Delta v = v_1 - v_0 = \sqrt{2v_0^2 + v_\infty^2} - v_0 \qquad (9.2.3)$$

As mentioned with Eq. (7.4.24), the parameter $C_3 := v_\infty^2$ is called **characteristic energy**. The eccentricity of the departure hyperbola is calculated with $v_1 = \mu(e+1)/h$ from Eq. (7.4.20) and $v_\infty = \mu\sqrt{e^2-1}/h$ from Eq. (7.4.23) and with Eq. (9.2.2) as

$$e = 1 + \frac{v_\infty^2}{v_0^2} = 1 + \frac{r_0 v_\infty^2}{\mu_\oplus} \qquad (9.2.4)$$

To determine the right timing angle θ_∞ for the injection burn, we use Eq. (7.4.21) and get

$$\theta_\infty = \arccos\frac{1}{e} = \arccos\frac{v_0^2}{v_0^2 + v_\infty^2} \qquad (9.2.5)$$

Figure 9.3 shows that this angle is measured relative to the direction of orbital movement of the Earth around the Sun.

Sensitivity Analysis

In practice, the injection burn can never be carried out exactly, but only with certain thrust errors $\delta v_\parallel$ and $\delta v_{\perp\odot}$ (see Sections 8.1.4 and 8.4.2). What is the impact of these thrust errors on the hyperbolic excess velocity with regard to its value as well as the asymptotic direction? Let's have a look at Eq. (9.2.3), which we rewrite as

$$v_\infty^2 + 2v_0^2 = (\Delta v + v_0)^2$$

If we differentiate and rearrange the result, we get

$$\frac{\delta v_\infty}{v_\infty} = \frac{v_1(v_1 - v_0)}{v_\infty^2}\frac{\delta v_\parallel}{\Delta v} \qquad (9.2.6)$$

We get two contributions to the error $\delta\theta_\infty$: the first, because of Eqs. (9.2.6) and (9.2.5) $\delta v_\parallel \to \delta v_\infty \to \delta\theta_\infty$; and the second because a perpendicular thrust error leads to a rotation of the line of apsides of the hyperbola, i.e. $\delta v_{\perp O} \to \delta \omega = \delta \theta_\infty$. The first contribution is

$$\delta \theta_\infty = -\frac{1}{e}\sqrt{\frac{e-1}{e+1}}\frac{\delta v_\infty}{v_\infty} = -\frac{v_0^2}{v_0^2+v_\infty^2}\frac{v_\infty}{v_1}\frac{\delta v_\infty}{v_\infty} = -\frac{v_0^2(v_1-v_0)}{v_\infty(v_0^2+v_\infty^2)}\frac{\delta v_\|}{\Delta v}$$

For the second we get according to Table 8.1 and because of $h = r_0 v_1 = \mu_\oplus v_1/v_0^2$:

$$\delta \theta_\infty = \delta \omega = -\frac{h}{e\mu_\oplus}\delta v_{\perp O} = -\frac{v_1}{ev_0^2}\delta v_{\perp O}$$

With Eqs. (9.2.2) and (9.2.4) we finally get

$$\delta \theta_\infty = -\frac{v_1(v_1-v_0)}{v_0^2+v_\infty^2}\frac{\delta v_{\perp O}}{\Delta v}$$

So, in total we have

$$\delta \theta_\infty = -\frac{v_1(v_1-v_0)}{v_0^2+v_\infty^2}\left(\frac{v_0^2}{v_1 v_\infty}\frac{\delta v_\|}{\Delta v} + \frac{\delta v_{\perp O}}{\Delta v}\right) \tag{9.2.7}$$

Example

For an Earth → Mars transit with $v_\infty = 3.040$ km s^{-1}, $v_0 = 7.76$ km s^{-1} (300 km parking orbit) and $v_1 = 11.34$ km s^{-1}, we get

$$\frac{\delta v_\infty}{v_\infty} = 4.4\frac{\delta v_\|}{\Delta v} \quad \text{and} \quad \delta \theta_\infty = -0.59\left(1.7\frac{\delta v_\|}{\Delta v} + \frac{\delta v_{\perp O}}{\Delta v}\right)$$

A 1% error in thrust direction and perpendicular to it leads to an error of $\delta v_\infty / v_\infty = 4.4\%$ and $\delta \theta_\infty = -0.91°$.

Departure Hyperbolas

In order to pass over into a heliocentric tangential Hohmann transfer orbit at the edge of the Earth SOI, the orbital plane of the departure hyperbola has to include the instantaneous velocity vector of the Earth. But apart from that, it may have a random orientation (see Fig. 9.4). So there are many possible departure hyperbolas. However, to include the Earth's velocity vector, the parking orbit has to be included as well. Without any additional propulsion demand, such a matching parking orbit can be reached only twice a day for a specific launch site. So there are only two launch windows per day for interplanetary flights. Once the S/C is in the right parking orbit, there is one injection burn opportunity per orbit. According to Fig. 9.3, for flights to the outer planets, i.e. for v_∞ parallel to $v_\oplus$ and for prograde orbits, this opportunity lies on the night side of the Earth, and for flights to inner planets, i.e. for v_∞ antiparallel to $v_\oplus$, on the day side of the Earth.

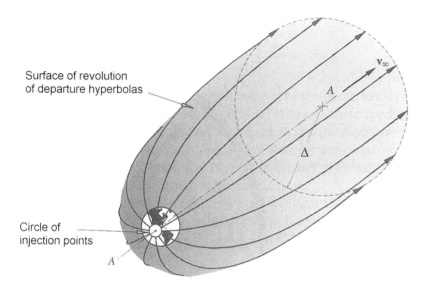

Figure 9.4 Possible orientations of the departure hyperbola.

9.3
Transit Orbits

9.3.1
Hohmann Transfers

We have now reached the edge of the Earth's SOI. From here, the trajectory is a heliocentric conic, so we use the heliocentric system of coordinates, which we indicate by a prime. In this primed coordinate system, the Earth has orbit velocity $v_\oplus$, and the velocity of our S/C is

$$v'_\infty = v_\infty + v_\oplus$$

As we have selected our entry conditions so that for flights to outer planets v_∞ is parallel to $v_\oplus$, and for inner planets v_∞ antiparallel to $v_\oplus$, this implies

$$v'_\infty = \begin{cases} v_\oplus + v_\infty & @ \to \text{outer planets} \\ v_\oplus - v_\infty & @ \to \text{inner planets} \end{cases} \quad (9.3.1)$$

We still have to determine the required transfer orbit, i.e. that excess velocity v_∞, to meet the target orbit. In Chapter 8 we learned that energetically a

Hohmann transfer would be most favorable. If we select a Hohmann transfer, the transfer injection burn to an outer or inner planet reads from Eq. (8.1.6) as

$$v_\infty = \Delta v_{\bullet \to H} = v_\oplus \left[\sqrt{\frac{a_{planet}}{a_H}} - 1 \right]$$

$$v_\infty = \Delta v_{\circ \to H} = v_\oplus \left[1 - \sqrt{\frac{a_{planet}}{a_H}} \right]$$

So for both types of missions we get with $a_H = (a_\oplus + a_{planet})/2$ (Mars: $a_H = 1.2618\, a_\oplus$),

$$v_\infty = v_\oplus \left| 1 - \sqrt{\frac{2 a_{planet}}{a_\oplus + a_{planet}}} \right| \qquad (9.3.2)$$

where a_H is the semi-major axis of the Hohmann transfer orbit, and $v_\oplus = 29.78\text{ km s}^{-1}$. For Mars, we get $v_\infty = 2.972\text{ km s}^{-1}$. Please note that this excess velocity is unprimed, so it is valid in the geocentric system, not in the primed heliocentric system. For the Hohmann transfer time we find from Eq. (8.1.3)

$$t_H = \pi \sqrt{\frac{a_H^3}{\mu}} = \frac{T_\oplus}{4\sqrt{2}} \left(\frac{a_\oplus + a_{planet}}{a_\oplus} \right)^{3/2} \qquad (9.3.3)$$

with $T_\oplus = 365.256\, d$ the period of the Earth orbit.

Orbit Phasing

With a Hohmann transfer, we would just touch the planetary target orbit. But to hit the target planet, the Earth and the target planet have to be in a specific mutual configuration. To determine this configuration, we make use of our assumption that both planetary orbits are circular and lie in the same plane (ecliptic plane). This restriction is insignificant, but it eases our explanation of the principle of configuration determination. According to Fig. 9.5a the angle $\theta_{\bullet\circ}$ between the initial configuration and the final configurations of the Earth and the target planet is

$$\theta_{\bullet\circ} = 180° - n_\circ t_H = 180° \left(1 - \sqrt{\frac{a_H^3}{a_\circ^3}} \right) \qquad (9.3.4a)$$

$$\theta_{\circ\bullet} = 180° - n_\bullet t_H = 180° \left(1 - \sqrt{\frac{a_H^3}{a_\bullet^3}} \right) \qquad (9.3.4b)$$

where $n_\odot$, $n_\bullet$ is the mean motion (= orbit angular velocity) of the outer (Mars) and inner planet (Earth). So for a Mars transit, we get: $t_H = 258.86\,d$, $\theta_{\bullet\odot} = 44.3°$, $\theta_{\odot\bullet} = -75.1°$.

When do these constellations come up again? This question is essential for interplanetary missions, because the narrow launch window is open only in this interval. With regard to a given direction in the ecliptic plane, let's say the vernal equinox (index 0), the orbit positions evolve according to Eq. (7.4.1b)

$$\theta_\bullet = \theta_{\bullet,0} + n_\bullet t$$

$$\theta_\odot = \theta_{\odot,0} + n_\odot t$$

and thus the constellation angle is

$$\theta_{\bullet\odot} = \theta_\odot - \theta_\bullet = \theta_0 + (n_\odot - n_\bullet)t$$

The constellation angle recurs after time $t = T_{syn}$, the so-called *synodic period*, when $\theta_{\bullet\odot} \to \theta_{\bullet\odot} - 2\pi$, i.e.

$$(n_\odot - n_\bullet)T_{syn} = -2\pi$$

From this follows

$$\frac{1}{T_{syn}} = \frac{n_\bullet}{2\pi} - \frac{n_\odot}{2\pi}$$

or

$$\frac{1}{T_{syn}} = \frac{1}{T_\bullet} - \frac{1}{T_\odot} \quad \text{synodic period} \tag{9.3.5}$$

and $T_\odot$, $T_\bullet$ are the orbital periods of the outer and inner planet, respectively. For Earth–Mars, we have $T_{syn} = 2.135$ yr.

In contrast to Moon missions, it is not possible to fly back to Earth at just any time. You have to wait for the right planetary constellation (see Fig. 9.5b), just as for the forward flight. How long do you have to wait until you get the right return flight constellation after you have arrived? The answer can be derived from similar phase considerations as above (exercise, Problem 9.1). It is

$$t_{wait} = T_{syn}\left(k - \left|1 - \frac{2t_H}{T_\oplus}\right|\right) \tag{9.3.6}$$

where k is any natural number for which $t_{wait} > 0$. The shortest waiting time therefore occurs with the smallest k, when $t_{wait} > 0$ for the first time. Finally,

9 Interplanetary Flight

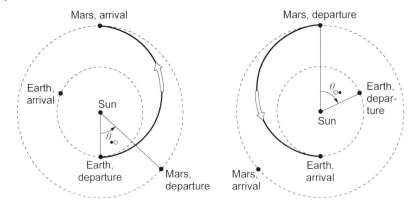

Figure 9.5 Relative configuration of Earth and Mars for forward (left) and return transits (right).

the minimum total flight time is calculated with forward and return flight time plus minimum waiting time:

$$t_{trip} = 2t_H + t_{wait} \tag{9.3.7}$$

Table 9.2 gives an overview of the derived characteristic times for the relevant planets in the solar system.

Table 9.2 The characteristic times for Hohmann transits from the Earth to the planets in the solar system.

Planet	T [yr]	T_{syn} [d]	t_H [d]	t_{wait} [d]	t_{trip} [yr]
Mercury	0.2408	115.9	105.5	66.9	0.761
Venus	0.6152	584.0	146.1	467.1	2.079
Earth	1.0000	—	—	—	—
Mars	1.8808	779.8	258.8	454.5	2.661
Jupiter	11.863	398.9	997.9	213.8	6.049
Saturn	29.414	378.0	2222	315.6	12.166

9.3.2
Non-Hohmann Transfers

Hohmann orbits may be the most favorable transfer orbits from an energetic point of view, but in Chapter 8 we already found out that Hohmann transfer orbits are very sensitive to initial thrust errors, and also that they take the longest time. So just a little more thrust would make sure that, with small thrust errors, the transfer orbit still intersects the target orbit, while the transfer time drastically decreases (cf. Fig. 9.6). But how does the crossing point and with it the transfer time change with the transfer excess velocity in the initial parking orbit?

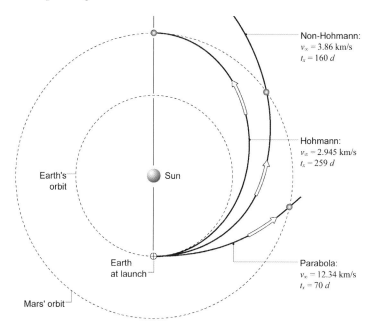

Figure 9.6 Various Earth–Mars transfer orbits.

We consider a *flight from an inner to an outer target orbit* (running example Earth → Mars), for which $v'_\infty = v_\oplus + v_\infty$ holds according to Eq. (9.3.1). We recall that we assumed coplanar and circular planetary orbits. An elliptic transfer orbit is of the form

$$r = \frac{a\left(1-e^2\right)}{1+e\cos\theta} \tag{9.3.8}$$

The ellipse touches the initial orbit with radius $a_\bullet$ at its periapsis. Therefore

$$a_\bullet = r_{per} = a\left(1-e\right)$$

and the initial speed at periapsis is

$$v'_\infty = \sqrt{\frac{\mu}{a}\frac{1+e}{1-e}} = \sqrt{\frac{\mu}{a_\bullet}(1+e)}$$

From this we derive the orbit number k (see Section 14.4.1)

$$1+e = \frac{a_\bullet v'^2_\infty}{\mu} =: k_\bullet \tag{9.3.9}$$

and

$$a = \frac{a_\bullet}{2 - k_\bullet} \quad (9.3.10)$$

The crossing (index ×) with the target orbit at $r = a_\odot$ determines the true anomaly of the crossing point via Eq. (9.3.8):

$$\cos\theta_\times = \frac{1}{e}\left[\frac{a(1-e)(1+e)}{a_\odot} - 1\right]$$

From this we find with Eqs. (9.3.9) and (9.3.10)

$$\cos\theta_\times = \frac{1}{k_\bullet - 1}\left(\frac{a_\bullet k_\bullet}{a_\odot} - 1\right) \quad @ \bullet \to \odot \quad (9.3.11)$$

According to Eq. (7.4.13c) we can translate this into an eccentric anomaly

$$\sin E_\times = \frac{\sqrt{1-e^2}\cdot\sin\theta_\times}{1+e\cos\theta_\times} \quad (9.3.12)$$

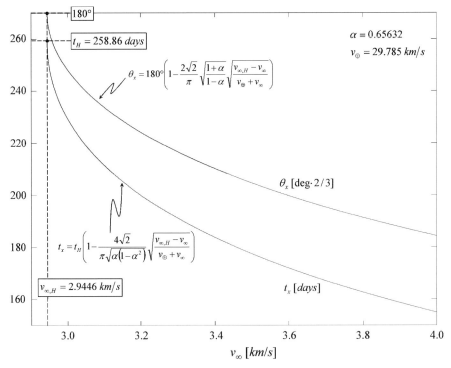

Figure 9.7 Dependence of the transition time $t_\times$ and the orbit crossing angle $\theta_\times$ as functions of the excess velocity v_∞ for an Earth → Mars transit.

to find from the Keplerian equation (7.4.16) the transit time

$$\sqrt{\frac{\mu}{a^3}} \cdot t_\times = E_\times - e \sin E_\times \tag{9.3.13}$$

Figure 9.7 depicts how, according to Eqs. (9.3.11) and (9.3.13), the transit time and the true anomaly of the crossing point decrease with increasing excess velocity, $k_\bullet = a_\bullet v'^2_\infty/\mu$, for an Earth → Mars transit.

For missions *from an outer to an inner planet*, for instance Earth → Venus, one has to substitute the indices $\bullet \leftrightarrow \circ$ in Eq. (9.3.9) to (9.3.11). Furthermore $\cos\theta_\times = (a_\circ k_\circ/a_\bullet - 1)/(1 - k_\circ)$ and $k_\circ = a_\circ v'^2_\infty/\mu = 1 - e$ holds. Otherwise Eqs. (9.3.12) and (9.3.13) remain unchanged.

Approximate Solution near Hohmann Transits

If the general elliptic transfer orbit passes into the specific Hohmann transfer orbit, the derivatives of the transit time and of the true anomaly diverge. These are the two drawbacks of a Hohmann transfer that we already mentioned in the sensitivity analysis in Section 8.1.2. So for practical purposes one will choose a transit orbit away from the Hohmann transfer so that, despite injection burn errors, an intersection with the target orbit is ensured, but otherwise close enough that one still profits from the energetic advantage of a Hohmann transfer.

Let's investigate the behavior of $t_\times$ and $\theta_\times$ near a Hohmann transfer. Specifically we want to trace the error chains $\Delta v_1 \to \Delta v_\infty \to \Delta a \to \Delta E_\times \to \Delta \theta_\times$ and $\Delta v_1 \to \Delta v_\infty \to \Delta a \to \Delta E_\times \to \Delta t_\times$ mathematically. For this we work in the chain backwards and start with $\Delta E_\times \to \Delta t_\times$. The time on the transfer ellipse is given by the above Keplerian equation. To determine the effects of tiny variations, we linearize the Keplerian equation

$$t_\times \cdot dn + n \cdot dt_\times = (1 - e \cos E_\times) dE - \sin E_\times \cdot de$$

Because $n = \sqrt{a^3/2\mu}$ and therefore $dn = 3/2\sqrt{a/\mu} \cdot da$, we obtain for small but finite variations of the orbit parameters and for $E_\times \approx \pi$

$$n \cdot \Delta t_\times = (1+e) \cdot \Delta E_\times - \frac{3}{2} t_\times \sqrt{\frac{a}{\mu}} \cdot \Delta a \tag{9.3.14}$$

We now have to find the chain link $\Delta a \to \Delta E_\times$ to arrive at $\Delta a \to \Delta t_\times$. Let's have a look at relation Eq. (7.4.13a)

$$r = a(1 - e \cos E)$$

We differentiate this equation by bearing in mind that the radius of the intersecting point remains constant, $r = a_{target} = const$, and that according to

Eq. (8.1.16) $de = (1 - e) \, da/a$. After some rearrangements we obtain

$$\frac{dE}{da} = -\frac{1}{a_H e_H} \tan \frac{E}{2}$$

Let us check. For $E < \pi$ we get with increasing semi-major axis a decreasing eccentric anomaly, as expected. Because we are close to the Hohmann transit $\tan E/2 \to \infty$. So in order to get finite variations, we have to integrate this equation by separating its variables

$$\int_\pi^{\pi - \Delta E_\times} \frac{dE}{\tan(E/2)} = -\frac{1}{e_H} \int_{a_H}^{a_H + \Delta a} \frac{da}{a}$$

In the left integral, we substitute $E = \pi - 2\delta \Rightarrow dE = -2 \cdot d\delta$, and we get

$$\int_\pi^{\pi - \Delta E_\times} \frac{dE}{\tan(E/2)} = -2 \int_0^{\Delta E_\times/2} \frac{d\delta}{\tan(\pi/2 - \delta)}$$

For $\delta \to 0$ the following is valid:

$$\frac{1}{\tan(\pi/2 - \delta)} = \frac{1}{\cot \delta} = \tan \delta \approx \delta$$

With this we get on one side of the integral

$$\int_\pi^{\pi - \Delta E_\times} \frac{dE}{\tan(E/2)} \approx -2 \int_0^{\Delta E_x/2} \delta \cdot d\delta = -\frac{(\Delta E)^2}{4}$$

and on the other side

$$-\frac{1}{e_H} \int_{a_H}^{a_H + \Delta a} \frac{da}{a} = -\frac{1}{e_H} \ln\left(1 + \frac{\Delta a}{a_H}\right) \approx -\frac{1}{e_H} \frac{\Delta a}{a_H}$$

In summary, we obtain for the chain link $\Delta a \to \Delta E_\times$

$$\Delta E_\times \approx -\frac{2}{\sqrt{e_H}} \sqrt{\frac{\Delta a}{a_H}}$$

So, any $\Delta E_\times$ decreases with the square root of Δa for $\Delta a \to 0$. Compared with this rapid variation, we can neglect the linear dependence Δa in Eq. (9.3.14) and obtain with $n = 2\pi/T \approx 2\pi/2t_H = \pi/t_H$

$$\frac{\Delta t_\times}{t_H} \approx -\frac{2}{\pi} \frac{1 + e_H}{\sqrt{e_H}} \sqrt{\frac{\Delta a}{a_H}}$$

We now work one more step backward in our error chain and look for the dependence $\Delta v_\infty \to \Delta a$. Because of Eq. (8.1.4) $e_H = (a_\bigcirc - a_\bullet)/(a_\bigcirc + a_\bullet)$ and since from Eq. (8.1.16) for a Hohmann transfer it follows that

$$\frac{da}{a_H} = 2\frac{a_\bigcirc}{a_\bullet}\frac{dv'_\infty}{v'_\infty} = 2\frac{a_\bigcirc}{a_\bullet}\frac{dv_\infty}{v_\oplus + v_\infty}$$

any excess velocity error Δv_∞ can be converted into a transit time error

$$\frac{\Delta t_x}{t_H} \approx -\frac{4\sqrt{2}}{\pi\sqrt{\alpha(1-\alpha^2)}}\sqrt{\frac{|v_{\infty,H} - v_\infty|}{v_\oplus \pm v_\infty}}$$

$$= -\frac{4\sqrt{2}}{\pi\sqrt{\alpha(1-\alpha^2)}}\sqrt{\frac{v_\infty}{v_\oplus \pm v_\infty}}\sqrt{\frac{\Delta v_\infty}{v_\infty}} \qquad (9.3.15)$$

with

$$\alpha := \frac{a_\bullet}{a_\bigcirc}$$

and $v_{\infty,H}$ the excess velocity for a Hohmann transfer. According to Eq. (9.3.1) the different signs denote the transit to an outer or an inner planet, respectively.

We now make the final step backward to derive $\Delta v_1 \to \Delta v_\infty$. From differentiating Eq. (9.2.2) we get

$$\frac{\Delta v_1}{v_1} = \frac{v_\infty^2}{v_1^2}\frac{\Delta v_\infty}{v_\infty}$$

This inserted into Eq. (9.3.15) finally delivers the transition time from the transplanetary injection burn to the intersection with the target orbit as a function of the injection burn error in the LEO parking orbit Δv_1. From $t_x = t_H \pm \Delta t_x$ we get

$$t_x \approx t_H\left[1 \mp \frac{4\sqrt{2}}{\pi\sqrt{\alpha(1-\alpha^2)}}\frac{v_1}{\sqrt{v_\infty(v_\oplus \pm v_\infty)}}\sqrt{\frac{\Delta v_1}{v_1}}\right] \qquad (9.3.16)$$

with $\alpha = a_\bullet/a_\bigcirc$. By the same token the true anomaly of the intersection point is calculated (exercise, Problem 9.4) to be

$$\theta_x \approx 180°\left[1 \mp \frac{2\sqrt{2}}{\pi}\sqrt{\frac{1+\alpha}{1-\alpha}}\frac{v_1}{\sqrt{v_\infty(v_\oplus \pm v_\infty)}}\sqrt{\frac{\Delta v_1}{v_1}}\right] \qquad (9.3.17)$$

In both equations the different signs in the first place denote the first and the second intersection of the transfer ellipse with the target orbit, respectively, and the second one the transit to an outer or inner planet, respectively. The divergences close to the Hohmann transfer are of the form $\sqrt{\Delta v_\infty} \propto \sqrt{\Delta v_1}$, in accordance with Fig. 9.7.

Example
For an Earth → Mars transit, $v_\infty = 2.972 \text{ km s}^{-1}$, $v_1 = 11.32 \text{ km s}^{-1}$ (300 km parking orbit), $v_\oplus = 29.78 \text{ km s}^{-1}$ and $\alpha = 0.656\,32$. From this we derive for the transit time to the first intersection point

$$t_x \approx t_H \left(1 - 3.380 \sqrt{\frac{\Delta v_1}{v_1}}\right)$$

Because for Mars $t_H = 259$ d, this means that for an injection burn error of $\Delta v_1 = 0.01 \text{ km s}^{-1}$ in LEO surplus to the regular Hohmann injection burn, $v_1 = 11.32 \text{ km s}^{-1}$ (that is a surplus of only 0.1%) we achieve a transition time reduction of $\Delta t = 26.0$ d, that is 10%! The first intersection angle is at

$$\theta_x \approx 180° \left(1 - 2.268 \sqrt{\frac{\Delta v_1}{v_1}}\right)$$

and the reduction of the intersection angle is $\Delta \theta_x = 12.1°$.

9.4
Arrival Orbit

After the heliocentric transition phase, the S/C enters the SOI of the target planet and crosses it on a hyperbola, which according to Eq. (9.2.4) has the eccentricity

$$e = 1 + \frac{r_{per} v_\infty^2}{\mu} \qquad (9.4.1)$$

For Hohmann transfers the approach hyperbola only has to include the line of movement of the target planet. Otherwise, and in analogy to departure (see Fig. 9.4), the orientation of the target orbit can be chosen randomly (see Fig. 9.8).

According to Eq. (7.4.22), the aiming radius, which we are free to choose by adjustment burns in the course of the heliocentric transit, is related to the minimum approach distance r_{per} to the target planet by

$$\Delta = r_{per} \sqrt{\frac{e+1}{e-1}} = r_{per} \sqrt{1 + \frac{2\mu}{r_{per} v_\infty^2}} \qquad (9.4.2)$$

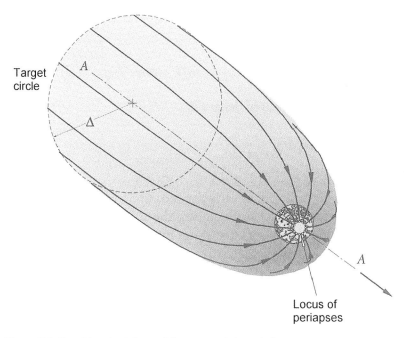

Figure 9.8 Possible orientations of the approach hyperbola.

Optimum Planetary Capture

It is now our goal to transfer the S/C with minimum delta-v effort from its unbounded hyperbolic approach (index h) orbit into an elliptically bound capture orbit with orbit parameters a_c, e_c. We know from Eq. (8.4.13) that the required injection burn to catch the S/C (to lower the infinite apoapsis to a finite value) preferably takes place at the periapsis where the hyperbolic speed is, according to Eqs. (7.4.19) and (7.4.20),

$$v_{h,per} = \sqrt{\frac{\mu(e_h + 1)}{r_{per}}} = \sqrt{v_\infty^2 + \frac{2\mu}{r_{per}}}$$

The latter equation results from Eq. (9.4.1). For the elliptical capture orbit the periapsis velocity is, according to Eqs. (7.4.10a) and (7.4.8a),

$$v_{c,per} = \sqrt{\frac{\mu(1 + e_c)}{r_{per}}}$$

So the required kick-burn at the joint periapsis is

$$\Delta v = \sqrt{v_\infty^2 + \frac{2\mu}{r_{per}}} - \sqrt{\frac{\mu(1 + e_c)}{r_{per}}}$$

Note that the delta-v only indirectly (via the periapsis) depends on the semi-major axis of the capture orbit. Now, if we want to minimize the Δv demand at a given eccentricity of the capture orbit, we can do that by determining the optimum periapsis distance. The result of this quite simple minimizing task (exercise) is

$$r_{per,opt} = \frac{2\mu}{v_\infty^2} \frac{1-e_c}{1+e_c} \tag{9.4.3}$$

so we get

$$\Delta v_{min} = v_\infty \sqrt{\frac{1-e_c}{2}} \tag{9.4.4}$$

as the minimum kick-burn. According to Eq. (9.4.2) the optimum periapsis distance is adjusted by the following optimum aiming radius

$$\Delta_{opt} = 2\sqrt{2} \frac{\sqrt{1-e_c}}{1+e_c} \frac{\mu}{v_\infty^2} \tag{9.4.5}$$

If aerobraking is possible, such as at Mars or Venus, then one first aims at a highly elliptical orbit $e \approx 1$, which is achieved by a very small injection burn (Eq. (9.4.4)) with a very low periapsis distance (Eq. (9.4.3)) (Of course one has to bear in mind that the aiming radius always has to be larger than the radius of the planet.) As we will see in Section 12.6.3, the captured S/C will then circularize because of the atmospheric drag at the periapsis, and finally after many orbits, depending on the drag force, it will be turned down to a circular orbit with radius $a = r_{per}$. This is the most effective way to achieve a circular capture orbit, and thus it is also regularly done on Mars missions, as for instance with NASA's Mars Reconnaissance Orbiter in March 2006.

9.5
Flyby Maneuvers

9.5.1
Basic Considerations

The exploration of the planets of our solar system is one of the most attractive undertakings in astronautics. However, the propulsion demands for missions to these planets are considerable. We had already seen that for Mars we would need $v_1 = 11.32 \text{ km s}^{-1}$. If we wanted to fly to Jupiter on a Hohmann transfer orbit, according to Eqs. (9.3.2) and (9.2.2) we would already need $v_1 = 14.04 \text{ km s}^{-1}$. If we even wanted to leave the solar system, we would need at least $v_\infty = (\sqrt{2} - 1)v_\oplus = 12.34 \text{ km s}^{-1}$ and therefore

$v_1 = 16.48 \text{ km s}^{-1}$. These propulsion demands are impossible to achieve with chemical propulsions, even with just a minimum payload. But despite this fact, the probes Voyager and Pioneer flew to the most remote planets of our solar system, already in the 1970s, and by now they have actually left our solar system. How did NASA manage to accomplish that? The trick is called "gravity-assist maneuver," also called "swing-by" or "flyby maneuver." Such a maneuver makes use of a near flyby past a planet to get additional momentum. You could compare it with a roller-blader, who just for a short time hitches up to a passing bus to gain speed. The flyby maneuver is not understandable in the two-body system. If an S/C enters the SOI of a planet it is deflected and leaves the SOI with exactly the same escape velocity it had when it entered the SOI. Only when the Sun is taken into account, and we watch the flyby in the heliocentric system, do we see that the S/C indeed picks up speed. The approached planet has an orbit velocity, and because of the gravitational "hitching up" of the S/C to the planet during flyby it is able to pick up part of its orbit velocity. It is a special feature of a multi-body system that its bodies can exchange energy and impulse. Note that momentum gain by flyby only works because the planet loses the same momentum the S/C gains. However, as the mass of a planet is a lot bigger than the mass of the S/C, the velocity change of the planet is insignificant and undetectable.

In fact, Pioneer and Voyager not only undertook one flyby maneuver, but several in a sequence. And with each flyby the S/C gained speed until the escape velocity of the solar system was exceeded. This trick is frequently used today to fly to other planets. Even when flying only to the next planet, Mars, you could save delta-v for Earth excess velocity if you pick up momentum before from Moon flybys (see Section 9.7). If you are clever, you could pick up this momentum and gain velocity even several times. But of course it all has its price, and here the price is time. Every flight to the Moon and back to the Earth requires a lot of time, and thus prolongs the flight to the real target, Mars. However, sometimes planets are constellated such that there is an intermediate planet on the path to flyby right to the target planet. In very rare cases, and they occur only once every few decades, there are several planets in a row to swing along to the outermost planets in our solar system. Voyager 2 flew along Jupiter, Saturn, Uranus and Neptune, a constellation that only occurs once every 189 years. This situation came along in the middle of the 1970s, and thus NASA took on this very tricky mission already at that early time of interplanetary space flight.

9.5.2
Flyby Framework

Let's assume a spacecraft approaches a planet from a large distance for a flyby and is just about to enter the planet's SOI (cf. Fig. 9.2) with incoming velocity v_{in} in the heliocentric reference system. The planet at that time may have velocity v_p in its orbital plane. Because we are only able to describe the flyby in the "planet–spacecraft" two-body system with their center of mass (which is essentially the planet) at rest, we have to map the incoming velocity into the planetocentric system, where $v_p = 0$. According to Fig. 9.9 the mapped velocity v_∞^-, with which the S/C enters the SOI, reads

$$v_\infty^- = v_{in} - v_p \tag{9.5.1}$$

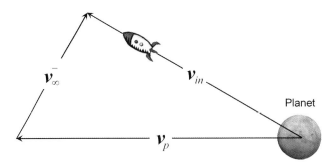

Figure 9.9 The velocity triangle in the heliocentric reference system at entry into the planet's SOI.

The subscript ∞ of the entry velocity denotes that it is a hyperbolic excess velocity. The superscript "−" means that we are before flyby. We are now in the SOI with the planet at rest. Figure 9.10 illustrates the most general situation for the flyby of a S/C within the SOI. The flyby takes place on a hyperbola in the flyby plane, which needs not to be coplanar with the planet's orbital plane. This non-coplanarity complicates the mathematical description somewhat, which we will however neglect for the time being, by assuming that the flyby plane is coplanar with the planet's orbital plane. This, of cause, implies that the incoming velocity v_{in} and the flyby trajectory is also in the planet's orbital plane.

The S/C then passes through the SOI, whereby it is deflected on a hyperbolic path, and leaves the SOI with exit velocity v_∞^+. As we are in a two-body system, the principle of conservation of energy holds, and according to the vis-viva Eq. (7.2.9) the following has to be valid for $r \to \infty$:

$$v_\infty^- = v_\infty^+ =: v_\infty \tag{9.5.2}$$

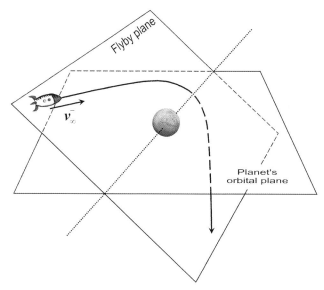

Figure 9.10 The hyperbolic flyby path of the S/C in the flyby plane relative to the planet's orbital plane.

The absolute value of the exit velocity of the S/C is thus identical to its entry velocity; only the flight direction changes. This velocity turn δ may be counterclockwise, $\delta > 0$, or clockwise $\delta < 0$, depending on whether the S/C intersects the planet's path "from inside" or "from outside" and whether it flies by "before" or "behind" the planet (see Fig. 9.11). "From inside" and "from outside" mean that the S/C intersects the planet orbit from inside and outside, respectively. "Before" and "behind" refer to the direction of movement of the planet. As seen, either a before or a behind flyby can cause clockwise and counterclockwise turns. But from Fig. 9.12 it can generally be said that, if the S/C flies by *behind* a planet, it is "carried along" the orbit by the planet, so it *gains speed*, $v_{out} > v_{in}$. The other way round is that it moves *before* the planet, then its flight momentum is "hampered" and the transferred momentum leads to a *reduction of its velocity*, $v_{out} < v_{in}$. This is not strictly true, but is mostly so. The latter case may seem to be academic, as usually you want to gain velocity. But this is only the case with flights to outer planets. To get to inner planets, one has to reduce the velocity of the S/C after it leaves the Earth. This can be done either by flybys to the Moon, or, if you want to go to Mercury, by flying by Venus – if the planetary constellation admits it.

It will be the main task in the next section to determine the exact deflection angle δ. Let's assume for a moment that we have found δ. Because $v_\infty^- = v_\infty^+ =: v_\infty$, the deflection can be described as a rotation of the velocity vector

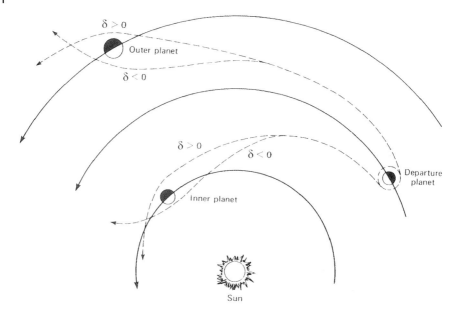

Figure 9.11 Before/behind flybys in the "from inside" case (above) and the "from outside" case (below). Note the different turn directions for these two cases.

(see Fig. 9.11) by angle δ

$$v_\infty^+ = \mathbf{R}_\delta v_\infty^- \tag{9.5.3}$$

with rotation matrix

$$\mathbf{R}_\delta = \begin{pmatrix} \cos\delta & -\sin\delta \\ \sin\delta & \cos\delta \end{pmatrix}$$

Once we know the exit vector v_∞^+ we can derive the outgoing vector v_{out} in the heliocentric system according to Fig. 9.12 and in analogy to Eq. (9.5.1) as

$$v_{out} = v_\infty^+ + v_p \tag{9.5.4}$$

So, to determine the final outgoing vector, everything hinges on the determination of the deflection angle, to which we turn now.

9.5.3
Flyby Analysis in the Planetocentric System

In this section we want to determine quantitatively the changes of flight direction and speed the S/C undergoes during flyby in the planetocentric coordinate system.

9.5 Flyby Maneuvers

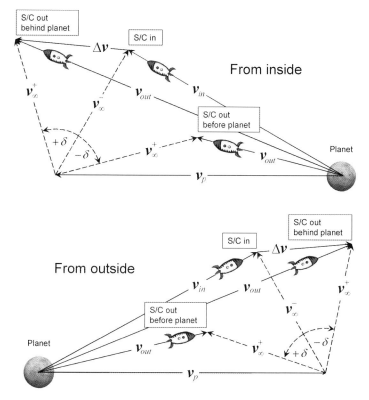

Figure 9.12 The vector diagram of a flyby from a heliocentric point of view (full vectors) for the "from inside" case (top) and the "from outside" case (bottom). The S/C enters the SOI of the planet with velocity v_{in}. During flyby behind or before the planet, it changes flight direction by angle δ and then leaves the SOI with velocity v_{out}.

Deflection Angle

We first calculate the deflection angle. According to Fig. 9.13 it is given by

$$\delta = \pi - 2\psi$$

The angle ψ in turn is, according to Fig. 7.11, the reverse angle of the escape angle θ_∞, for which $\cos\theta_\infty = -1/e$ was valid (see Eq. (7.4.21)). So

$$\tan\psi = \tan(\pi - \theta_\infty) = -\tan\theta_\infty = \sqrt{e^2 - 1} = \frac{\Delta}{-a}$$

where the latter follows from Eq. (7.4.22). While the aiming radius is our freely adjustable parameter, the hyperbolic parameter $-a > 0$ is determined from the vis-viva equation (7.2.10) by the entry condition $r \to \infty$ at the SOI

$$v_\infty^2 = -\frac{\mu_p}{a}$$

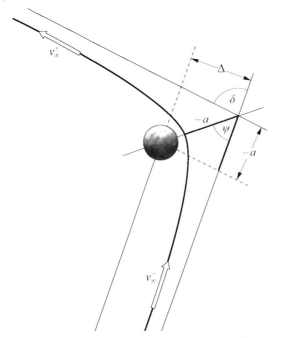

Figure 9.13 Flyby parameters in the planetocentric reference system.

Inserting this and $\delta = \pi - 2\psi$ into the above equation leads to the result

$$\delta = \pi - 2\arctan\frac{\Delta \cdot v_\infty^2}{\mu_p}$$

or

$$\tan\frac{\delta}{2} = \frac{\mu_p}{\Delta \cdot v_\infty^2} \tag{9.5.5}$$

So the deflection depends on the adjustable aiming radius Δ and the given entry velocity v_∞. Observe that for $\Delta \to 0$ the deflection angle tends to $\delta \to 180°$, which is the limiting turn the S/C theoretically would do if it could fly at zero distance past the planet's center of mass.

Aiming Radius

The aiming radius is a key parameter to control a flyby maneuver. How is it determined when entering the SOI? At the edge of the planet's SOI the position vector from the planet to the entering S/C is

$$\bar{r}_\infty = \bar{r}_{S/C} \bar{r}_p \tag{9.5.6}$$

So we find from Fig. 9.14 for the aiming radius vector

$$\Delta = r_\infty^- + \hat{v}_\infty^- \left(-\hat{v}_\infty^- \cdot r_\infty^-\right) = r_\infty^- - \hat{v}_\infty^- \left(\hat{v}_\infty^- \cdot r_\infty^-\right) \quad (9.5.7)$$

where $\hat{v}_\infty^-$ is the unit vector of v_∞^-. With this expression we can mathematically define before/behind flybys:

before flyby $\Delta \cdot v_p > 0$

behind flyby $\Delta \cdot v_p < 0$

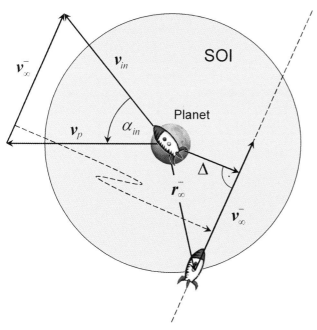

Figure 9.14 Entry triangles. The left triangle is the velocity triangle from Fig. 9.9. Its entry velocity is taken at the SOI entry point (right triangle) to determine the aiming radius vector.

Turn Direction

To determine the direction in which the S/C turns during flyby, we have to consider the relation between the r_∞^- and Δ (see Fig. 9.14). The S/C will make a positive turn if Δ points to the right of the planet (as seen from the S/C towards the planet). This orientation can be expressed by the orientation of the vector cross-product relative to the plane's normal vector n. Of the two possible normal vectors n, we choose that vector which is parallel to the angular momentum of a counterclockwise turn (right-hand rule):

$$\mathrm{sgn}(\delta) = \mathrm{sgn}\left[(\bar{r}_\infty \times \boldsymbol{\Delta}) \cdot \boldsymbol{n}\right] = \mathrm{sgn}\left[(\bar{r}_\infty \times \boldsymbol{\Delta})_z\right] = \mathrm{sgn}\left(\bar{r}_{\infty,x}\Delta_y - \bar{r}_{\infty,y}\Delta_x\right)$$

The vector components x, y lie in the flyby plane and z is vertical along n. Rather than assigning this orientation to the deflection angle δ, we prefer to assign it to the value of Δ, because the deflection can then be described in terms of only two parameters, v_∞ and Δ_p. This can be done, because our deflection Eq. (9.5.5) is sensitive to the sign of Δ. So

$$\Delta_p = \mathrm{sgn}\left(\bar{r}_{\infty,x}\Delta_y - \bar{r}_{\infty,y}\Delta_x\right) \frac{v_p^2}{\mu_p}|\Delta| \quad \text{normalized aiming radius} \quad (9.5.8)$$

and

$$\delta = 2\arctan \frac{v_p^2}{\Delta_p \cdot v_\infty^2} \quad \text{deflection angle} \quad (9.5.9)$$

where the aiming radius is normalized to a dimensionless form for reasons we will see later. These two equations describe the wanted amount and direction of the deflection angle.

Velocity Change

From Eq. (9.5.3) with Eq. (9.5.9) we immediately find for the velocity change vector

$$\Delta v = v_\infty^+ - v_\infty^- = \boldsymbol{R}_\delta v_\infty^- - v_\infty^-$$

To determine its value let's have a look at the velocity triangle depicted in Fig. 9.15. As it is an isosceles triangle, the following is valid:

$$\sin\frac{\delta}{2} = \frac{\Delta v/2}{v_\infty}$$

Inserting this into Eq. (9.5.5) yields

$$\frac{\Delta v}{v_\infty} = \frac{2}{\sqrt{1 + (\Delta \cdot v_\infty^2/\mu_p)^2}} \quad \text{flyby delta-}v \quad (9.5.10)$$

This is the wanted delta-v of the flyby maneuver, which of course and according to Fig. 9.12 is the same in the heliocentric reference system.

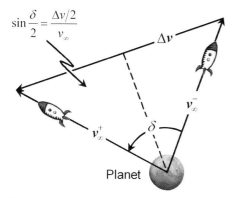

Figure 9.15 The velocity transfer triangle from a planet point of view.

Conclusions

It is remarkable that there is only one lumped parameter $\Delta \cdot v_\infty^2/\mu_p$, which is decisive for the amount of the deflection as well as for the amount of the velocity change. The smaller it is, the bigger is Δv and δ, and the other way round. In summary we can say that:

> The **delta-**v and the **deflection angle** of a flyby **increase** with
> - decreasing entry velocity v_∞^2
> - decreasing aiming radius Δ
> - increasing mass of the target planet $\mu_p = m_p G$

So, a sharp flyby with a small entry velocity past a massive planet such as Jupiter changes velocity and flight direction far more than a flyby at a large distance and high entry velocity past a small planet. This is intuitively clear, but it's good to know how it works out quantitatively.

The key parameter to tune a flyby obviously is the aiming radius Δ. Because it is usually the goal of a flyby to maximize the delta-v (consistently with the change of the other orbital parameters), Δ is to be chosen as small as possible. Quite naturally $r_{per} = R_p$ is the limit, in which case the S/C would scrape the surface of the planet with radius R_p. According to Eq. (9.4.2) this limit translates into the constraint

$$r_{per} = \sqrt{\left(\frac{\mu_p}{v_\infty^2}\right)^2 + \Delta^2} - \frac{\mu_p}{v_\infty^2} > R_p$$

for the aiming radius, from which we derive by rearrangement

$$\Delta > \sqrt{\left(R_p + \frac{\mu_p}{v_\infty^2}\right)^2 - \left(\frac{\mu_p}{v_\infty^2}\right)^2} \tag{9.5.11}$$

Even this limit should not be exhausted, firstly because the true aiming radius might deviate from the determined one by measurement errors, and secondly because nearly all planets possess an atmosphere. If the S/C dives too deeply into it, the drag forces might annihilate the anticipated delta-v gain. In most cases, though, not only is the delta-v gain alone decisive, but also the outgoing direction is important, because it must match the direction to the target. Therefore, for detailed mission planning, the above constraint plus drag have to be taken into account just as side constraints.

9.5.4
Flyby Analysis in the Heliocentric System

Flyby in the Orbital Plane

Now that we know the upshot of a flyby in the planetocentric coordinate system, it is quite easy to determine the wanted outgoing velocity v_{out} of the S/C in the heliocentric reference system. We only need to map the incoming velocity v_{in} into the planetocentric system to obtain v_∞^-. We then turn this vector by the angle δ to receive v_∞^+, which we map back into the heliocentric system to finally get v_{out}. If these steps are performed with Eq. (9.5.1), (9.5.3), and (9.5.4) and applying some trigonometric relations we get

$$v_{out} = \mathbf{R}_\delta (v_{in} - v_p) + v_p \tag{9.5.12}$$

where

$$\mathbf{R}_\delta = \begin{pmatrix} \cos\delta & -\sin\delta \\ \sin\delta & \cos\delta \end{pmatrix} = \frac{1}{1+\chi^2}\begin{pmatrix} \chi^2 - 1 & -2\chi \\ 2\chi & \chi^2 - 1 \end{pmatrix}$$

with

$$\delta = 2\arctan\frac{1}{\chi}$$

$$\chi := \Delta_p \left(\frac{v_{in} - v_p}{v_p}\right)^2$$

Generalized Flyby

We finally generalize the calculations for v_{out} to a flyby, which does not take place in the planet's orbital plane, see Fig. 9.10. Let v_{in} be any incoming vector, which hits the planet's SOI at S/C position $r_{S/C}$ as measured from the heliocentric origin, the Sun. It maps into the planetocentric system as the entry vector $\bar{v}_\infty = v_{in} - v_p$. According to Fig. 9.14 and Eq. (9.5.7) $\Delta = \bar{r}_\infty - \hat{v}_\infty (\hat{v}_\infty \cdot \bar{r}_\infty)$ still holds. Now that the flyby plane's normal vector n is no longer in the direction of the z axis, we have from the above the general expression

$$\Delta_p = \text{sgn}\left[(\bar{r}_\infty \times \Delta) \cdot n\right] \frac{v_p^2}{\mu_p} |\Delta| \quad (9.5.13)$$

The next step would be the deflection described by the rotation matrix. Because this is given only in the flyby plane spaned by $(\hat{v}_\infty, \hat{\Delta})$ we have to transform the vectors into a coordinate system in this plane. Let $\boldsymbol{T}_{P \to F}$ be the matrix made up from two Euler rotations, which transforms the planetocentric reference system into the flyby plane (the first rotation along z axis to bring the x axis along the nodal line of the intersecting planes (see Fig. 9.10), and the second rotation along this new x axis to bring the z axis along $\hat{v}_\infty \times \hat{\Delta}$). Then the outgoing vector can be written as

$$v_{out} = \boldsymbol{T}_{P \to F}^{-1} \boldsymbol{R}_\delta \boldsymbol{T}_{P \to F} (v_{in} - v_p) + v_p \quad (9.5.14)$$

In conclusion, the following calculation scheme for a general flyby can be given:

Flyby Calculation Scheme

1. For the S/C at the edge of the SOI of the flyby planet, determine planet's r_p and S/C position $r_{S/C}$, the instantaneous orbital velocity of the planet v_p, and the incoming velocity v_{in} of the S/C.

2. Determine:
$$\bar{r}_\infty = r_{S/C} - r_p$$
$$\bar{v}_\infty = v_{in} - v_p \quad \Rightarrow \quad v_\infty = \sqrt{\bar{v}_\infty \bar{v}_\infty} \quad \text{and} \quad \hat{v}_\infty = \bar{v}_\infty / v_\infty$$

3. Determine the aiming radius vector and the normalized aiming radius
$$\Delta = \bar{r}_\infty - \hat{v}_\infty (\hat{v}_\infty \cdot \bar{r}_\infty)$$
From this follows
$$\Delta_p = \text{sgn}\left[(\bar{r}_\infty \times \Delta) \cdot n\right] \frac{v_p^2}{\mu_p} \sqrt{\Delta \Delta} \quad \text{(dimensionless)}$$

4. Determine the rotation matrix
$$\boldsymbol{R}_\delta = \frac{1}{1 + \chi^2} \begin{pmatrix} \chi^2 - 1 & -2\chi & 0 \\ 2\chi & \chi^2 - 1 & 0 \\ 0 & 0 & 1 + \chi^2 \end{pmatrix} \quad \text{with} \quad \chi = \Delta_p \frac{v_\infty^2}{v_p^2}$$

5. Determine transformation matrix $\mathbf{T}_{P \to F}$, which by two Euler rotations transforms the planetocentric reference system into the flyby plane $[\hat{v}_\infty^-, \hat{\Delta}]$. For flybys in the planet's orbital plane, $\mathbf{T}_{P \to F} = \mathbf{1}$ holds.

6. Determine $v_{out} = \mathbf{T}_{P \to F}^{-1} \mathbf{R}_\delta \mathbf{T}_{P \to F} (v_{in} - v_p) + v_p$

Numerical Calculations

To illustrate the effects of a flyby, we assume a flyby in the planet's orbital plane and choose v_p as a basis for the reference frame, $v_p = v_p(1,0)$, relative to which the incoming and outgoing velocity vector is measured, $v_{in} = v_{in}(\cos \alpha_{in}, \sin \alpha_{in})$, $v_{out} = v_{out}(\cos \alpha_{out}, \sin \alpha_{out})$, cf. Fig. 9.14. The results are shown in Figs. 9.16 to 9.18 for $v_{in}/v_p = 1$. Because at flyby a S/C has about the same heliocentric orbital speed as the planet, we chose $v_{in}/v_p = 1$ as a good ballpark figure. A special situation occurs when $\alpha_{in} = 60°$ and $\Delta_p = \sqrt{3} = 1.732$, which is elaborated in Fig. 9.19. In this case $\Delta v = v_p = v_{in}$ (see Fig. 9.20). But because $\delta = -60°$ as well, it follows that $v_{out} = 0$: the S/C comes to a full stop in the heliocentric system. The outgoing angle α_{out} therefore becomes undefined and hence also the deflection angle $\Delta \alpha = \alpha_{out} - \alpha_{in}$.

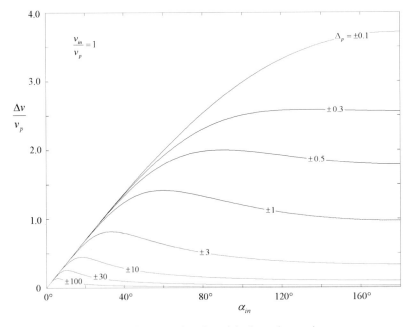

Figure 9.16 Normalized delta-v as a function of the incoming angle α_{in} in the heliocentric system for $\Delta_p < 0$ and $\Delta_p > 0$.

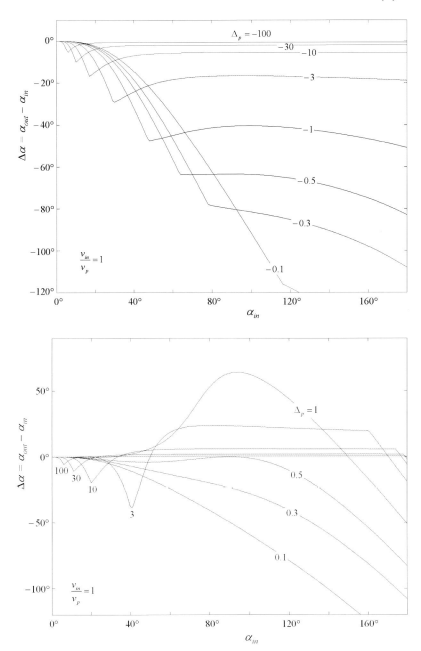

Figure 9.17 Deflection angle $\Delta\alpha$ as a function of the incoming angle α_{in} in the heliocentric system for a flyby with $\Delta_p < 0$ (above) and $\Delta_p > 0$ (below).

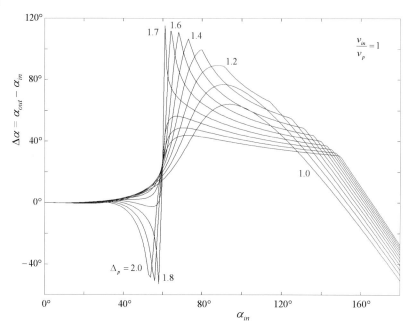

Figure 9.18 Deflection angle $\Delta\alpha$ as a function of the incoming angle α_{in} in the critical interval $\Delta_p = 1.0$ to 2.0.

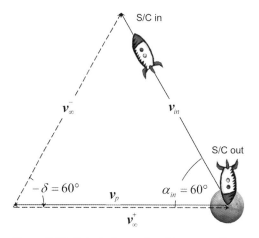

Figure 9.19 Flyby before a planet for $v_{out} = 0$.

9.5.5
Change of Orbital Elements

We finally want to know how the orbital elements are modified by a flyby maneuver. If the received Δv is small, we can formally consider the induced momentum transfer as a kick-burn, $\delta v = v_{out} - v_{in} = (\delta v_\|, \delta v_{\perp\circ}, \delta v_{\perp\perp})$, for which the orbital element changes in the heliocentric reference system are given by Eq. (8.4.3). If the momentum transfer is quite big, then we have to determine the orbital elements through the outgoing state vector (r_{out}, v_{out}), where r_{out} is the position vector at the time the S/C leaves the SOI. This procedure is described in detail in Section 14.4.1.

At any rate, if the flyby is in the planet's orbital plane, then the S/C receives a Δv with components $\Delta v_{\parallel}$ and $\Delta v_{\perp O}$ only. According to Table 8.1 in this case only the orbital elements a, e, ω are modified. If the flyby is not in the orbital plane then the S/C receives also a $\Delta v_{\perp\perp}$ and the orbital elements i, Ω are also effected. If the flyby plane is perpendicular to the planet's orbital plane, then only i, Ω, ω are effected. So the decisive reason for in-plane flybys is to alter the flight direction via ω and/or the semi-major axis, which because of $\varepsilon = -\mu/2a$ corresponds to a change in orbital energy. This maneuver was chosen for the Voyager and Pioneer space probes at Jupiter to escape from the solar system. Out-of-plane flybys make use of the inclination change. The Ulysses space probe, for instance, flew in February 1992 right above Jupiter to be propelled on a polar orbit around the sun with inclination $i = 80.2°$. Any of these maneuvers would otherwise require a lot of delta-v.

Tisserand Relation

According to Section 11.3 and 11.4, a flyby past a planet can also be interpreted as a trajectory in the restricted three-body system Sun–planet–S/C. We want to apply those results to a planetary flyby in the heliocentric inertial (sideral) system. We therefore transform the energy conservation equation (11.4.1) in the synodic system of the revolving planet into the heliocentric inertial system by a simple Eulerian rotation. By doing so, the centrifugal term transforms as $\omega^2 \left(x^2 + y^2\right)/2 \to h\omega$ (see Murray and Dermott (1999, p. 71)) and we obtain

$$\frac{E_{tot}}{m} = -\left(\frac{\mu_{sun}}{\Delta r_{sun}} + \frac{\mu_p}{\Delta r_p}\right) - h\omega + \frac{1}{2}v^2 = const \tag{9.5.15}$$

Here $\omega = const$, with $\omega^2 = G\left(m_{sun} + m_p\right)/a_p^3 \approx \mu_{sun}/a_p^3$, is the constant angular velocity (mean motion), of the planet's circular orbit around the Sun with orbital radius a_p; Δr_{sun} and Δr_p are the distances of the S/C to the sun and the planet, respectively; and h and v is its angular momentum and speed in the heliocentric system.

Note: *Equation (9.5.15) expresses an energy conservation, which is quite peculiar. Other than the kinetic and gravitational energy, the new term $-h\omega$ shows up, which we can interpret physically as a flyby potential. As long as the S/C is outside the planetary SOI, $E_{flyby} = -h\omega = const$ and therefore according to Newton's third law (Eq. (7.1.11)), $-dE_{flyby}/dr = F_{flyby} = 0$. So, the S/C does not experience any deflection force outside the SOI, in accordance with our expectation. When the S/C dives into the SOI, $h = r \times v$ changes (the angular* momentum is not constant in the heliocentric system!) and thus the flyby energy changes as well, which can be interpreted as a deflection force. The energy transferred by a complete flyby to the S/C is

$$\Delta E_{flyby} = -(h_{out}\omega - h_{in}\omega) = -[\mathbf{r} \times (\mathbf{v}_{out} - \mathbf{v}_{in})]\,\boldsymbol{\omega} = -(\mathbf{r} \times \Delta\mathbf{v})\,\boldsymbol{\omega}.$$

We now apply Eq. (9.5.15) to a position external to the SOI of the planet. In this case $m_{planet}/r_{planet} \ll m_{sun}/r_{sun}$, so we are effectively in a two-body system where the vis-viva Eq. (7.2.10), $v^2 = \mu_{sun}(2/r_{sun} - 1/a)$, holds. Therefore, and because $h\omega = h\omega\cos i$, with orbital inclination i, we find

$$-\frac{\mu_{sun}}{r_{sun}} - h\sqrt{\frac{\mu_{sun}}{a_p^3}}\cos i + \frac{\mu_{sun}}{r_{sun}} - \frac{\mu_{sun}}{2a} = \text{const}$$

so

$$\frac{h}{\sqrt{\mu_{sun}a_p^3}}\cos i + \frac{1}{2a} = \text{const}$$

We finally apply this equation to the S/C orbit before and after the flyby at the planet. With Eq. (7.3.7), $h^2 = \mu_{sun}a(1-e^2)$, we arrive at

$$\frac{a_p}{a_{in}} + 2\cos i_{in}\sqrt{\frac{a_{in}}{a_p}(1-e_{in}^2)} = \frac{a_p}{a_{out}} + 2\cos i_{out}\sqrt{\frac{a_{out}}{a_p}(1-e_{out}^2)} \quad (9.5.16)$$

$$= T_p = \text{const}$$

This is the so-called **Tisserand Relation** (Tisserand 1896) with Tisserand's parameter T_p. As a flyby invariant it relates the unknown orbital elements of the exiting S/C to those known of the entering S/C. For a deflection in the orbital plane, $i_{in} = i_{out} = 0$. The Tisserand relation holds for any flyby, also for flights of unknown comets and asteroids by planets. Because their orbital elements change during flyby, it would be hard to decide whether a newly observed celestial body is identical to a known one or not. The Tisserand relation is here a helpful decision criterion.

9.6 Earth–Moon Orbits

9.6.1 Rapprochement Orbits

In Section 9.1 we already mentioned that the patched-conics method cannot be employed for Earth–Moon transits. In this case the transfer orbits neither are approximations to Keplerian orbits, nor can they even be described analytically. This is why they have to be determined numerically. There are numerical solutions that are interesting from a general, but also from a practical, point of view. Of general interest are the highly symmetrical periodic orbits embracing Earth and Moon. French mathematicians extensively studied these types of orbits, and that is why such orbits are called "rapprochement orbits."

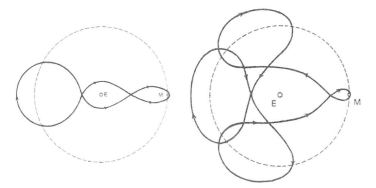

Figure 9.20 Two typical rapprochement orbits in the Earth–Moon system.

Figure 9.20 shows two typical rapprochement orbits. The transfer body nicely shuttles on a symmetrically closed curve between Earth and Moon.

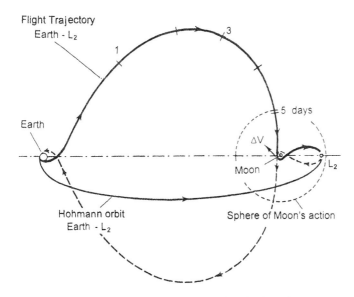

Figure 9.21 A Farquhar transfer orbit to the L_2 point based on a rapprochement orbit with a very low propulsion demand.

There is an immense number of rapprochement orbits. Apart from their periodicity, they have another important feature in common: they are unstable. Any tiny orbital perturbation amplifies until the orbit is no longer symmetric or periodical, but takes on a chaotic course.

There are however variants of these orbits that are also interesting from a practical point of view, e.g. the low energetic transfer orbits to the libration points (see Section 11.3) between the Earth and the Moon. Figure 9.21 shows such a low-energy orbit to libration point L_2, which was studied by Farquhar and coworkers. It takes advantage of a special flyby past the Moon to swing by to L_2.

9.6.2
Free-Return Trajectories

When mission planning for the first manned US missions to the Moon was at issue, the safety of the crew in case of a main engine failure played a crucial role. Thus a trajectory was selected which assured the return of the astronauts to the Earth even with a total main engine failure. This special trajectory was called a "free-return" trajectory and it is depicted in Figure 9.22. It is a symmetric and periodic trajectory, and thus is a rapprochement orbit. It passes the

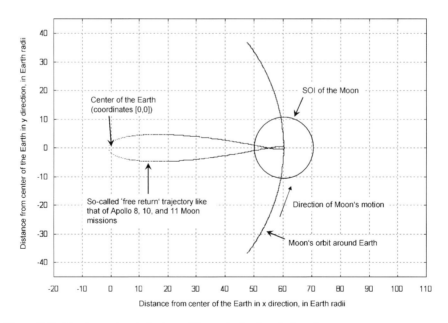

Figure 9.22 The free-return trajectory of the early Apollo missions to the Moon displayed in the geocentric coordinate system co-rotating with the Moon.

surface of the Moon with a minimum distance of 111 km, which corresponds exactly to the periselene altitude of the missions Apollo 8, 10, and 11. Later moon-landing missions entered into a circular Moon orbit at this point, from which it was then possible to descent to the Moon's surface. If the mission had to be aborted for any reason, the engine ignition required for braking behind the Moon into the circular Moon orbit would not have happened, and the astronauts would automatically have returned to the Earth. This indeed happened with Apollo 13.

The trajectory shown in Figure 9.22 was calculated by numerical integration of the equation of motion for the restricted three-body system Earth–Moon–S/C. In the geocentric reference frame co-rotating with the Moon it is fully symmetric with regard to the Earth–Moon connecting line. The actual trajectories of the Apollo missions were slightly asymmetric, such that on return the spacecraft would touch the Earth's atmosphere in order to guarantee an automatic reentry. This asymmetry was achieved by a slight shift of the position of the periselene.

During the outward as well as the return flight the free-return trajectory is clearly elliptical in the proximity of the Earth, as the gravitational influence of the Earth is dominant. Approaching the edge of the Moon's SOI, the trajectory becomes more and more a straight line. Here the orbit velocity has already decreased quite a lot, and the gravitational influence of the Earth and the Moon plus centrifugal force just cancel each other out. In this area the real trajectory deviates the most from the patched-conics approximation. In the surroundings of the Moon the trajectory is bent into a hyperbola. All this is not obvious from Fig. 9.22 because the reference frame co-rotates.

9.7
Weak Stability Boundary Transfers

Apart from flyby maneuvers there is another possibility to reduce the propulsion demand of an interplanetary mission in a multi-body system: the so-called weak-stability-boundary transfer. A weak stability boundary, often abbreviated by WSB, refers to the transfer area between two gravitational fields, i.e. the area around the edge of an SOI. For a WSB transfer, you inject an S/C into a WSB, as illustrated in Fig. 9.23 with the example of the Sun–Earth–Moon system for the Sun–Earth WSB, such that it arrives there with $v \approx 0$ as measured in the co-rotating system. Under these circumstances the energy demand (equals propulsion demand) $\Delta\varepsilon \approx v \cdot \Delta v$ (see Eq. (8.1.1)) for any Δv, which carries the vehicle into a transfer orbit to the target, becomes arbitrarily small. So, effectively, a WSB transfer is a bi-elliptic transfer (see Section 8.3), except that $v \approx 0$ is not achieved after an indefinite period of time at indefinite distance, but quite conveniently after a reasonable period of time and at a distance that is not too far away. If the bi-elliptic transfer is already advantageous to the Hohmann-Transfer with regard to delta-v, the WSB transfer is even more favorable than the bi-elliptic transfer, as the gravitational saddle point (see Fig. 11.8) between the two celestial bodies – also called L_1 point – is energetically lower than at infinite distance. In summary, one saves time and propulsion effort as compared to the bi-elliptic transfer.

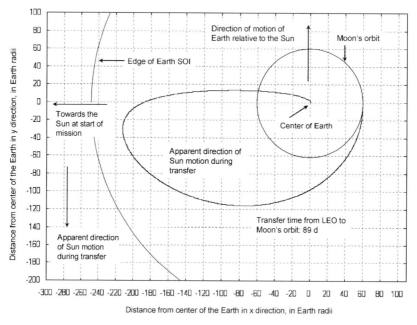

Figure 9.23 Earth → Moon transfer orbit exploiting solar gravitation at Earth–Sun WSB. (Perigee altitude: 400 km above the Earth's surface. Perigee velocity: 1.410 times the circular orbit velocity.)

The WSB transfer from Earth to Moon, depicted in Figure 9.23, was calculated by numerical integration of the equation of motion in the limited four-body system Sun–Earth–Moon–S/C. It also makes use of another feature of this multi-body system. After the S/C has been transferred from a 400 km LEO to a highly elliptical transfer orbit to the Earth–Sun WSB, it remains there for quite a long period of time because of $v \approx 0$. During that time the position of the Sun changes because of the orbital movement of the Earth, and thus the gravitational force of the Sun affects the S/C more and more from a lateral direction. This gives the S/C the correct and necessary Δv to transfer it into a transfer orbit to the Moon, where it approaches its surface at a minimum 111 km altitude with very low relative velocity without engine ignition. Because of the combination of these three-body effects – WSB transfer plus solar acceleration – the delta-v of the trajectory shown in Figure 9.23 is more than 200 m s^{-1} less than for the Apollo flight path. Instead, the transfer time increases from three days to 90 days, which is however of minor importance for unmanned missions.

Lowest Thrust Transfers

Flyby maneuvers and WSB transfers are the ingredients for ultimate low-thrust missions to the Moon and to the planets in our solar system. The Japanese mission HITEN demonstrated how to get to the Moon with as little propellant as possible. It first made use of a Moon flyby (see Fig. 9.24) to carry the spacecraft to the WSB between the Earth and the Sun. Then, without engine ignition, just by acceleration of the Sun, it was transferred via the L_5 libration point to the Moon, where the Moon caught the S/C in a highly elliptical orbit by ballistic capture. If you take a closer look, you will realize that this trajectory is impossible with a patched-conics approximation, as it would violate energy conservation; in a three- or four-body system it is, however, possible.

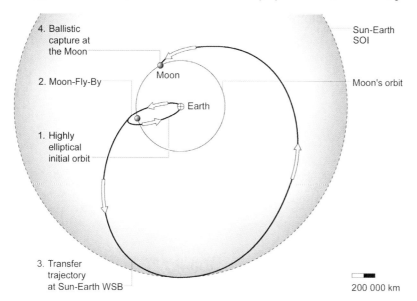

Figure 9.24 The Moon flyby and WSB transfer trajectory of the Japanese Moon probe HITEN.

A similar dexterous sequence of maneuvers in the Earth–Moon region was conceived for the Japanese NOZOMI mission to Mars. Figure 9.25 shows the maneuvers preceding the injection into the transfer orbit. Just as with HITEN, the S/C was first supposed to reach a highly elliptical transfer orbit to the Moon. Then a Moon flyby was used to get to the WSB between the Earth and the Sun where the acceleration of the Sun turned it back to the Moon.

250 | 9 Interplanetary Flight

A second Moon flyby brought the S/C back to Earth, where at the closest approach distance a trans-Mars injection burn initiated the transfer flight to Mars.

That was at least the mission planning. However, due to an unfortunate construction, the propellant froze in the pipeline to the main engine, and thus this journey could not be carried out as planned. Consequently, Japanese mission control postponed the trans-Mars injection for two years to wait for the next favorable Earth–Mars constellation. During that time the NOZOMI spacecraft gained enough impetus by repeated Moon flybys and WSB transfers that its weak reaction control engines sufficed to give it the final small injection burn to Mars.

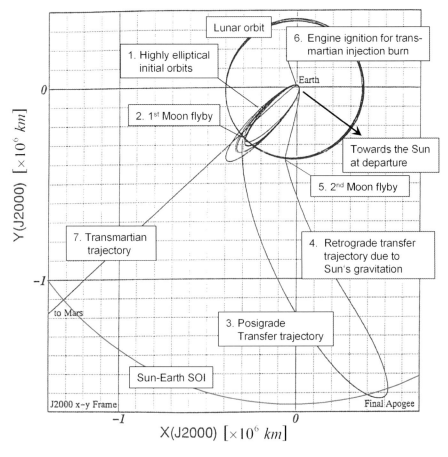

Figure 9.25 The sequence of Moon flyby and WSB maneuvers of the Japanese Mars probe NOZOMI preceding the trans-Mars injection in the co-rotating Earth–Moon system. (Courtesy: ISAS.)

Problems

Problem 9.1 Planet Waiting Time
Prove Eq. (9.3.6) from basic considerations.
  Hint: Prove first that for flights to outer planets

$$t_{wait} = T_{syn} \left(k + 1 - \frac{2t_H}{T_\oplus} \right) \quad @ \oplus \leftrightarrow \bigcirc$$

and for flights to inner planets

$$t_{wait} = T_{syn} \left(k - 1 + \frac{2t_H}{T_\oplus} \right) \quad @ \oplus \leftrightarrow \bullet$$

Problem 9.2 Forgetful Asteroid Prospector
Consider two asteroids that move in circular coplanar heliocentric orbits with the following elements:

  Asteroid A: $a = 2.0$ AU, $\omega + \Omega = 139°$, $t_0 = 2025$, January 1.0
  Asteroid B: $a = 3.5$ AU, $\omega + \Omega = 271°$, $t_0 = 2025$, January 1.0

An absent-minded asteroid prospector working on A decides on 2025, January 1.0, to move his ship, with the greatest economy in fuel, to B.

(a) Show that the first available take-off time is 2026 January 26.8.
(b) When he arrives at B, he discovers that he has left his Geiger counter on A and has to go back for it. Show that his minimum waiting time on B if the return journey is also made under the fuel economy condition is 1.930 years (neglect the asteroids' gravitational fields.)

Adopted from (Roy, 2005, problem 12.10).

Problem 9.3 Flyby Distances
Prove that the distances indicated with $-a$ in Fig. 9.13 indeed have this value.

Problem 9.4 Near Hohmann Transfer, I
Prove Eq. (9.3.17).

Problem 9.5 Near Hohmann Transfer, II
In Section 9.3.2 we found for near-Hohmann transfers for the dependence for the chain link $\Delta a \to \Delta E_\times$ in lowest approximation

$$\Delta E_\times = -\frac{2}{\sqrt{e_H}} \sqrt{\frac{\Delta a}{a_H}} =: -2\sqrt{\alpha}$$

Show that in the next higher approximation in Δa

$$\Delta E_x = -2\sqrt{\alpha}\left(1 - \frac{\alpha}{4}\right)$$

with corresponding impacts on $\Delta t_\times$ and $\Delta \theta_\times$.

10
Reentry

10.1
Introduction

10.1.1
Thermal Problem Setting

The reentry of a spacecraft is subject to the same aerodynamic and physical laws and equations as ascent (see Eqs. (6.2.6) and (6.2.7)), and you might therefore think that the conditions of both events are the same. But they actually differ very much, for instance with regard to their heat load. So the difference is not due to the equations. It is due to the initial conditions. During launch, $v = 0$ and the flight path angle is $\gamma = 90°$, while at reentry it is exactly the other way round $v \approx 7.9 \text{ km s}^{-1}$ and $\gamma \approx 0°$. Prior to reentry, the S/C in LEO possesses a high amount of energy of approximately 33 MJ kg^{-1}. This energy has to be annihilated during reentry in a controlled way and in a relatively short period of time, such that the S/C is not to be damaged.

The following rough estimate shows that this is not easy to accomplish. The total orbital energy of an S/C in LEO is $E_{tot} \approx mg_0R/2$. A capsule in a so-called ballistic reentry (see Section 10.3.2) converts this into frictional heat, which is released as a heat flow rate $\dot{Q} = E_{tot}/\tau$ within typically $\tau \approx 0.5$ min. Usually 99.9% of that is released via heat convection to the air flow. The rest, which is described by the so-called *Stanton number St*, which has roughly the universal value of $St \approx 0.1\%$, is taken up by the surface of the S/C. The thermal shield of the S/C with an area of A and with emissivity of typically $\varepsilon \approx 0.85$ then radiates away the absorbed heat with a heat flux $\dot{q}_{S/C}$, which according to the Stefan–Boltzmann law is related to the shield temperature as

$$T^4 = \frac{\dot{Q}}{\varepsilon \sigma A} = \frac{\dot{q}_{S/C}}{\varepsilon \sigma} \tag{10.1.1}$$

with the Stefan–Boltzmann constant $\sigma = 5.6705 \times 10^{-8}$ Wm^{-2}K^{-4}. When the thermal shield is in thermal balance, we get the following from the above considerations:

$$St\frac{E_{tot}}{\tau} = St\frac{mg_0 R}{2\tau} = A\varepsilon\sigma T^4 \qquad (10.1.2)$$

Example 1
The Mercury capsule had a mass of $m = 1\,450$ kg and a thermal shield with an area $A = 2.8$ m². From Eq. (10.1.2) we derive that during reentry a shield temperature of roughly $T \approx 1\,890$ K $= 1\,610$ °C occurred. These temperatures are at the limit today's heat insulations can withstand.

Example 2
If a Space Shuttle with a mass $m \approx 100\,000$ kg and an effective stagnation point area of $A \approx 5$ m² would also reentry on a ballistic flight path, it would not be able to survive at theoretically $T \approx 3\,680$ °C, as the thermal tiles at the nose and at the front edge of the shuttle wings are designed for a maximum of $1\,750$ °C. The shuttle actually reduces the maximum heat load by drastically increasing the reentry time through the atmosphere and thus also the time of maximum heat generation to about 10 min (equals blackout time) because of its lift. The so-called angle-of-attack (see Section 10.8.3) also displaces part of the generated heat to the lower side, which we will consider in the effective stagnation point area by the factor 2. According to Eq. (10.1.2) these measures reduce the maximum temperature to $T \approx 1\,600$ °C.

These results are only estimates. To get more accurate results, we have to calculate the time-dependent heat flux as well as its maximum. The total heat flux of the inflowing air is calculated as

$$\dot{q}_{tot} = \frac{\dot{Q}_{tot}}{A} = \frac{\dot{m}H_{tot}}{A} = \frac{\rho v A}{A}\frac{v^2}{2} = \frac{1}{2}\rho v^3$$

where H_{tot} is the totally generated enthalpy, which equals the kinetic energy of the air, which impinges with velocity v. The portion which is transferred to the S/C is

$$\dot{q}_{S/C} = St \cdot \dot{q}_{air} = St\frac{1}{2}\rho v^3$$

This is the *theoretical* result for the heat flux of a flat plate with area A. Wind tunnel tests show that the heat flux at curved surfaces obeys a somewhat different relation because it is not distributed evenly. It becomes maximal where the curvature is greatest: at the stagnation point. So the heat flux transferred to the S/C at the stagnation point is *empirically* derived to be

$$\dot{q}_{S/C} = St \sqrt{\rho \rho_q \frac{R_0}{R_n}} \cdot v^3 \qquad (10.1.3)$$

with $\rho_q = 0.0291 \text{ kg m}^{-3}$ @ $St = 0.001$, $R_0 = 1$ m and R_n the radius of curvature of the (nose) surface at which the stagnation point occurs.

This heat flux at the stagnation point achieves its maximum at a given entry velocity profile $v = v(\rho)$ for $d\dot{q}/dv = 0$. This is the conditional equation which we will use in this chapter to calculate the maximum heat flux, which in turn can be inserted into Eq. (10.1.1) to derive the maximum surface temperature of the S/C.

10.1.2
Entry Interface

Let's have a closer look at the reentry process into the atmosphere. Reentry formally commences at the so-called entry interface at which the orbit parameters take on the values r_e, v_e, γ_e.

> According to international agreements, the entry interface is located at an altitude of 400 000 ft = 122 km, i.e. at the border between heterosphere and homosphere.

The atmosphere of course does not abruptly set in at this altitude, but drag and lift start there to have an influence on the entering vehicle. For the homosphere below 120 km the atmospheric density obeys the barometric formula (see Eq. (6.1.5)):

$$\rho(r) = \rho_0 \exp\left(-\frac{r-R}{H}\right) = \rho_0 \exp\left(-\frac{h}{H}\right),$$

with $\rho_0 = 1.752 \text{ kg m}^{-3}$ and $H = 6.7 \pm 0.1$ km scale height for the Earth.

10.1.3
Deorbit (Phase A)

Before we study the reentry into the atmosphere, we need to know how the S/C gets from its preceding trajectory to the entry interface. Usually the starting point is a circular Earth orbit, the radius of which we denote by r_i (initial). Deorbit is initiated by a deorbit burn at a certain position on this orbit, which transfers the S/C onto an entry ellipse with a low-lying perigee, which intersects the entry interface at a predetermined flight path angle γ_e (see Fig. 10.1).

10 Reentry

The deburn position has to be chosen such that the entry point is at the right distance from the anticipated landing site. We now want to evaluate three questions: What is the required delta-v for the deorbit? At which position is the entry interface attained? What is the entry velocity at entry interface? It is now our objective to determine these three values.

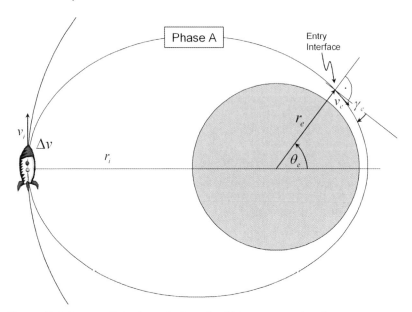

Figure 10.1 Reentry flight phase A: From deorbit burn to entry interface.

At deorbit burn the S/C is positioned at apogee of the entry ellipse, which has the still to be determined orbital elements a, e. If this ellipse is supposed to intersect the entry interface at position r_e, θ_e with entry velocity v_e, the ellipse is unequivocally determined because Eqs. (7.3.5), (7.3.10) and (7.3.12) state that the following holds:

$$\cos \gamma_e = \frac{1 + e \cos \theta_e}{\sqrt{1 + 2e \cos \theta_e + e^2}} \tag{10.1.4a}$$

$$v_e = \frac{\mu}{h}\sqrt{1 + 2e \cos \theta_e + e^2} \tag{10.1.4b}$$

$$r_e = \frac{a(1 - e^2)}{1 + e \cos \theta_e} = R + h_e = 6500 \text{ km} \tag{10.1.4c}$$

The orbital elements are not independent from each other because at the deorbit position

$$r_i = a(1 + e) \tag{10.1.5}$$

From this condition and from Eq. (10.1.4c) we derive

$$e \cos \theta_e = \frac{a\left(1-e^2\right)}{r_e} - 1 = \alpha\left(1-e\right) - 1 \tag{10.1.6}$$

with

$$\alpha := \frac{r_i}{r_e} = \frac{r_i}{6500 \text{ km}} > 1 \tag{10.1.7}$$

and

$$h = \sqrt{\mu a\left(1-e^2\right)} = \sqrt{\mu r_i\left(1-e\right)}$$

We insert these results back into Eq. (10.1.4a) and (10.1.4b) and find after some trivial steps

$$\cos \gamma_e = \alpha \sqrt{\frac{1-e}{2\alpha - 1 - e}}$$

$$v_e = v_i \sqrt{2\alpha - 1 - e}$$

$$e \cos \theta_e = \alpha\left(1-e\right) - 1$$

$$\Delta v = \sqrt{\frac{\mu}{r_i}} - \sqrt{\frac{\mu}{a}\frac{1-e}{1+e}} = v_i\left(1 - \sqrt{1-e}\right)$$

with

$$v_i = \sqrt{\frac{\mu}{r_i}}$$

the orbital velocity of the initial orbit and Δv the propulsion demand for the deorbit burn.

These equations have to be interpreted as follows. At a given r_i, r_e, γ_e the first equation delivers the eccentricity e of the entry ellipse. With this we find from the other equations the wanted entry velocity and the propulsion demand for deorbit. This is exactly what we are going to do now.

From the first equation we derive after some rearrangements

$$e = \frac{\alpha^2 - (2\alpha - 1)\cos^2 \gamma_e}{\alpha^2 - \cos^2 \gamma_e} \tag{10.1.8}$$

We insert this into the other three equations and find

$$v_e = \alpha v_i \sqrt{\frac{2\alpha_-}{\alpha^2 - \cos^2 \gamma_e}} = \sqrt{\frac{2\mu}{r_e} \frac{\alpha \alpha_-}{\alpha^2 - \cos^2 \gamma_e}} \qquad (10.1.9a)$$

$$\Delta v = v_i \left(1 - \cos \gamma_e \sqrt{\frac{2\alpha_-}{\alpha^2 - \cos^2 \gamma_e}}\right) = v_i - \frac{\cos \gamma_e}{\alpha} v_e \qquad (10.1.9b)$$

$$\cos \theta_e = \frac{\alpha_-^2 \cos^2 \gamma_e - \alpha^2 \sin^2 \gamma_e}{\alpha_-^2 + \alpha^2 \sin^2 \gamma_e} \qquad (10.1.9c)$$

with

$$\alpha := \frac{r_i}{r_e} = \frac{r_i}{6500 \text{ km}}, \quad \alpha_- := \alpha - 1 = \frac{r_i}{r_e} - 1, \quad v_i = \sqrt{\frac{\mu}{r_i}}$$

These are the wanted expressions. There are only two cases of practical interest: $\alpha \approx 1$ and $\alpha \to \infty$, which we will investigate now.

Interplanetary Reentry ($\alpha \to \infty$)

The S/C approaches Earth from infinity, for instance from the Moon or from a planet. In this case we simply get

$$v_e \approx \sqrt{\frac{2\mu}{r_e}\left(1 - \frac{1}{\alpha}\right)} \approx \sqrt{\frac{2\mu}{r_e}}$$

$$\Delta v \approx v_i \left(1 - \cos \gamma_e \sqrt{\frac{2}{\alpha}}\right) \approx v_i \approx 0 \qquad (10.1.10)$$

$$\theta_e = 2\gamma_e \quad @ \; \gamma_e^2 \ll 1$$

As expected, the entry velocity is just the second cosmic velocity (see Eq. (7.4.7)), and the deorbit burn effort becomes arbitrarily small. Of course the burn cannot be performed so precisely as to hit exactly the entry interface. Therefore the entry path usually must be adjusted several times during approach to pass through the so-called entry corridor. Figure 10.2 displays the entry corridor for Apollo 11. Three adjusting maneuvers were planned for this Moon mission to hit the corridor.

LEO Reentry ($\alpha \to 1$)

The S/C initially is in a LEO. For $\gamma_e \leq 10°$ the trigonometric functions can be approximated and because then $\gamma^4 \ll 20 \cdot \alpha_-$, Eqs. (10.1.9a) and (10.1.9b) can

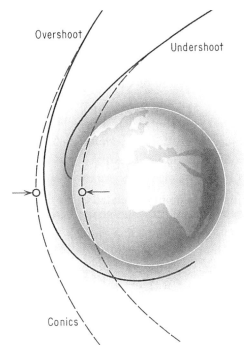

Figure 10.2 Limiting paths of the Apollo reentry corridor.

be written as (exercise, Problem 10.1)

$$v_e = v_i \left(1 + \frac{3}{4}\alpha_- - \frac{\gamma_e^2}{8\alpha_-}\right)$$

$$\Delta v = \frac{v_i}{4}\left[\alpha_- + \frac{\gamma_e^2}{2\alpha_-}\right]$$

(10.1.11)

From this follows

$$v_e < v_i \quad @ \quad \gamma_e > \sqrt{6} \cdot \alpha_-$$
$$v_e > v_i \quad @ \quad \gamma_e < \sqrt{6} \cdot \alpha_-$$

For θ_e no simpler expression can be derived than that in Eq. (10.1.9c).

Example

After undocking from the ISS at 400 km altitude ($\alpha = 1.0428$, $v_i = 7.669 \text{ km s}^{-1}$) the Space Shuttle should deorbit such that it encounters the entry interface with a standard flight path angle $\gamma_e = 1°$. According to Eqs. (10.1.9c) and (10.1.11) we obtain: $\theta_e = 46.1°$, $v_e = 1.031 \times v_i = 7.91 \text{ km s}^{-1}$, and $\Delta v = 0.0116 \times v_i = 88.8 \text{ m s}^{-1}$ (cf. Section 10.8.1).

On the other hand the Soyuz capsule after undocking from the ISS usually acquires a flight path angle of $\gamma_e = 3°$ at entry interface. We find from Eqs. (10.1.9c) and (10.1.11): $\theta_e = 76.2°$, $v_e = 1.024 \times v_i = 7.85 \text{ km s}^{-1}$, and $\Delta v = 0.0187 \times v_i = 143 \text{ m s}^{-1}$.

Note: For all flat entries from LEO the entry velocity is $v_e \approx 7.9 \text{ km s}^{-1} = v_0$, while for entries from infinity (celestial bodies) $v_e \approx 11.1 \text{ km/s} = \sqrt{2} \cdot v_0$. Therefore the assumptions, which we are going to be used in the following reentry investigations, $\varepsilon_e = v_e^2/v_0^2 = 1$ and $\varepsilon_e = v_e^2/v_0^2 = 2$, respectively, are excellent assumptions for these two cases.

10.2
Equations of Motion

We are now at entry interface, where atmospheric reentry commences. As reentry is subject to the same physical laws as ascent, we adopt the general orbit equations (6.2.6) to (6.2.12). But in contrast to ascent, no propulsion is required, and this is why we set thrust and also the mass change rate $\dot{m} = 0$ to zero.

According to our current definition a declining S/C has a negative flight path angle. But to be in line with standard conventions it is to be positive. So we formally apply the transformation $\gamma \to -\gamma$, and $\dot{\gamma} \to -\dot{\gamma}$ to Eqs. (6.2.6) to (6.2.12), which results in

$$\dot{v} = -\frac{D}{m} + g \sin \gamma \qquad (10.2.1)$$

$$v\dot{\gamma} = -\frac{L}{m} + \left(g - \frac{v^2}{r}\right) \cos \gamma \qquad (10.2.2)$$

$$\dot{r} = \dot{h} = -v \sin \gamma \qquad (10.2.3)$$

with

$$L = \frac{1}{2}\rho(r)v^2 C_L(\alpha, Re, v) A_\perp \qquad (10.2.4)$$

$$D = \frac{1}{2}\rho(r)v^2 C_D\left(\alpha^2, Re, v\right) A_\perp \qquad (10.2.5)$$

$$g(r) = g_0 \frac{R^2}{r^2}, \quad R = \text{radius of the planet} \qquad (10.2.6)$$

Drag D and lift L, as already described for ascent, depend on the angle of attack (AOA) α, on the Reynolds number Re, and to a lesser extent on the hypersonic velocity v. We want to neglect however this dependency in the following. Remember that from now on every flight path angle γ is to be understood as a positive angle (see Fig. 10.3).

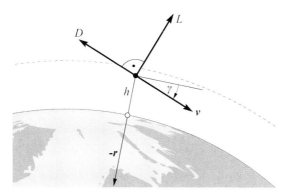

Figure 10.3 Definition of the reentry variables.

10.2.1
Normalized Equations of Motion

The most common method to study the behavior of reentry is to solve numerically the equations of motion in their time-dependent representation. Equations (10.2.1) to (10.2.3) however are not suitable for that. For numerical treatments in general it is advantageous to make the involved variables and thus the equations of motion themselves dimensionless. Dimensions with unit 1 is the most natural and hence optimum choice to describe the physics of nature. In addition it makes the problem of choosing the optimum spatial, time, and mass units dispensable. We do so by expanding Eqs. (10.2.4) and (10.2.5) for drag and lift with the barometric formula Eq. (6.1.5).

$$D = \frac{1}{2}\rho(r)v^2 C_D A_\perp = mv^2 \frac{\kappa_D}{H} e^{-\frac{h}{H}} \tag{10.2.7}$$

$$L = \frac{1}{2}\rho(r)v^2 C_L A_\perp = mv^2 \frac{\kappa_L}{H} e^{-\frac{h}{H}} \tag{10.2.8}$$

where we have introduced the *dimensionless drag and lift coefficients*

$$\kappa_D := \frac{C_D A_\perp H}{2 m} \rho_0 \tag{10.2.9}$$

$$\kappa_L := \frac{C_L A_\perp H}{2 m} \rho_0 = \kappa_D \frac{C_L}{C_D} = \kappa_D \frac{L}{D} \tag{10.2.10}$$

Typically $\kappa_D \approx 25$ and $L/D = 0.3$–2.5 apply, the former value for a capsule (Apollo 4: $\langle L/D \rangle = 0.368$) and the latter for a winged body at low altitudes.

After some trivial transformations (exercise, Problem 10.2) one gets the following form for the normalized equations of motion

$$\mu' = -\mu^2 \kappa_D \frac{R}{H} e^{-\eta} + \sin\gamma$$

$$\gamma' = -\mu \kappa_D \frac{R}{H} \frac{L}{D} e^{-\eta} + \frac{1-\mu^2}{\mu} \cos\gamma \qquad (10.2.11)$$

$$\eta' = -\mu\alpha^2 \frac{R}{H} \sin\gamma$$

$$\chi' = \mu\alpha^2 \frac{R}{H} \cos\gamma$$

with

$v = v_e \mu$

γ [rad]

$h = H\eta$ $\qquad \alpha = \dfrac{v_e}{v_0}$

$x = H\chi$ $\qquad v_0 = \sqrt{g_0 R_\oplus} = 7.905 \text{ km s}^{-1}$

$t = \alpha \cdot t_0 \cdot \tau \qquad t_0 = \sqrt{R_\oplus/g_0} = 807.2 \text{ s}$

$(\ldots)' = \dfrac{d}{d\tau}$

The first three equations are intertwined. At first glance this was not obvious from Eqs. (10.2.1) to (10.2.3), as both lift and drag depend exponentially on the altitude via the atmospheric density such that the ratio L/D is a constant. This is why we use it in the following as a convenient constant to characterize the S/C. We also added the normalized non-intertwined equation $\dot{x} = v\cos\gamma$, which describes the downrange x. We did this because this enables us to derive via the time-dependent solutions $h(t)$, $x(t)$ the spatial path of the entry $h = h(x)$. The equations of motion in form of Eq. (10.2.11) are therefore optimally adapted to be coded and solved numerically, such as by a Runge–Kutta method. For specific problems more elaborate equations without the approximations (see beginning of Section 10.3) made here are used. The relatively simple equations above however show quite well the general entry behavior, so we will limit ourselves to them.

Numerical Solutions

To get a first overview of the entry behavior, Figs. 10.4–10.6 with reentry trajectories in the upper part describe the reentry of a capsule with a typical $L/D = 0.3$ for three different entry angles: a steep $\gamma_e = 45°$, a medium

$\gamma_e = 10°$, and a very flat entry with $\gamma_e = 2°$. They were calculated with a step-size controlled Runge–Kutta method with the normalized system of equations above. The two parts of each diagram have the same entry angle. The time-dependent velocities, altitudes, and decelerations are shown in the lower panels. The different entry profiles as a function of the entry angle first attracts attention.

Note the quite different scales of the downrange x axes, so visually the depicted entry angles are steeper than in reality. Only in Fig. 10.4a the x and y scales are the same, so the depicted profiles have accurate contours.

The most serious effects of flat entry angles are the increasing ranges and entry durations: An S/C with $\gamma_e = 45°$ hits the ground only 150 km downrange of the entry point, with $\gamma_e = 10°$ this comes to approximately 1000 km, and with $\gamma_e = 2°$ approximately 4000 km, with correspondingly increased entry durations. Apart from that the entry profiles are the same at the beginning. With an altitude of down to 60 km, with steeper ones even down to 40 km, the entry body moves on a straight line. Only then do lifting forces become significant. Lifts with $L/D > 0.2$ have dramatic consequences. Instead of going down continuously, the S/C literally rebounds off the atmosphere, falls back, dips in again and then for large lifts and very flat entries, slightly rises again, until it completely goes down. If the excursions are small, they are called "reflections", and they are known as "skips" for larger excursions. They are typical for capsules with $L/D \approx 0.3$, and they are very pronounced for winged bodies with $L/D > 1$ if there are no countermeasures. These effects, which are of quite practical interest, will be treated in detail in Section 10.6. Winged bodies with their more fragile structures should not enter with an angle that is too steep, because the decelerations already attain 20 g at 10°. Neither the crew nor the system would be able to endure that. These winged bodies have to enter with a very flat angle with typically 2° (Fig. 10.6), so that during the second reflection they only have to endure a maximum deceleration of merely 3 g, which is tolerable. The reason for the reduction of the deceleration is that the reduction of the velocity, i.e. the acceleration, is spread over a longer period of time, and thus it is reduced at any point in time. The critical accelerations with steeper entries $\gamma_e \geq 10°$ are always just before the reflection, and they drastically increase with an increasing entry angle. For $\gamma_e = 45°$ and $L/D = 0.3$ it already amounts to 118 g!

10.2.2
Reduced Equations of Motion

Numerical solutions are imperative for real missions, but they furnish no insight into why the entry profiles are as they are. For our goal of understanding, we need to find solutions, or at least partial solutions, that mathematically

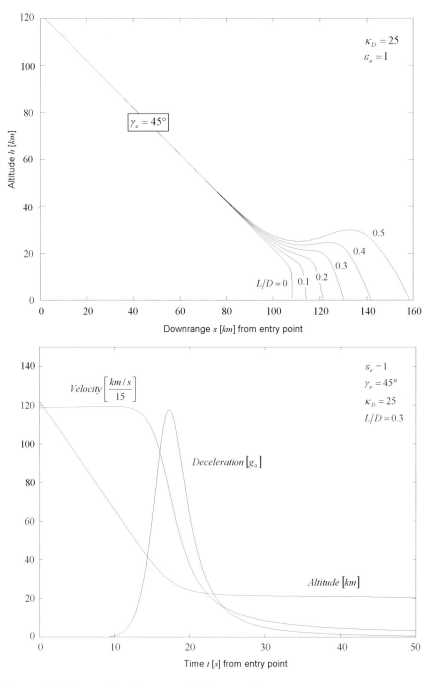

Figure 10.4 Entry profiles of a spacecraft with $\gamma_e = 45°$.

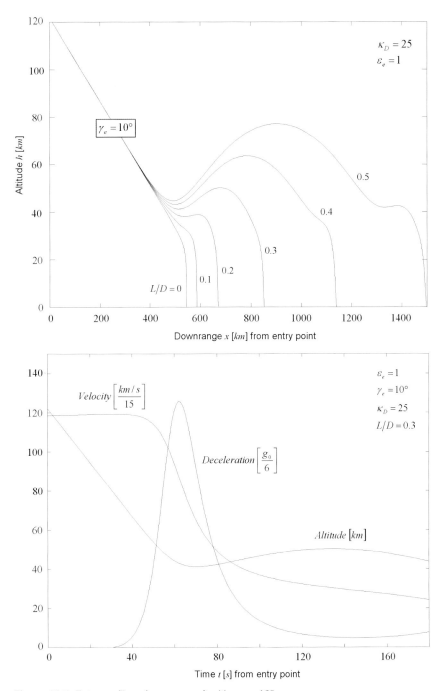

Figure 10.5 Entry profiles of a spacecraft with $\gamma_e = 10°$.

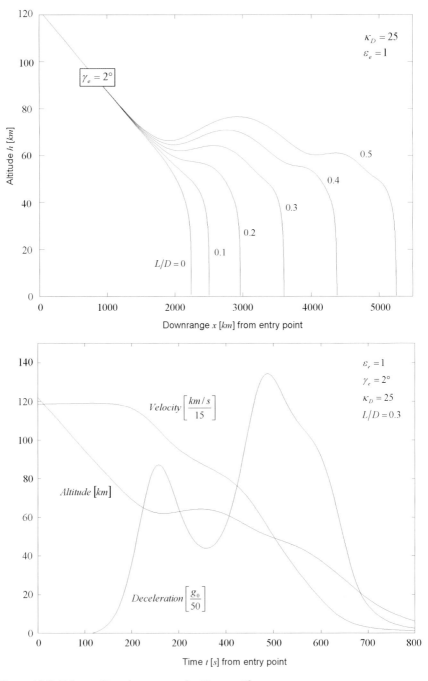

Figure 10.6 Entry profiles of a spacecraft with $\gamma_e = 2°$.

relate entry parameters and entry trajectory and thus show up the limits of applicability. The above dimensionless equations of motion are still too complicated for analytical solutions, so we are looking for simplifications. The third differential equation relates the altitude linearly with time. This linearity allows us to substitute the time variable by the altitude variable. Thereby we get rid of the time dependence of the entry trajectory and obtain the entry variables as a function of altitude $v(h), \gamma(h)$. This is exactly what we want. So we are looking for differential equations with h as the independent variable. We again make use of Eqs. (10.2.7) and (10.2.8); this time however we substitute the entry velocity by the new dimensionless variable

$$\varepsilon := \frac{v^2}{v_0^2} = \frac{E_{kin}(v)}{E_{kin,0}} \quad \text{with} \quad v_0^2 = g_0 R \qquad (10.2.12)$$

which is the instantaneous kinetic energy of the S/C with regard to the orbital kinetic energy at deorbit burn (see Eq. (7.4.4)), which for a circular orbit is $v_0^2 = g_0 R$. We are now set to replace the time variable t by the altitude variable. But we do so by introducing a dimensionless altitude variable λ, which quite naturally occurs in the differential equations

$$\lambda := \frac{2\kappa_D}{\sin \gamma_e} e^{-\frac{h}{H}} \qquad (10.2.13)$$

Altogether this allows us (exercise, Problem 10.3) to transform the equations of motion into the following dimensionless form of only two intertwined differential equations

$$\frac{d\varepsilon}{d\lambda} = -\frac{\sin \gamma_e}{\sin \gamma}\varepsilon + \frac{2H}{\lambda R} \qquad \varepsilon \text{ equation} \qquad (10.2.14a)$$

$$\frac{d(\cos \gamma)}{d\lambda} = \frac{\sin \gamma_e}{2}\frac{L}{D} - \left(\frac{1}{\varepsilon} - 1\right)\frac{H \cos \gamma}{\lambda R} \qquad \gamma \text{ equation} \qquad (10.2.14b)$$

Let's have a closer look at these equations. They describe the state changes of the entry body (velocity, kinetic energy ε, and flight path angle γ) as a function of the instantaneous altitude λ. If we compare them with Eqs. (10.2.1) and (10.2.2) we see the following. On the right side of Eq. (10.2.14a) the first term is the modified drag term, and the second term is the modified gravitational term. On the right side of Eq. (10.2.14b) we have the modified lift term as the first term, and the gravitational term ($1/\varepsilon$) as the second reduced by the centrifugal force (-1), the so-called *reduced gravitational term*. To be able to distinguish between the two equations later, we call the first one ε *equation* and the second one γ *equation*.

Equations (10.2.14) make clear that, apart from the required state variables $\varepsilon(\lambda)$ and $\gamma(\lambda)$ there are two characteristic planet constants, R and H, and two characteristic S/C constants, κ_D and κ_L or κ_D and L/D. The latter are actually not constant, but they weakly depend on the Mach number and Reynolds number.

Equations (10.2.14) provide the possibility to directly derive the deceleration, which is an important figure for both humans and vehicle structure. By considering $\dot{h} = -v \sin \gamma$ from Eq. (10.2.3) we get

$$a = \frac{dv}{dt} = \frac{1}{d(\ln \varepsilon)/dv} \frac{d(\ln \varepsilon)}{d\lambda} \frac{d\lambda}{dh} \frac{dh}{dt}$$

$$= \frac{\varepsilon v_0^2}{2v} \left[-\frac{\sin \gamma_e}{\sin \gamma} + \frac{2H}{\varepsilon \lambda R} \right] \left(-\frac{\lambda}{H} \right) (-v \sin \gamma)$$

from which because of $v^2 = \varepsilon v_0^2 = \varepsilon g_0 R$ follows

$$a = -\frac{v^2 \lambda}{2H} \sin \gamma_e + g_0 \sin \gamma = -g_0 \left(\frac{\varepsilon \lambda R}{2H} \sin \gamma_e - \sin \gamma \right) \quad (10.2.15)$$

So, except for a short period of time after entry, when the S/C accelerates due to gravitation and negligible drag ($\lambda \ll R/H$), the expression in the brackets is positive and the vehicle decelerates. This is a quite practical equation, which later we will use a lot.

10.3
Preliminary Considerations

From the above numerical solutions of the complex entry profiles it is evident that there are no global analytical solutions. We will therefore focus on certain entry phases, which allow approximate analytical solutions, in particular in the critical deceleration phase.

General Approximations

To do so we have to make some gentle approximations to simplify the equations:

- Gravitation $g = g_0 = const.$ According to Eq. (10.2.6), this assumption entails an error of $(6370/6248)^2 \leq 4.0\%$. If one chooses for g the mean value of $\langle g \rangle_h = 9.62 \text{ m s}^{-2}$ at an altitude of 61 km, the error is even reduced to $(6370/6309)^2 \leq 2.0\%$
- $\frac{v^2}{r} \approx \frac{v^2}{R} \approx \frac{v^2}{r_e}$. This assumption entails an error of $6370/6248 \leq 2.0\%$.
- κ_D is assumed to be constant during the whole reentry process. The actual deviations from this constant for a S/C with a constant angle of

attack are no more than ±10%. If the angle of attack slightly changes, the important parameter L/D is still within a ±10% range.

The first two errors are negligible with regard to the third assumption, with regard to all prior assumptions (e.g. a constant scale height for the barometric formula (6.2.12), the assumption that the Earth is a non-rotating inertial system), and with regard to the other qualitative assumptions we will make later on.

10.3.1
Reentry – Phase B

The S/C is now at the entry interface at 122 km altitude with state vector (r_e, γ_e). Reentry is roughly divided into two different phases, as illustrated in Fig. 10.7. Directly after entry, the aerodynamic drag is so low that drag can practically be neglected. So the body descends with the entry angle almost in free fall in the direction to Earth. The motion equations (10.2.1) to (10.2.3) therefore, and because of $v^2 \approx gR^2/r \approx gr$ (circular orbit velocity), reduce to

$$\dot{v} \approx g \sin \gamma$$

$$\dot{\gamma} \approx 0$$

$$\dot{h} = -v \sin \gamma$$

in this reentry phase. Integration results in

$$\gamma = \gamma_e \qquad (10.3.1a)$$

$$v = v_e + g \sin \gamma_e \cdot t \approx const \qquad (10.3.1b)$$

$$h = h_e - v_e \sin \gamma_e \cdot t \qquad (10.3.1c)$$

This is exactly the behavior we observe in the numerical results of phase A in Figs. 10.4–10.6. As v_e is still very big, altitude decreases strongly while the velocity increases only slightly.

Note: *The reason for the absolutely straight reentry trajectory, rather than one which is bent downwards due to gravity, is the centrifugal force, which like in a circular orbit still counterbalances the gravitational force.*

Below approximately 80 km, the impinging air behaves like a free molecular flow with a rapidly increasing aerodynamic drag. This is where the crucial phase C starts. The transition between the two phases takes place when $d\varepsilon/d\lambda = 0$ and is characterized by the onset of a deceleration. According to Eq. (10.2.14a), this implies that

$$\frac{\sin \gamma_e}{\sin \gamma} = \frac{2H}{\varepsilon \lambda R}$$

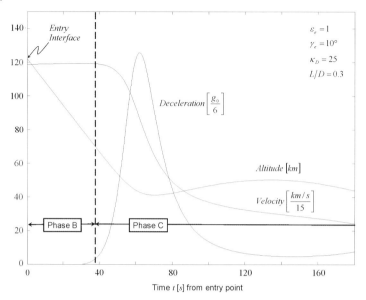

Figure 10.7 Definition of reentry phases B and C.

Because of Eq. (10.2.13) and $\varepsilon \approx 1$ and $\gamma \approx \gamma_e$, this determines the altitude of transition between the two phases to be

$$h_{B \to C} = H \ln \left(\frac{\kappa_D R}{\sin \gamma_e H} \right) \tag{10.3.2}$$

It mainly depends on γ_e and reaches for flat entries, $\gamma_e = 2° - 10°$, $h_{B \to C} = 85\text{–}95$ km, and for steep entries, $\gamma_e = 10°\text{–}45°$, $h_{B \to C} = 75 - 85$ km (cf. Figs. 10.4–10.6)

10.3.2
Ballistic Reentry Without Perturbations

The above numerical calculations show that the entry profiles in phase C may be quite different and complex. So it would be foolish to tackle the equations of motion head on. First of all, we need to understand the basic behavior of the solutions without the perturbing lift and gravitational terms L/D and H/R. This determines our approach: step by step, from simple approximations of equations to the more difficult ones. So we start out by not considering the perturbation terms L/D and H/R to find solutions for a non-disturbed reentry, and later, we will also take into account lift and gravitational perturbations.

We first assume that the S/C does not have any lift, $L/D = 0$. This is the so-called ballistic reentry. The expression "ballistic" refers to "like the flight of a ball", which does not have any lift. Except for truly spherical reentry bodies, which do not exist in practice, $L = 0$ is only valid for axially symmetric bodies with absolutely no angle of attack (AOA, angle between the axis of symmetry and flight direction). In reality even small AOAs produce differences in the position of the center of mass and the center of aerodynamic pressure and therefore significant lift. If this is not desired – because skipping can easily occur (see Section 10.6), and then it is more difficult to determine the landing site – so if a true ballistic flight path is wanted, one can slowly roll the S/C to average out such lifts. Mercury for instance had a nominal roll rate of 15° per second.

Large Drag Assumption

In addition we assume that, compared to drag, all other forces acting on the vehicle are negligible. This implies that we assume that the gravitational term in the ε equation is negligibly small

$$\frac{H}{\lambda R} \ll \frac{\sin \gamma_e}{\sin \gamma} \frac{\varepsilon}{2} \leq \frac{\varepsilon}{2}$$

With Eq. (10.2.13) this results in

$$e^{h/H} \ll \varepsilon \frac{\kappa_D R}{\sin \gamma_e H}$$

At high altitudes $\varepsilon \approx 1$ and therefore

$$h \leq H \ln \frac{0.1 \cdot \kappa_D R}{\sin \gamma_e H} = h_{B \to C} - 15.4 \text{ km} \approx 70 \text{ km}$$

where the condition "much smaller than" was quantitatively replaced by the value 0.1. At low altitudes we get with $\varepsilon = v^2/v_0^2$ and $\gamma_e \leq 45°$

$$v \gg 0.0055 \cdot e^{h/2H} v_0$$

As we will we later in Eq. (10.3.7) $v = v_e e^{-\lambda/2} \approx v_0 e^{-\lambda/2}$ and therefore

$$\frac{\kappa_D}{\sin \gamma_e} e^{-\frac{h}{H}} < 5.2 + \frac{h}{2H}$$

For $\gamma_e > 2°$ and a typical $\kappa_D \approx 25$ this implies $h > 30$ km, which will later be verified in Fig. 10.9.

As the gravitational term in the γ equation is of the same form, we get the same result for it also here. In addition, because at high altitudes the centrifugal term counterbalances the gravitational term, $1/\varepsilon - 1 \approx 0$, for high altitudes we can also neglect the entire reduced gravitational term. We therefore arrive at the conclusion that the assumption that there are no further forces acting on the vehicle holds for the important altitude range 30 km $< h <$ 70 km.

Because the decisive action of reentry happens in this altitude range 30 km $< h <$ 70 km we adjust the scale height of the barometric formula for further calculations to this altitude, which is

$$\boxed{H \approx 7.6 \text{ km}} \quad @ \text{ } 30 \text{ km} < h < 70 \text{ km} \quad (10.3.3)$$

Solution of the Equation of Motion

Setting those negligible terms to zero, the corresponding differential equations are derived from Eq. (10.2.14) as

$$\frac{d(\ln \varepsilon)}{d\lambda} = -\frac{\sin \gamma_e}{\sin \gamma} \quad (10.3.4a)$$

$$\frac{d(\cos \gamma)}{d\lambda} = 0 \quad (10.3.4b)$$

This set of equations is easily solved. The second equation directly yields

$$\boxed{\cos \gamma = \cos \gamma_e = const} \quad (10.3.5)$$

and therefore

$$\sin \gamma = \sin \gamma_e = const$$

Because the reentry body is subject just to drag, it reentries on a straight line (cf. Figs. 10.4–10.6)) and decelerates. This deceleration is described by the term on the right side of the first equation (10.3.4a). We can solve this equation as well by inserting Eq. (10.3.5)

$$\int_{\varepsilon_e}^{\varepsilon} d(\ln \varepsilon') = -\int_{\lambda_e}^{\lambda} d\lambda' \quad \Rightarrow \quad \ln \frac{\varepsilon}{\varepsilon_e} = -(\lambda - \lambda_e)$$

From this follows that

$$\boxed{\varepsilon = \varepsilon_e e^{-(\lambda - \lambda_e)} \approx \varepsilon_e e^{-\lambda}} \quad (10.3.6)$$

The latter holds because $\lambda_e \approx 10^{-7}$. The kinetic energy hence decreases exponentially with λ, and because it itself decreases exponentially with increasing altitude the kinetic energy decreases double-exponentially with decreasing altitude. With Eq. (10.2.12) we find from Eq. (10.3.6)

$$\boxed{v = v_e e^{-\lambda/2}} \quad (10.3.7)$$

As we will see later, this dependence is a good description for any early entry phase where drag exceeds lift and drag forces, which is why these results, despite their simplicity, are of wide significance even for very flat reentries (cf. Section 10.6).

10.3.3
Maximum Heating for Ballistic Reentries

As stated in our thermal problem setting in Section 10.1.1, it is our goal to determine the maximum heat flux during reentry. The heat flux on the S/C is (Eq. (10.1.3))

$$\dot{q}_{S/C} = St \sqrt{\rho \rho_q \frac{R_0}{R_n}} \cdot v^3$$

To find the maximum $\dot{q}_{S/C}(v)$ we need to have the dependence of ρ on the speed. From Eq. (10.2.13)

$$\lambda = \frac{2\kappa_D}{\sin \gamma_e} e^{-h/H} = \frac{2\kappa_D}{\sin \gamma_e} \frac{\rho}{\rho_0}$$

We know from the numerical calculations that the maximum deceleration occurs where the body first deviates from the straight reentry path. Because we expect the maximum heating around maximum deceleration, we apply the simple solution (10.3.7), $v = v_e e^{-\lambda/2}$. The atmospheric density as a function of v then is determined to be

$$\rho = \rho_0 \frac{\sin \gamma_e}{\kappa_D} \ln \frac{v_e}{v}$$

Hence

$$\dot{q}_{S/C} = St \sqrt{\rho_0 \rho_q \frac{R_0}{R_n} \frac{\sin \gamma_e}{\kappa_D} \cdot \sqrt{\ln \frac{v_e}{v}} \cdot v^3}$$

As $\dot{q}$ is monotonous in v, this results in the condition equation for a maximum $\dot{q}$:

$$\frac{1}{\dot{q}} \frac{d\dot{q}}{dt} \propto \frac{1}{\dot{q}} \frac{d\dot{q}}{dv} = \frac{1}{v \ln(v/v_e)} \left(\frac{1}{2} + 3 \ln \frac{v}{v_e} \right) = 0$$

From this it follows that

$$v_{\max \dot{q}} = \frac{v_e}{e^{1/6}} = 6.3 \text{ km s}^{-1} \cdot \sqrt{\varepsilon_e} \approx 6.3 \text{ km s}^{-1} \quad (10.3.8)$$

So the S/C experiences its maximum heating even long before its critical acceleration at $v_{crit} = 4.5$ km s^{-1} (see Eq. (10.5.8)). This validates our expectation that the simple Eq. (10.3.6) can be applied even better than we thought. The altitude where the maximum heating is reached is derived from $v = v_e e^{-\lambda/2}$ as $\lambda_{\max \dot{q}} = 1/3$ and because of Eq. (10.2.13)

$$h_{\max \dot{q}} = H \ln \frac{6\kappa_D}{\sin \gamma_e} \tag{10.3.9}$$

For $\gamma_e = 3°$–$10°$ maximum heating therefore occurs at an altitude of about 55 km and for high entry angles, $\gamma_e \approx 45°$, at 41 km. Finally the maximum heat flux at the S/C stagnation point is found to be

$$\dot{q}_{S/C \max} = St \cdot v_e^3 \sqrt{\frac{\rho_0 \rho_q \sin \gamma_e}{6e \cdot \kappa_D} \frac{R_0}{R_n}} = \dot{q}_{bal} \cdot \left(\frac{v_e}{v_0}\right)^3 \sqrt{\frac{\sin \gamma_e}{\kappa_D} \frac{R_0}{R_n}} \tag{10.3.10}$$

with

$$\dot{q}_{bal} = St \cdot v_0^3 \sqrt{\frac{\rho_0 \rho_q}{6e}} = 2.762 \times 10^7 \frac{\text{kg}}{\text{s}^3}, \quad v_0 = 7.905 \text{ km s}^{-1} \quad \text{and} \quad R_0 = 1 \text{ m}$$

With this and Eq. (10.1.1) the maximum temperature at this point is calculated to be

$$T_{\max}^4 = \frac{T_{bal}^4}{\varepsilon} \cdot \left(\frac{v_e}{v_0}\right)^3 \sqrt{\frac{\sin \gamma_e}{\kappa_D} \frac{R_0}{R_n}} \tag{10.3.11}$$

with

$$T_{bal} = \left(\frac{\dot{q}_{bal}}{\sigma}\right)^{1/4} = 4698 \text{ K}$$

and $\varepsilon \approx 0.85$ the emissivity of the heat shield.

10.3.4
Reentry with Lift

We now make a step forward to solve the basic differential equations (10.2.14) for reentry by allowing for a lift of the reentry vehicle, but still neglecting gravitational and centrifugal forces via $H/R = 0$. The differential equations then read

$$\frac{d (\ln \varepsilon)}{d\lambda} = -\frac{\sin \gamma_e}{\sin \gamma}$$

$$\frac{d (\cos \gamma)}{d\lambda} = \frac{\sin \gamma_e}{2} \frac{L}{D}$$

From Eqs. (10.2.4) and (10.2.5) we would expect $L/D \approx \text{const}$. In fact, L and D depend somewhat differently on speed, so that L/D slightly depends on speed. Within our approximations (see Section 10.3) $L/D = \text{const}$, however, is a good assumption, which we will adopt from now on. This allows us to solve the second equation directly by separating the variables

$$\cos \gamma = [1 + b(\lambda - \lambda_e)] \cos \gamma_e \tag{10.3.12}$$

where

$$b := \frac{\tan \gamma_e}{2} \frac{L}{D} \tag{10.3.13}$$

describes the lift power (buoyancy). What does this quite important equation tell us? The vehicle entries the atmosphere at the entry interface $\lambda = \lambda_e$ with $\gamma = \gamma_e$. As it descends, λ increases. According to Eq. (10.3.12) a positive lift, $b > 0$, decreases the flight path angle steadily until $\cos \gamma = 1$, when a horizontal flight with $\gamma = 0$ is attained. Of course, lift continues to act on the vehicle, which leads now to an increase in altitude and hence a decreasing λ, which via Eq. (10.3.12) implies $\cos \gamma < 1$, but this time with a negative flight path angle $\gamma < 0°$. In total, the positive lift results in a steadily upward curved path (cf. the numerical calculations). If lift is negative, that is if the reentry body turns upside down, then the path steadily turns down. From Eq. (10.3.12) we find after some minor trigonometric conversions

$$\sin \gamma \approx \sin \gamma_e \sqrt{1 - 2c(\lambda - \lambda_e)} \tag{10.3.14}$$

with

$$c := b \cot^2 \gamma_e = \frac{\cot \gamma_e}{2} \frac{L}{D}$$

where we have neglected the term of order $b^2 \lambda^2$ in the root, which is equivalent to $\lambda L/D \ll 4 \cot \gamma_e$, and which typically holds for altitudes $h > H \ln(100 L/D)$. We insert this expression into the first differential equation to find

$$d(\ln \varepsilon) = -\frac{\sin \gamma_e}{\sin \gamma} d\lambda = -\frac{d\lambda}{\sqrt{1 - 2c(\lambda - \lambda_e)}}$$

We can solve this equation analytically to arrive at

$$\ln \frac{v}{v_e} = \frac{1}{2c} \left(\sqrt{1 - 2c(\lambda - \lambda_e)} - 1 \right) \quad @ \quad \frac{\lambda L}{D} \ll 4 \cot \gamma_e \tag{10.3.15}$$

This solution holds for any L/D values, even for a high-lift reentry, as long as $\lambda L/D \ll 4\cot\gamma_e$. There even exists a fully analytical solution without the approximation $\lambda L/D \ll 4\cot\gamma_e$. But because this is much more complex and because it does not help to understand, we pursue it in an exercise (see Problem 10.5). As expected, Eq. (10.3.15) passes over into Eq. (10.3.6) for $L/D \propto c \to 0$.

10.4
Second-Order Solutions

After these introductory considerations we now take the final step forward in solving the reentry equations of motion by allowing for the perturbation terms of gravitation and centrifugal forces

$$\frac{d(\ln\varepsilon)}{d\lambda} = -\frac{\sin\gamma_e}{\sin\gamma} + \frac{2H}{\varepsilon\lambda R}$$

$$\frac{d(\cos\gamma)}{d\lambda} = \frac{\sin\gamma_e}{2}\frac{L}{D} - \left(\frac{1}{\varepsilon} - 1\right)\frac{H\cos\gamma}{\lambda R}$$

Because $H/R \approx 0.001$ we assume the H/R-terms are perturbations of first order with respect to the terms considered so far. We will take these perturbations fully into account. However, because they are small, it will suffice to apply for $\cos\gamma$ and ε the undisturbed terms of Section 10.3.2 in these perturbation terms, i.e. we will not consider perturbations of perturbations. We are looking for solutions with this so-called second-order perturbation analysis. These solutions will not be globally exact (we already know that there are no globally exact solutions), but they will be applicable for a quite extended region of λ.

10.4.1
Flight Path Angle

For $\cos\gamma$ and ε we insert the unperturbed expressions from Eqs. (10.3.5) and (10.3.6). The γ equation then reads

$$d(\cos\gamma) = \left[\frac{\sin\gamma_e}{2}\frac{L}{D} - \left(\frac{e^\lambda}{\varepsilon_e} - 1\right)\frac{H}{\lambda R}\cos\gamma_e\right]d\lambda$$

from which by direct integration follows that

$$\cos\gamma = \left[1 + b(\lambda - \lambda_e) - \frac{H}{\varepsilon_e R}\int_{\lambda_e}^{\lambda}\frac{e^x - \varepsilon_e}{x}dx\right]\cos\gamma_e$$

The integral, which comprises the perturbation, can be solved analytically

$$\int_{\lambda_e}^{\lambda} \left(\frac{e^x}{x} - \frac{\varepsilon_e}{x} \right) dx = Ei(\lambda) - Ei(\lambda_e) - \varepsilon_e \ln \frac{\lambda}{\lambda_e}$$

We find from any special formulary that the exponential integral function $Ei(x)$ can be expressed as

$$Ei(\lambda) - Ei(\lambda_e) = f^\lambda - f^{\lambda_e} + \ln \frac{\lambda}{\lambda_e} \approx f^\lambda + \ln \frac{\lambda}{\lambda_e} \qquad (10.4.1)$$

where we have defined the exponential-like function

$$f^x := \int_0^x \frac{e^y - 1}{y} dy = \sum_{n=1}^\infty \frac{x^n}{nn!} \qquad (10.4.2)$$

Due to the globally converging series expansion this function can be easily calculated numerically and is depicted in Fig. 10.8. We therefore find with

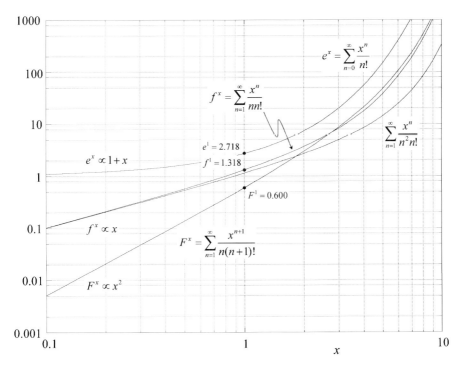

Figure 10.8 Representation of the functions f^x, F^x as defined in the text and e^x.

Eq. (10.2.13)

$$\delta(\lambda, \varepsilon_e) := \int_{\lambda_e}^{\lambda} \frac{e^x - \varepsilon_e}{x} dx = f^\lambda - (\varepsilon_e - 1) \ln \frac{\lambda}{\lambda_e} = f^\lambda - (\varepsilon_e - 1) \frac{h_e - h}{H} \quad (10.4.3)$$

The solution for the flight path angle therefore is

$$\cos \gamma = [1 + b(\lambda - \lambda_e) - p\delta(\lambda, \varepsilon_e)] \cos \gamma_e \quad \text{FPA equation} \quad (10.4.4)$$

where

$$p := \frac{H}{\varepsilon_e R}$$

describes the strength of the perturbation. The FPA (flight path angle) equation (10.4.4) is of high significance for the understanding of reentry. It indirectly describes the course of the reentry path with altitude for any entry condition. Because of these properties, we will make use of it for equilibrium reentries.

Let's have a closer look at reentry with the FPA equation. For $\varepsilon_e = 1$, at high altitudes $\delta(\lambda, 1) \approx \lambda \ll 1$. Therefore $\cos \gamma \approx \cos \gamma_e$: the entry vehicle descends on a straight line. With decreasing altitude, λ and hence $\delta(\lambda, 1)$ increase exponentially. Since b is in the range 0.01–0.1 while $p = 0.001$, first the lift term $b\lambda$ is significant. It increases $\cos \gamma$, so γ decreases: the vehicle will slowly deviate upwards from the straight line. At lower altitudes, $\lambda > 1$, δ increases exponentially with λ (therefore double-exponentially with decreasing h), so the gravity term quickly becomes significant. The specific value of $b = \tan \gamma_e L/(2D)$ depends on the entry angle and the lift. If it is quite substantial, the right side of Eq. (10.4.4) at some point becomes unity, where the vehicle flies horizontally. But because the lift continuously lifts the vehicle, it begins to ascent ($\gamma < 0$), implying a decreasing $\cos \gamma$, because also λ decreases. So, if b is sufficiently big, the vehicle may reverse the flight path angle and ascent before gravity overcomes this excursion. If b is too small, then there is just an indication of an upturn, but gravity will beat it soon. This is exactly what we see in the numerical calculations Fig. 10.3–10.6.

10.4.2
Critical Deceleration

The critical deceleration a_{crit}, that is the maximum deceleration during reentry, can be derived from Eq. (10.2.15). We rearrange this equation to

$$a = -g_0 \frac{R \sin \gamma_e}{2H} \left(\varepsilon \lambda - \frac{2H \sin \gamma}{R \sin \gamma_e} \right)$$

We now apply second-order perturbation analysis. Because H/R is a perturbation, $\sin\gamma \approx \sin\gamma_e$, and hence

$$a = -g_0 \sin\gamma_e \left(\frac{\varepsilon\lambda R}{2H} - 1\right) \tag{10.4.5}$$

This expression is no longer exact, but because even for extremely flat entry angles $\varepsilon_{crit}\lambda_{crit} \geq 0.1$ and because $2H/R = 0.0021$, this still is a very good approximation for all practical purposes. From the maximum condition $da/d\lambda = 0$ we derive from Eq. (10.4.5) in a few steps the critical λ to be

$$\lambda_{crit} = \left[1 + \frac{2H}{\varepsilon_{crit}R}\right]\frac{\sin\gamma_{crit}}{\sin\gamma_e} \tag{10.4.6}$$

We could now insert this into Eq. (10.4.5) to find the wanted a_{crit}. However, in principle this is not permissible because $\varepsilon_{crit}, \gamma_{crit}$ in Eq. (10.4.6) themselves depend on λ_{crit}. But for the upcoming special cases this will be not a problem, and Eq. (10.4.6) will therefore be of great value.

10.5
Low-Lift Reentry (First-Order Solutions)

Up to this point we have solved the equations of motion including gravitational forces in second-order perturbation analysis. Now that we move on to solve also the ε equation, this is no longer possible. If we still want to take gravitation into account, we can do so only by applying first-order perturbation analysis. So the following solutions will apply only for more restricted altitudes.

10.5.1
Velocity

The problem is that we can no longer exactly integrate the expression $1/\sin\gamma = 1/\sqrt{1-\cos^2\gamma}$ with $\cos\gamma = [1 + b(\lambda - \lambda_e) - p\delta(\lambda,\varepsilon_e)]\cos\gamma_e$. We can do so only with linear approximations by assuming

$$c^2\lambda^2 \ll 1 \quad \text{and} \quad q^2\delta^2(\lambda,\varepsilon_e) \ll 1$$

If these approximations are applied we obtain

$$\sin\gamma \approx \sin\gamma_e\sqrt{1 - 2c\lambda + 2q\delta} \tag{10.5.1}$$

with

$$q := p\cot^2\gamma_e = \frac{H}{\varepsilon_e R}\cot^2\gamma_e$$

from which it follows that
$$\frac{\sin \gamma_e}{\sin \gamma} \approx \frac{1}{\sqrt{1 - 2c\lambda + 2q\delta}} \approx 1 + c\lambda - q\delta$$

We insert this result into the ε equation and separate the variables
$$d(\ln \varepsilon) = \left(-\frac{\sin \gamma_e}{\sin \gamma} + \frac{2H}{\varepsilon \lambda R} \right) d\lambda$$

Because the second term on the right side is the perturbation, we can adopt for ε the unperturbed expression from Eq. (10.3.6). With Eq. (10.5.1) and the approximations $c^2\lambda^2 \ll 1$ and $q^2\delta^2(\lambda, \varepsilon_e) \ll 1$ we then derive

$$\ln \frac{\varepsilon}{\varepsilon_e} = -\int_{\lambda_e}^{\lambda} (1 + c\lambda - q\delta)\, d\lambda + 2p \int_{\lambda_e}^{\lambda} \frac{e^x}{x} dx$$

$$\approx -\lambda - \frac{1}{2}c\lambda^2 + q \int_0^{\lambda - \lambda_e} \delta(x)\, dx + 2p\left[Ei(\lambda) - Ei(\lambda_e) \right]$$

So we finally find with Eq. (10.4.1)
$$\ln \frac{v}{v_e} \approx -\frac{\lambda}{2} - \frac{1}{4}c\lambda^2 + \frac{q}{2}\Delta(\lambda, \varepsilon_e) + p\delta(\lambda, 0) \tag{10.5.2}$$

with
$$\Delta(\lambda, \varepsilon_e) := \int_0^{\lambda} \delta(x, \varepsilon_e) \cdot dx = F^{\lambda} - (\varepsilon_e - 1)\lambda \left(\ln \frac{\lambda}{\lambda_e} - 1 \right)$$

where we have introduced the function
$$F^x := \int_0^x f^y dy = \sum_{n=1}^{\infty} \frac{x^{n+1}}{n(n+1)!}$$

Which like f^x can easily be calculated numerically by series expansion. It is also depicted in Fig. 10.8.

10.5.2
Entry Trajectory

Within the first-order perturbation analysis, that is for $c^2\lambda^2 \ll 1$ and $q^2\delta^2(\lambda, \varepsilon_e) \ll 1$, it can be shown easily (exercise, Problem 10.4) that the course of the trajectory with altitude can be described analytically by

$$x \approx \cot \gamma_e \left\{ (h_e - h) + \frac{H}{\sin^2 \gamma_e} [b\lambda - p \cdot \Theta(\lambda, \varepsilon_e)] \right\} \tag{10.5.3}$$

with

$$\Theta(\lambda, \varepsilon_e) := \int_0^{\lambda - \lambda_e} \frac{\delta(x, \varepsilon_e)}{x} dx = \sum_{n=1}^{\infty} \frac{\lambda^n}{n^2 n!} - \frac{1}{2}(\varepsilon_e - 1) \ln^2 \frac{\lambda}{\lambda_e}$$

Here x is the downrange distance relative to the entry point. This dependence is illustrated in Fig. 10.9 for an entry with $\gamma_e = 45°$. We recognize the straight entry line $x = \cot \gamma_e (h_e - h)$. The actual entry trajectory deviates from this for positive lift by an upturn and for negative lift by a downturn. The trajectory representation ends where $c^2 \lambda^2 \ll 1$ is no longer valid. The numerical solutions of the full equations of motion for this case show that, for $L/D = 0.3$, in the further course of the trajectory the vehicle flies horizontally at 21 km altitude for a moment and then finally descends. For $L/D = 0.4$, there is a reflection point at 23.5 km altitude, a maximum at 24.6 km altitude, and thereafter a final descend. For $L/D = 0.5$ the reflection point is at 25 km and the maximum at 30 km altitude.

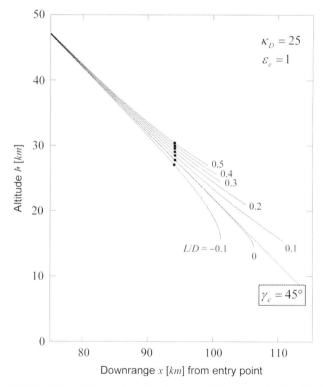

Figure 10.9 Reentry trajectories for $\gamma_e = 45°$ as given by Eq. (10.5.3) for different lift coefficients. On the x axis is the downrange distance from the entry point. The full dots denote the critical altitudes.

10.5.3
Critical Deceleration

To determine λ_{crit} we employ the approximate solutions of $\varepsilon(\lambda)$, $\sin\gamma(\lambda)$ for the unperturbed reentry from Eqs. (10.3.6) and (10.3.14) and obtain for this first-order perturbation analysis

$$\lambda_{crit} = (1+2ep)\sqrt{1-2c+2q\delta\left(1,\varepsilon_e\right)}$$

with

$$\delta\left(1,\varepsilon_e\right) = 1.318 - (\varepsilon_e - 1)\ln\frac{1}{\lambda_e}$$

and

$$\ln\frac{1}{\lambda_e} = 18.2 - \ln\frac{2\kappa_D}{\sin\gamma_e}$$

Because $2ep = 2eH/\varepsilon_e R < 0.0065$, this term is negligible and therefore we find with c from Eq. (10.3.14)

$$\begin{aligned}\lambda_{crit} &= \sqrt{1-2c+2p\delta\left(1,\varepsilon_e\right)} \\ &= \sqrt{1-\cot\gamma_e\frac{L}{D}+2\cot^2\gamma_e\frac{H}{\varepsilon_e R}\delta\left(1,\varepsilon_e\right)}\end{aligned} \quad (10.5.4)$$

Note: *Because we have assumed $c, p \ll 1$, essentially $\lambda_{crit} \approx 1$. Lift and gravity cause only minor variations from this value.*

The critical altitude at which the critical deceleration happens is determined from Eq. (10.2.13) as

$$h_{crit} = H\ln\frac{2\kappa_D}{\lambda_{crit}\sin\gamma_e} \quad (10.5.5)$$

Together with Eq. (10.5.4) this equation describes how the critical altitude changes as a function of lift and entry angle. It increases with increasing lift (see Fig. 10.9) and with decreasing entry angle. For $L \approx 0$ and $\gamma_e = 3°-10°$ critical altitudes are about 43–52 km.

For the critical deceleration we apply Eq. (10.4.5) to the critical point. Inserting Eq. (10.5.4) yields for $c^2\lambda^2 \ll 1$ and $q^2\delta^2(\lambda,\varepsilon_e) \ll 1$

$$a_{crit} = -\frac{v_e^2\sin\gamma_e}{2eH}\left[1-\frac{1}{4}\cot\gamma_e\frac{L}{D}+\frac{H}{\varepsilon_e R}(\xi-2e)\right] \quad (10.5.6)$$

10.5 Low-Lift Reentry (First-Order Solutions)

with

$$\xi(\varepsilon_e) := 2.636 + (\varepsilon_e - 0.400) \cdot \cot^2 \gamma_e - \left[\cot^2 \gamma_e (\varepsilon_e - 1) - 2\right] \ln \frac{1}{\lambda_e}$$

For $\varepsilon_e = 1$ we have

$$\xi - 2e = 0.600 \cdot \cot^2 \gamma_e + 33.6 - 2\ln \frac{2\kappa_D}{\sin \gamma_e} \approx 0.600 \cdot \cot^2 \gamma_e + 21.0$$

and therefore

$$a_{crit} = -\frac{v_0^2 \sin \gamma_e}{2eH}\left(1.025 - \frac{1}{4}\cot \gamma_e \frac{L}{D} + 6.30 \times 10^{-4} \cot^2 \gamma_e\right) \quad @ \; \varepsilon_e = 1$$

Because

$$\frac{v_0^2}{2eH} = \frac{R}{2eH}g_0 = 154 \cdot g_0$$

we finally obtain

$$\boxed{a_{crit} = -154 \cdot g_0 \sin \gamma_e \left(1.025 - \frac{1}{4}\cot \gamma_e \frac{L}{D} + 6.30 \times 10^{-4} \cot^2 \gamma_e\right)} \quad @ \; \varepsilon_e = 1 \qquad (10.5.7)$$

For decreasing entry angles the critical deceleration deviates more and more from the simple relationship $a_{crit} = -154 \cdot g_0 \varepsilon_e \sin \gamma_e \cdot 1.025$ (see Fig. 10.10) to larger values. This is counteracted by a positive lift. In Fig. 10.10 the critical deceleration is plotted according to Eq. (10.5.7) for $\varepsilon_e = 1$ for different L/D.

Example
For manned missions the reentry trajectory is chosen such that the critical deceleration never exceeds the maximum tolerable value of 10 g. From Eq. (10.5.7) it follows that for ballistic entries, $L = 0$, from LEO $\gamma < 2.4°$. Of course, it is quite difficult to adjust the entry angle exactly to such small values. If, however, one linearly extrapolates our results in Fig. 10.10 beyond our approximations to an entry vehicle with a lift of $L/D = 0.3$, which is typical for capsules with a heat shield, such as for Apollo and Gemini, then one could presume that it would be possible to increase the entry angle to $\gamma_e > 6°$ yet still having $a_{crit} < 10g$. This is corroborated by numerical calculations, which remarkably show $a_{crit} = 10.1 \; g_0$ for $L/D = 0.3$, $\gamma_e = 6°$. So our perturbation analysis seems to hold even beyond the theoretical limits.

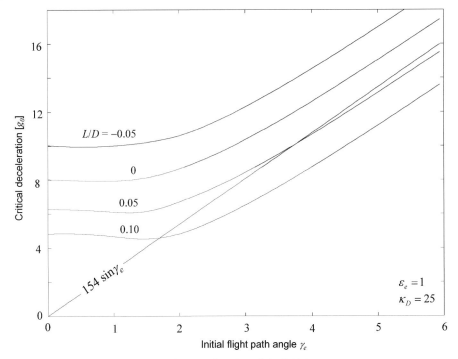

Figure 10.10 The critical deceleration as a function of the initial flight path angle (entry angle) and lift. Values for $\gamma_e > 1.5°$ are from Eq. (10.5.7), and for $\gamma_e \leq 1.5°$ numerical solutions.

For $\gamma_e \leq 1.5°$ our approximations definitely no longer apply, because $\cot \gamma_e$ diverges. On the other hand, $\sin \gamma_e \to 0$. One could presume that in all $a_{crit}(\gamma_e \to 0)$ would converge. Numerical calculations, which continue the analytical solutions for $\gamma_e \leq 1.5°$ (see Fig. 10.10) corroborate this supposition. With a semi-analytical ansatz, V.A. Yaroshevsky could even show that for $L = 0$ the critical deceleration converges to (see Table 6.1)

$$a_{crit}(\gamma_e \to 0) = 0.273 \cdot g_0 \sqrt{R/H} = 7.91 \cdot g_0$$

This is in excellent agreement with our numerical calculations. This limiting case, however, is of no practical interest, since for $\gamma_e \to 0$ the downrange distance becomes infinite. It is just the other way round that, to determine precisely the landing site of a capsule, the downrange distance should be as small, and hence the entry angle as large, as possible. These contradictory requests can only be resolved by a capsule with lift.

We close our considerations for low-lift reentries by determining from Eq. (10.5.2) the critical velocity at which the deceleration becomes maximal

$$v_{crit} = \frac{v_e}{\sqrt{e}} \left[1 - \frac{\cot \gamma_e}{8} \frac{L}{D} + \frac{H}{2\varepsilon_e R} \zeta(\varepsilon_e) \right] \approx \frac{v_e}{\sqrt{e}} = 4.5 \text{ km s}^{-1} \cdot \sqrt{\varepsilon_e} \quad (10.5.8)$$

For the latter we have chosen $v_e = 7.44 \text{ km s}^{-1}$, which is more realistic, because due to the Earth's rotation the entry velocity with respect to the atmosphere is effectively reduced.

10.6 Reflection and Skip Reentry

10.6.1 Reflection

From the discussion of the FPA equation (10.4.4) we saw that it nicely reproduces the upturn of the reentry trajectory for positive lift. In fact, we can take the FPA equation to determine the point – the reflection point – where the vehicle turns back to increasing altitudes. From the reflection condition $\cos \gamma = 1$ we derive from Eq. (10.4.4) for the reflection altitude λ_r

$$1 + \frac{\tan \gamma_e}{2} \frac{L}{D} \lambda_r - \frac{H}{\varepsilon_e R} \delta(\lambda_r, \varepsilon_e) - \frac{1}{\cos \gamma_e} = 0 \tag{10.6.1}$$

Reflections typically take place at $h_r > 20 \text{ km} \to \lambda_r < 3$, where according to Fig. 10.8 $\delta(3, \varepsilon_e \approx 1) = f^3 \approx 10$. Therefore $H\delta/\varepsilon_e R \ll 1$, which implies that Eq. (10.6.1) can be reduced to

$$1 + \frac{\tan \gamma_e}{2} \frac{L}{D} \lambda_r - \frac{1}{\cos \gamma_e} = 0 \tag{10.6.2}$$

From this and with Eq. (10.2.13) it follows for the reflection altitude that

$$h_r = H \ln \left(\frac{L}{D} \frac{\kappa_D}{1 - \cos \gamma_e} \right) \tag{10.6.3}$$

Figure 10.11 shows the reflection altitudes as a function of the entry angle for a given L/D and entry angle as calculated from Eq. (10.6.1). The results are almost identical to those from Eq. (10.6.3) except for $L/D < 0.2$ and $\gamma_e < 20°$, because then the gravity term is no longer negligible compared to the lift term.

Note: *The reflection altitude is independent from the entry velocity! One would have expected that it decreases with increasing entry speed because the higher entry momentum defies the ability to turn the vehicle up. But, on the other hand, the lifting force, which does the turn, increases quadratically with speed (cf. Eq. (10.2.4)), which just compensates the higher inertia of the vehicle.*

What are the entry parameters for which a reflection occurs? The condition derived from Eq. (10.6.1) reads

$$\frac{L}{D} \geq \left(\frac{1}{\cos \gamma_e} - 1 \right) \frac{1}{\lambda_r} + \frac{H}{\varepsilon_e R} \frac{\delta(\lambda_r, \varepsilon_e)}{\lambda_r} \tag{10.6.4}$$

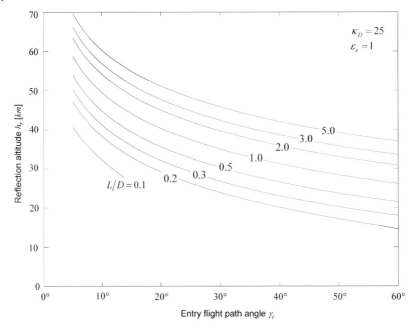

Figure 10.11 Reflection altitudes as a function of the entry angle and lift for $\varepsilon_e = 1$ (as derived from Eq. (10.6.1)).

This equation states that the vehicle reflects at given λ if L/D satisfies this equation. We are now seeking that minimal L/D for which reflection sets in at altitude $\lambda_{r,hor}$, that is, where it just flies horizontally for the first time. Minimizing the term on the right side of Eq. (10.6.4) delivers the condition

$$\left(\frac{1}{\cos \gamma_e} - 1\right) + \frac{H}{\varepsilon_e R}\left[e^{\lambda_{r,hor}} - \varepsilon_e - \delta\left(\lambda_{r,hor}, \varepsilon_e\right)\right] = 0 \tag{10.6.5}$$

The root of this equation for a given entry angle, which can be determined numerically, delivers $\lambda_{r,hor}$ and hence $h_{r,hor}$. Inserting it into Eq. (10.6.4) one obtains the wanted minimal L/D for a given entry angle. This dependence is displayed in Fig. 10.12 for $\varepsilon_e = 1$ and $\varepsilon_e = 2$.

If reflection is not desirable at all, then $L/D < 0.1$ has to be ensured. This can be achieved either by a continuous rotation of the capsule (which was done with Mercury), which ensures $\langle L/D \rangle_t = 0$ on a time average, or by turning the capsule upside down such that the lift vector points down, implying $L/D < 0$.

In the course of the path after reflection, the vehicle speed quickly diminishes so that gravitation, $p\delta \approx 1$, outweighs the lift and therefore the S/C descends. This second entry phase, induced by the relentless gravitation, is not reflected by the term δ, though, because we assumed in the perturbation

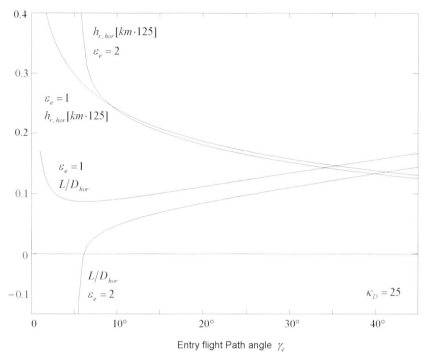

Figure 10.12 Minimum values for L/D and corresponding altitudes at which reflection of a reentering vehicle sets in, for $\varepsilon_e = 1$ and $\varepsilon_e = 2$.

term of the γ equation just $\varepsilon = \varepsilon_e \exp(-\lambda)$. What in fact happens is as follows. Because in the ε equation $d\varepsilon \propto -d\lambda / \sin\gamma$, the velocity always decreases, independently of whether the S/C ascends ($d\lambda < 0$, $\sin\gamma < 0$) or descends ($d\lambda > 0$, $\sin\gamma > 0$), and therefore also its energy $\varepsilon \propto v^2$ decreases steadily. Therefore the term $-H\cos\gamma/\varepsilon\lambda R$ in the γ equation steadily increases, and at some time becomes of order unity, independently of the flight path angle. It then dominates the lift, and after reflection at negative flight path angles $\gamma < 0$ it forces the vehicle to turn down, $d(\cos\gamma) \propto -d\lambda > 0$, and when it begins to descend again, $\gamma > 0$, to turn down even more rapidly, $d(\cos\gamma) \propto -d\lambda < 0$. This behavior, which we have derived from the equations of motion, just reflects the fact that lift declines quadratically with decreasing speed so that gravity takes over and makes the vehicle sink in the long run. In order to verify this behavior in detail from the equations of motions, one has to solve the time-dependent form (Eq. (10.2.11)) numerically. The λ-dependent form (Eq. (10.2.14)) applied for the analysis here is of no use for this, because λ as the independent variable has to progress per definition. But after reflection λ decreases. So the λ-dependent form cannot provide us with a trajectory after reflection.

Moderate reflections are usually desirable for capsule reentries because they decrease speed without an increase of deceleration. We will now see how reflections can be driven to the extreme to utilize them for achieving moderate reentry decelerations for manned missions even at very high entry speeds such as from interplanetary missions.

10.6.2
Skip Reentry

A skip reentry is the intelligent use of reflections to purposefully reduce the critical acceleration. It was used for the first time for the reentries of Apollo capsules after return from the Moon. Here the entry velocity roughly corresponded to the second cosmic velocity of 11 km s^{-1}, and thus the entry energy was $\varepsilon_e = 2$. So the mission managers encountered the big problem to reduce the double amount of kinetic energy, as compared to the preceding Mercury and Gemini LEO missions, and this seemed to be a special challenge for thermal protection and critical acceleration. The problem was solved by skip reentry. For reentry $\gamma_e \approx 6°$ was chosen and the design of the capsule was laid out such that the center of mass of the capsule did not coincide with the center of pressure to obtain a $L/D \approx 0.3$. From Fig. 10.12 it can be seen that for $\varepsilon_e = 2$ and $\gamma_e \approx 6°$ reflections occur for $L/D > 0$. Therefore $L/D = 0.3$ was a sure choice. Such an intentional reflection is called a "skip". By means of skipping, the initial speed can be reduced to such an extent that in a second dip reentry the deceleration forces are tolerable (see Figs. 10.5 and 10.6). The key purpose of skipping is a recurring stepwise speed reduction.

Exit Velocity

How big is the speed reduction brought about by one skip? To determine it we examine the equations of motions (10.2.1) and (10.2.2). From the discussion about approximations in Section 10.3.2 we know that at skipping altitudes 30 km $\leq h_{skip} \leq$ 70 km, when drag is about maximum, the gravitational term can be neglected.

Note: *In the γ equation the reduced gravitational term has for $1 \leq \varepsilon_e \leq 2$ a more sustainable effect because of the surplus centrifugal force of the approaching S/C. This centrifugal force effectively increases the lift, resulting in higher reflection altitudes, as will be found from the following calculations. So we are on the safe side.*

10.6 Reflection and Skip Reentry

Neglecting the gravitational terms we derive from Eqs. (10.2.1) and (10.2.2)

$$\dot{v} = -\frac{D}{m}$$

$$v\dot{\gamma} = -\frac{L}{m}$$

In order to determine the speed reduction we need the dependence $v(\gamma)$ to apply the symmetric condition $\gamma_{out} = -\gamma_{in}$ at the reflection point. So we need to get rid of the time dependence. We do this by dividing the above equations, yielding

$$\frac{\dot{v}}{\dot{\gamma}} = \frac{dv}{d\gamma} = v\frac{D}{L}$$

From this follows

$$\int_{v_e}^{v} \frac{dv}{v} = \frac{D}{L} \int_{\gamma_e}^{\gamma} d\gamma$$

and

$$v = v_e \exp\left[(\gamma - \gamma_e)\frac{D}{L}\right]$$

As skipping means an almost symmetrical flight path (see Fig. 10.4), during the exit from the atmosphere obviously the reflection condition

$$\gamma_{out} = -\gamma_{in} = -\gamma_e$$

has to be valid. So, for the reduced exit velocity from the atmosphere, we get

$$v_{out} = v_e \exp\left(-2\gamma_e \frac{D}{L}\right) \qquad (10.6.6)$$

Example

For the Apollo missions to the Moon the return velocity, and so the entry velocity of the capsule, was $v_e = 11.0 \text{ km s}^{-1}$. The selected entry angle was $\gamma_e = 6.5°$ and the L/D ratio was $L/D = 0.3$. Therefore, and according to Eq. (10.6.6), the exit velocity after the skip was $v_{out} = 0.470\, v_e = 5.2 \text{ km s}^{-1}$. With approximately this velocity, the capsule was dipped again for a double dip (see below).

"Double Dip" Reentry with Apollo Flights

For reentries from outside LEO, there exists a so-called entry corridor, which for Apollo was only $5.0° < \gamma_e < 7.0°$ wide (see Figs. 10.13 and 10.2). For $\gamma_e > 7.0°$ the maximum admissible deceleration of 12 g would have been exceeded during the skips. For $\gamma_e < 5.0°$ the splash down point would have been too imprecise due to extended recoils from the atmosphere, or the risk of not or too weakly grazing the atmosphere for a skip would have been to large. This narrow corridor required a very precise approach from the Moon. In order to better determine the landing point for the Apollo and the Soviet Zond Moon flights, a so-called "double dip" reentry was used during the return process with a lift reversion: During the first reentry the lift vector was flown upwards, as described above. After the skip, however, the capsule was rotated so that the lift vector pointed downwards, so that the negative lift kept the flight altitude roughly at the reflection altitude. With this maneuver it was possible to avoid bouncing back and the increasing inaccuracy of the landing point coming with this. In addition, the deceleration at lower altitudes could be kept to a more constant level, which altogether led to a safer landing.

Figure 10.14 depicts the original entry trajectory of Apollo 11. Judged against a comparable reentry trajectory from LEO (see Fig. 10.3) we recognize that the maximum after the first reflection is less developed. This is just the result of the negative lift. The reflection altitude of $185\,000$ ft $= 56.5$ km can easily be verified by applying Eq. (10.6.3) with the entry flight inclination

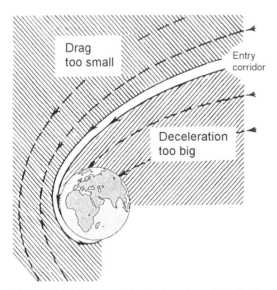

Figure 10.13 Entry corridor for the return of the Apollo capsules from the Moon.

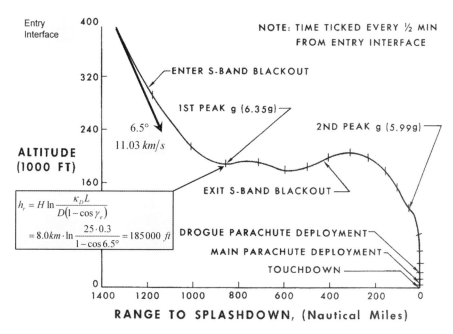

Figure 10.14 The original reentry trajectory of Apollo 11 with an entry flight inclination angle of 6.5° and reflection altitude of 185 000 ft.

angle of 6.5° and scaleheight $H = 8.0$ km at that altitude. This proves that the reflection altitude formula Eq. (10.6.3) is reliable and indeed is independent from the entry speed.

10.6.3
Phygoid Modes

It should now be obvious to drive skipping to the extreme and to skip not only once but again and again at a very shallow entry angle to slowly but steadily reduce the speed. This indeed would be possible. Any reentry body with $L > 0.5$ at $\varepsilon_e = 1$ and $\gamma_e < 2°$ will experience multiple or continuous skipping. This can be easily verified numerically and is done and displayed in Fig. 10.15 for $L/D = 1.3$ and $\gamma_e = 1.2°$, which are typical entry parameters for the Space Shuttle. Such shallow continuous skipping ups and downs are called *hypersonic phygoid modes*. These are oscillatory variations of altitude, where the flight path angle periodically oscillates with decreasing amplitude around zero.

The characteristic feature of a phygoid motion is that the S/C at very high speed and at very flat flight path angle, i.e. $\cos \gamma \approx 1$ and $\sin \gamma \approx \gamma$, oscillates around a mean flight path angle, $\gamma_D = const$, so $\langle \dot\gamma \rangle_t \approx 0$. We therefore can approximate the equations for a phygoid motion from the general equations of motion (10.2.1), (10.2.2), and (10.2.3) as

$$\dot{\gamma} = -\frac{L/v}{m} + \left(\frac{g}{v} - \frac{v}{r}\right)$$

$$\dot{v} = -\frac{D}{m} + g\gamma$$

$$\dot{h} = -v\gamma$$

The oscillations should be noticeable in variations of the flight path angle. In seeking a differential equation for that, we differentiate the above γ equation with respect to time

$$\ddot{\gamma} = -\frac{1}{m}\left(\frac{\partial(L/v)}{\partial h}\dot{h} + \frac{\partial(L/v)}{\partial v}\dot{v}\right) + 0 \cdot \dot{h} - \left(\frac{g}{v^2} + \frac{1}{r}\right)\dot{v}$$

where we have assumed that the oscillations take place at about a constant altitude, $r \approx \text{const}$. From Eqs. (10.2.4) and (10.2.5) it follows that

$$\frac{\partial(L/v)}{\partial h} = -\frac{(L/v)}{H} \quad \text{and} \quad \frac{\partial(L/v)}{\partial v} = \frac{L}{v^2}$$

and therefore

$$\ddot{\gamma} = -\frac{1}{m}\left[\frac{L}{H}\gamma + \frac{L}{v^2}\left(g\gamma - \frac{D}{m}\right)\right] - \left(\frac{g}{v^2} + \frac{1}{r}\right)\left(g\gamma - \frac{D}{m}\right)$$

from which follows

$$\ddot{\gamma} = -\left[\frac{L}{mH}\gamma + \frac{gL}{mv^2}\left(\gamma - \frac{D}{mg}\right)\right] - g\left(\frac{g}{v^2} + \frac{1}{r}\right)\left(\gamma - \frac{D}{mg}\right)$$

and

$$\ddot{\gamma} = -\left[\frac{L}{mH} + g\left(\frac{g}{v^2} + \frac{1}{r} + \frac{L}{mv^2}\right)\right]\gamma + \frac{D}{m}\left(\frac{g}{v^2} + \frac{1}{r} + \frac{L}{mv^2}\right)$$

Because it follows from the γ equation with $\langle\dot{\gamma}\rangle_t \approx 0$ and $\cos\gamma \approx 1$ that $v^2 \approx \langle v\rangle_t^2 = r(g - L/m)$, we find

$$\frac{g}{v^2} + \frac{1}{r} + \frac{L}{mv^2} \approx \frac{g}{r(g-L/m)} + \frac{1}{r} + \frac{L}{r(mg-L)}$$

$$= \frac{1}{r}\frac{mg + mg - L + L}{mg - L} = \frac{1}{r}\frac{2mg}{mg - L}$$

Hence

$$\ddot{\gamma} = -\left(\frac{L}{mH} + \frac{2g}{r}\frac{mg}{mg - L}\right)\gamma + \frac{2D}{mr}\frac{mg}{mg - L} \tag{10.6.7}$$

This is the differential equation of a somewhat odd linear oscillator which we can write in the general form

$$\ddot{\gamma} = -\left(\omega_0^2 + \omega_1^2\right)\gamma + c$$

Obviously there are two contributions to the phygoid oscillation with angular frequencies

$$\omega_0 = \sqrt{\frac{L}{mH}} \quad \text{and} \quad \omega_1 = \sqrt{\frac{2g}{r}\frac{mg}{mg-L}} \tag{10.6.8}$$

If we trace back these contributions in the derivation, we see that the ω_1 oscillation is caused by the causal chain (mode): declining S/C $\to$ increasing speed $\to$ increasing centrifugal force + increasing lift $\to$ decreasing flight path angle $\to$ upturn. On the other hand, the ω_0 oscillation stems from the chain: declining S/C $\to$ decreasing altitude $\to$ exponentially increasing atmospheric pressure $\to$ strongly increasing lift $\to$ quickly decreasing flight path angle $\to$ immediate upturn. We could interpret the latter process also as a bouncing off the atmosphere. These two oscillations happen on quite different time scales. Because

$$\frac{L}{mg} \approx \frac{RL}{mv^2} = \kappa_D \frac{L}{D}\frac{R}{H}e^{-h/H} \approx 0.2$$

it follows that

$$\frac{\omega_1^2}{\omega_0^2} \approx \frac{2g}{r}\bigg/\frac{L}{mH} = 2\frac{mg}{L}\frac{H}{r} \approx 2 \times 0.2 \times 0.001 = 0.0004$$

The short-term ω_0 mode therefore is the more forceful mode by orders of magnitude, which is why we can neglect ω_1. We therefore can simplify Eq. (10.6.7) to

$$\ddot{\gamma} = -\omega_0^2 \gamma + 2\frac{D}{mr}\frac{mg}{mg-L} = -\omega_0^2(\gamma - \gamma_D)$$

with

$$\gamma_D = 2\frac{D}{L}\frac{H}{r}\frac{mg}{mg-L} \approx 0.1° \tag{10.6.9}$$

To solve the equation we substitute $x := \gamma - \gamma_D$. This implies $\ddot{\gamma} = \ddot{x}$ we therefore get the differential equation $\ddot{x} = -\omega_0^2 x$ with the solution $x = x_0 \cos(\omega_0 t + \varphi)$. By resubstitution and because $\gamma(t=0) = \gamma_e$ we finally get

$$\gamma = \gamma_D + (\gamma_e - \gamma_D)\cos\omega_0 t \tag{10.6.10}$$

The drag-induced offset $\gamma_D \approx 0.1°$ is the time-averaged value of the FPA. It determines the long-term decline of the mean altitude of the vehicle and is easily recognized as such in Fig. 10.15. The period of the phygoid motion is determined via Eqs. (10.2.8) and (10.2.10) to be

$$T = \frac{2\pi}{\omega_0} = 2\pi\sqrt{\frac{mH}{L}} = 2\pi\frac{H}{v}e^{\frac{h}{2H}}\sqrt{\frac{D}{\kappa_D L}} \quad \textbf{phygoid period} \quad (10.6.11)$$

So the period decreases exponentially with the altitude at which the phygoid motion takes place. This exponential dependence is nicely depicted in Fig.10.15. Although the mean flight altitude decreases only slightly with the phygoid motion, the phygoid period (and the mode amplitude) decreases quickly. For a phygoid motion at an altitude of typically $h = 80$–90 km ($H \approx 5.8$ km, see Table 6.2) and for $\kappa_L = \kappa_D L/D = 32.5$, the period is $T = 15$–35 min.

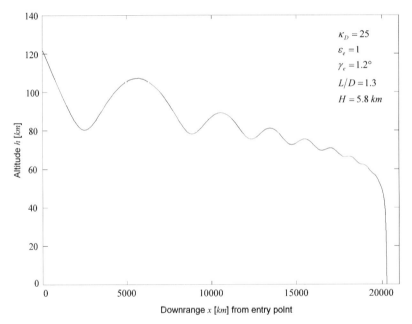

Figure 10.15 The phygoid oscillation which would result from an uncontrolled lift reentry of a Space Shuttle ($\gamma_e = 1.2°$, $L/D = 1.3$). The scale height $H = 5.8$ km is adjusted to the altitude where the phygoid oscillation happens.

Permanent skipping may seem the ideal, because gentle, velocity reduction at a first glance. However, this has a serious drawback: without any lift control, the downrange distance (and with it the landing site) virtually cannot be determined. Note that in the case of Fig. 10.15 the distance traveled until landing is about 20 000 km – that's halfway around the globe! This is not acceptable either for winged bodies or even less for capsules (which have to land on the ground or in the water with a good accuracy to pick them up), so phygoid modes have to be avoided at any rate. But one can turn the objective upside down. According to an idea of the famous Austrian space engineer Eugen Sänger (1905–1964), it would be possible to design an intercontinental transport high-lift vehicle with, say, $L/D \approx 2.5$ without any propulsion by accelerating it to LEO speeds and then using wave-like gliding along the surface of the atmosphere to reach a given target point anywhere on the globe with a controllable lift in the late phase. The time required to orbit the Earth would then be about 1 h 45 min. The critical acceleration of the first skip would be only 0.1 g, increasing for the following skips until it was about only 0.4 g in the final phase.

10.7
Lifting Reentry

We have seen that for $\gamma < 2°$ the approximations for a ballistic entry are no longer valid. For heavy S/Cs, however, reentries at such flat angles are the only way to keep the maximum heat load below tolerable values by spreading the deceleration and hence friction over a much longer time period. This can be achieved by an increased L/D of the S/C. This is why all the larger S/Cs are so-called winged bodies, such as the Space Shuttle.

We will now analyze this limiting case of a lifting reentry, where an adjustable lift is utilized to maintain a very small and constant flight path angle. So

$$\gamma = \gamma_e \approx 1° = const \quad \textbf{lifting reentry conditions} \tag{10.7.1}$$

Hence $\dot\gamma = 0$. Because of this condition we have: $\sin\gamma = \sin\gamma_e \approx \gamma_e$ and $\cos\gamma \approx 1$. Therefore the left sides of Eqs. (10.2.2) and (10.2.14b) vanish. In order also to have the right sides zero, the following must hold (approximation: $g \approx g_0, r \approx R$)

$$\frac{L}{m} = g - \frac{v^2}{r} \approx g_0\left(1 - \frac{v^2}{v_0^2}\right) \quad \text{with} \quad v_0^2 = g_0 R \tag{10.7.2}$$

This equation implies that the lift has to be constantly adjusted according to the entry conditions. Thereby we have two forces pulling "up" (adjusted lift plus centrifugal force) which just balance the gravitational force pulling "down". This is where the expression *equilibrium glide* (a.k.a. *lifting reentry* or *gliding reentry*) comes from.

Uncontrolled Lifting Reentry

We already came across a lifting reentry when we investigated phygoid modes. In that case, however, the FPA was not constant at any point in time, but constancy was achieved only on time average:

$$\langle \gamma \rangle_t = \gamma_D = 2 \frac{D}{L} \frac{H}{r} \frac{mg}{mg - L} = const$$

Furthermore the FPA oscillated around this mean value, because we assumed an unregulated constant L/D ratio.

Controlled Lifting Reentry

Phygoid modes are undesirable for a steady lifting reentry. Therefore, a reentry vehicle is required with $L/D_{max} \geq 0.7$ and for which lift and drag can be adjusted separately. Lift must be adjustable to such an extent that the phygoid oscillation can be compensated and therefore $\gamma = \gamma_e = const$ can be enforced at any time. This indeed is possible via the so-called *angle-of-attack* (AOA) α and the so-called *roll angle* (a.k.a. *bank angle*) μ. Figure 10.16 (see later) displays for the Space Shuttle how drag and lift can be varied simultaneously with the AOA. Thereby the L/D ratio changes in a way characteristic for the S/C. Of course, for $\alpha = 90°$, $L/D = 0$ always holds, because vertical lift must vanish at that angle.

Independent from the AOA, a bank angle can be set. The cosine of the bank angle determines the lift component, which points into the z direction, i.e. upwards. It is only this component which balances the gravitational and centrifugal force and which is decisive in our equations. If the bank angle is $\mu = 90°$, then the S/C is tilted fully sideways and no upward lift is generated. In summary:

> For a controlled lifting reentry the angle of attack and bank angle of the vehicle are adjusted such that the generated upward lift plus the centrifugal force exactly compensate the gravitational force such that $\gamma = const$.

10.7.1
Equations of Motion and L/D Control Law

How must L/D be adjusted at a given altitude such that Eq. (10.7.2) is fulfilled? Since we do not know the velocity as a function of altitude for a lifting reentry, Eq. (10.7.2) is not a control law for L/D in itself – rather, we have to solve the equation of motion. Applying the condition (10.7.1) to the basic equations of motion (10.2.14) we find

$$\frac{d(\ln\varepsilon)}{d\lambda} = -1 + \frac{2H}{\varepsilon\lambda R}$$

$$\frac{d(\cos\gamma)}{d\lambda} = \frac{\sin\gamma_e}{2}\frac{L}{D} - \left(\frac{1}{\varepsilon}-1\right)\frac{H\cos\gamma}{\lambda R}$$

equations of motion (10.7.3)

We already know the solution of the γ equation. It is Eq. (10.4.4). The solution of the decoupled ε equation can be easily derived. We first separate the variables

$$d(\ln\varepsilon) = \left(-1 + \frac{2H}{\varepsilon\lambda R}\right)d\lambda$$

This equation of motion is the same as the one in Section 10.5.1 except that here $\gamma = \gamma_e = const \Rightarrow c = q = 0$. Therefore we derive from Eq. (10.5.2) the second-order solutions

$$\ln\frac{v}{v_e} = -\frac{\lambda}{2} + \frac{H}{\varepsilon_e R}\delta(\lambda,0) \tag{10.7.4a}$$

$$\cos\gamma = \left[1 + \frac{\tan\gamma_e}{2}\frac{L}{D}\lambda - \frac{H}{\varepsilon_e R}\delta(\lambda,\varepsilon_e)\right]\cos\gamma_e \tag{10.7.4b}$$

The second equation is the FPA equation (10.4.4), which also holds for lifting reentry as long as $\tan\gamma_e \cdot \lambda L/(2D) \ll 1$, that is $\exp(h/H) \gg L\kappa_D/D$, i.e. down to about $h = 45$ km. With the FPA equation we have the answer to the question of how to adjust the lift. Because, if for an equilibrium glide $\cos\gamma = \cos\gamma_e = const$ must hold, then the last two terms in the square bracket must cancel each other. This condition provides the L/D control law:

$$\frac{L}{D} = \frac{2H}{\varepsilon_e R \tan\gamma_e}\frac{\delta(\lambda,\varepsilon_e)}{\lambda} \qquad \textbf{L/D control law} \tag{10.7.5}$$

At very high altitudes, that is for $\lambda_e \ll \lambda < 1$, and with $\varepsilon_e = 1$, in Eq. (10.7.5) we can approximate $\delta(\lambda,1) = f^\lambda \approx \lambda < f^1 = 1.318$ and therefore with $H \approx 7.6$ km we find

$$\frac{L}{D} \approx \frac{2H}{\tan\gamma_e R} \approx 0.10\text{–}0.14$$

This value is much smaller than the typical lift of a winged body, such as the Space Shuttle, which has $L/D \approx 1.3$. Drag has to be as big as possible at very high altitudes, but not to an extent that the flight attitude becomes unstable. Therefore the AOA is set to the limiting value $\alpha \approx 45°$. The bank angle μ has to be adjusted that the vertical lift component achieves the value $\cos\mu \cdot L/D = 0.10$–0.14. So

$$\mu \approx \arccos(0.12/1.3) = 85° \tag{10.7.6}$$

Because this is too close to the critical value $\mu = 90°$, where the Space Shuttle would plunge down, NASA limits the bank angle to $\mu = 80°$.

10.7.2
Critical Deceleration Parameters

Having found the solutions of the equations of motion, it is now straightforward to determine the critical deceleration and those trajectory parameters, namely critical altitude and critical speed, at which it is achieved. Because $\gamma = \gamma_e$, Eq. (10.4.5) is exact and therefore also Eq. (10.4.6). This is true for any entry phases down to the smallest velocities as long as $\gamma = \gamma_e = \gamma_{crit}$. From this we derive the critical altitude

$$\lambda_{crit} = 1 + \frac{2H}{\varepsilon_{crit} R} \quad @ \gamma = \gamma_e = const$$

How big is ε_{crit}? From Eq. (10.7.4) we find that for $H \cdot \delta(\lambda_{crit} \approx 1.0)/R \approx 10^{-3}$ it follows that $\varepsilon_{crit} = \varepsilon_e \exp(-\lambda_{crit})$. So for $\lambda_{crit} \approx 1$ we have $\varepsilon_{crit} = \varepsilon_e/e$ and therefore

$$\lambda_{crit} = 1 + \frac{2eH}{\varepsilon_e R} \approx 1 \tag{10.7.7}$$

Thereby we have proven self-consistently that Eq. (10.7.7) is correct. According to Eq. (10.5.5) we find for the critical altitude

$$h_{crit} = H \ln \frac{2\kappa_D}{\sin\gamma_e} \approx 60 \text{ km} \tag{10.7.8}$$

With Eq. (10.7.4a) the critical velocity is found to be

$$\ln \frac{v_{crit}}{v_e} = -\frac{1}{2} + \frac{H}{\varepsilon_e R}\delta(1,0)$$

Because

$$\frac{H}{\varepsilon_e R}\delta(\lambda,0) \approx \frac{H}{R} \cdot \frac{h_e - h}{H} = \frac{h_e - h}{R} \approx 0.01$$

it follows that

$$v_{crit} = \frac{v_e}{\sqrt{e}}\left(1 + \frac{H}{\varepsilon_e R}\delta(1,0)\right)$$

$$\approx \frac{v_e}{\sqrt{e}} = 4.5 \text{ km s}^{-1} \cdot \sqrt{\varepsilon_e} \approx 4.5 \text{ km s}^{-1} \tag{10.7.9}$$

From Eq. (10.4.5) we finally derive the critical deceleration

$$a_{crit} = -\frac{v_0^2 \sin \gamma_e}{2H}\left[\varepsilon_{crit}\lambda_{crit} - 2\frac{H}{R}\right] \approx -\frac{v_0^2 \sin \gamma_e}{2H}\varepsilon_{crit}\lambda_{crit} \approx -\frac{v_0^2 \sin \gamma_e}{2H}\frac{\varepsilon_e}{e}$$

from which follows

$$a_{crit} \approx -\frac{R}{2eH}\sin \gamma_e \cdot g_0 \approx -154 \cdot \sin \gamma_e \cdot g_0 \qquad (10.7.10)$$

This equation states that the critical deceleration may take on virtually any value by adjusting the entry angle accordingly. This is depicted in Fig. 10.10 by the straight line. For $\gamma_e = 1.0°$ the deceleration amounts to a modest $a_{crit} = -2.7 \cdot g_0$.

10.7.3
Maximum Heat Load

The structural load capacity of the Space Shuttle is about $|a_{crit}| \approx 5\, g_0$. Other than this the maximum heat load, which via the heat emission corresponds to the maximum surface temperature of the thermal tiles, is another critical parameter that has to be taken care of. To determine the maximum heat flux we consider Eq. (10.7.2) by applying Eq. (10.2.8)

$$\rho = \rho_0 \frac{H}{\kappa_L R}\left(\frac{v_0^2}{v^2} - 1\right)$$

So, the heat flux onto the vehicle surface is

$$\dot{q}_{S/C} = St\sqrt{\rho\rho_q \frac{R_0}{R_n}} \cdot v^3 = St\sqrt{\rho_0\rho_q \frac{R_0}{R_n}\frac{H}{\kappa_L R}}\sqrt{\frac{v_0^2}{v^2} - 1} \cdot v^3$$

From the maximum condition

$$\frac{1}{\dot{q}}\frac{d\dot{q}}{dv} = \frac{v_0^2}{v^3\left[v_0^2/v^2 - 1\right]}\left(3\frac{v^2}{v_0^2} - 2\right) = 0$$

we finally derive the critical velocity

$$v_{\max \dot{q}} = v_0\sqrt{\frac{2}{3}} = 6.1 \text{ km s}^{-1} \qquad (10.7.11)$$

To find the altitude at which the maximum heat load is achieved, we consider Eq. (10.7.4a). Since $v_{\max \dot{q}}$ is about the same as for the ballistic reentry, we expect the maximum heat load at about 50 km altitude at which λ is of order unity. Because from the above $H\delta(\lambda,0)/(\varepsilon_e R)$ is negligible, $v = v_e e^{-\lambda/2}$ at these altitudes. We therefore get for a lifting reentry from LEO, $v_e \approx v_0$, $\lambda_{\max \dot{q}} = -\ln(2/3)$, and because of Eq. (10.2.13)

$$h_{\max \dot{q}} = H \ln \frac{4.933 \cdot \kappa_D}{\sin \gamma_e} \quad @ \; v_e \approx v_0 \tag{10.7.12}$$

For typically $\gamma_e = 1°$ and $\kappa_D = 25$ we find $h_{\max \dot{q}} \approx 65$ km. For the wanted maximum heat flux to the S/C we thus find

$$\dot{q}_{S/C,\max} = St \cdot v_0^3 \frac{2}{3} \sqrt{\frac{\rho_0 \rho_q}{3}} \sqrt{\frac{R_0}{R_n} \frac{H}{\kappa_L R}} = \dot{q}_{lift} \sqrt{\frac{R_0}{R_n} \frac{HD}{\kappa_D RL}} \tag{10.7.13}$$

with

$$\dot{q}_{lift} = St \cdot \frac{2}{3} v_0^3 \sqrt{\frac{\rho_0 \rho_q}{3}} = 4.293 \times 10^7 \; \frac{kg}{s^3} \quad \text{and} \quad R_0 = 1 \; m$$

With this and Eq. (10.1.1) the maximum temperature load at the stagnation point of the S/C can be determined to be

$$T^4_{\max} = \frac{T^4_{lift}}{\varepsilon} \cdot \sqrt{\frac{R_0}{R_n} \frac{HD}{R L} \frac{1}{\kappa_D}} \tag{10.7.14}$$

with

$$T_{lift} = \left(\frac{\dot{q}_{lift}}{\sigma}\right)^{1/4} = 5246 \; K$$

and $\varepsilon \approx 0.85$ the emissivity of the heat shield.

10.8
Space Shuttle Reentry

Now that we know the general behavior of a winged body reentering with equilibrium glide, we want to exemplify this case for the Space Shuttle. The reentry profile described hereafter is typical for any kind of winged body, but it sticks to the real Space Shuttle profile. The data of this specific example are taken from the reentry of my mission STS-55 from May 6, 1993.

10.8.1
From Deorbit Burn to Entry Interface

Typically, the shuttle is in an initial circular LEO orbit at altitude $h_i = 300$ km. 60 min and 40 s before touchdown, and with it 170° west of the touchdown

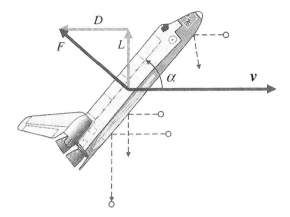

Figure 10.16 Drag and lift of a Space Shuttle at a typical AOA $\alpha = 40°$.

point, it executes a deorbit burn of 176 s duration to reduce the orbit velocity of 7.73 km s^{-1} by just 0.0885 km s^{-1} (that is, only 1.1%! – see example in Section 10.1.3). It thus declines on an entry ellipse that will cross the entry interface after 28 min at an altitude of 122 km with $\gamma_e = 1.2°$. As at this point the drag is still almost zero, the speed of the S/C increases on its further way according to Eq. (10.3.1b) as

$$v = v_e + g \sin \gamma_e \cdot t$$

up to approximately $v_0 =$ Mach 28. During that time, the goal is to have a drag as big as possible, so that the velocity does not exceed $v \approx v_0$. To accomplish this, the shuttle is placed at an AOA (angle of attack) of about $\alpha = 40°$ (see Fig. 10.16). The air molecules can be considered ballistic at these altitudes. They collide with the large bottom side of the Shuttle, transferring more momentum and hence more desired drag. But at the same time they cause increased lift. This is undesirable as in this phase of almost free fall the gravitational force and the centrifugal force balance each other. So an upward lift would bring the equilibrium glide out of balance and skipping would result. This is why the S/C is temporarily tilted sideways so that the bank angle is $\mu = 80°$ at an unaltered AOA, so that the shuttle is in a sloping lateral position with very low vertical lift. This is in line with our L/D Control Law from Eq. (10.7.6). NASA limits the bank angle to $\mu \leq 80°$, because any uncontrollable slight increase beyond 80° would lead to the Space Shuttle plunging straight down. The technical term for this roll maneuver is "roll reversals" or "bank reversals" because they are carried out alternately to the left and right side. Roll reversals allow the Space Shuttle to be steered laterally, which at this very early reentry phase increases the cross-track range a lot. This is highly desirable to ensure that the Shuttle reaches the landing site and still has sufficient range capability to properly align with the runway heading (see TAEM phase in Section 10.8.4). Roll maneuvers are continually carried out during reentry right down to Mach 2.5, but below 80 km with a slowly decreasing bank angle down to 45° in order that at decreasing speed the glide equilibrium is maintained by an increased lift.

10.8.2
From Entry Interface down to 80 km

With this attitude the Shuttle reaches the entry interface with an entry angle of $\gamma_e \approx 1.2°$. Here, the atmospheric density has increased to a degree that the aerodynamic drag decelerates the velocity to a bit less than v_0, and the right side of Eq. (10.7.2) becomes slightly positive. Now and for the following, the flight path angle can be maintained at $\gamma \approx 1°$ so as not to exceed the maximum admissible heat load. In this configuration, the Shuttle flies at a constant velocity (see Fig. 10.6) down to an altitude of about 80 km. So it merely reduces its potential energy by drag.

In this reentry phase the already mentioned skipping might occur. To avoid this, the flight path angle must be controlled carefully to maintain $\gamma \approx 1°$. This is difficult to achieve, because during the bank reversals, which can last up to one minute, the lift may become very large, in particular when the bank angle is about zero. This would immediately induce a skipping action. To avoid this, the AOA is simultaneously increased during the bank reversal so that lift increases not too much. Thereby the drag increases insignificantly (see Fig. 10.17).

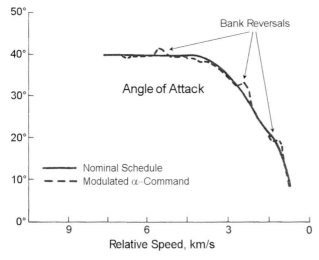

Figure 10.17 AOA profile as a function of entry velocity.

10.8.3
Blackout Phase

In the following flight phase at altitudes 80 km → 50 km the shuttle retains its AOA = 40° as well as its orbit angle $\gamma = 1°$. At these altitudes the thermal shield of the shuttle heats up so much that the impacting air around the shuttle ionizes and makes it impermeable for electromagnetic waves. For about 15 min there is no radio communication with Mission Control. That's why this phase is called the "blackout phase". The drag is now so strong that velocity is fiercely reduced.

Apart from the altitude-dependent flight profile, which we had discussed in Section 10.4.1, for this flight phase it is possible to provide a time-dependent expression for the key trajectory parameters. To do so we consider the equations of motion in the time-dependent form as given by Eq. (10.2.1) with $\cos \gamma \approx 1$, $\sin \gamma \approx 0$, and Eq. (10.7.2)

$$\dot{v} = -\frac{D}{m} \tag{10.8.1}$$

$$\frac{L}{m} = g_0 \left(1 - \frac{v^2}{v_0^2}\right)$$

We eliminate m from both equations, and get

$$\frac{\dot{v}}{g_0} = -\frac{D}{L}\left(1 - \frac{v^2}{v_0^2}\right) \tag{10.8.2}$$

Note that $v < v_0$ is mandatory, as for $v = v_0$ we get $\dot{v} = 0$, i.e. there would not be any deceleration. Separating the variables from Eq. (10.8.2) results in

$$\int_{v_e/v_0}^{v/v_0} \frac{dx}{1-x^2} = \operatorname{arctanh} \frac{v}{v_0} - \operatorname{arctanh} \frac{v_e}{v_0} = -\frac{g_0}{v_0} \int_0^t \frac{D}{L} dt = -\frac{Dg_0}{Lv_0} t$$

Because of the constant AOA, $D/L = const$ is valid during that flight phase, and we can extract this term from the integral. After some modifications, we get

$$v = v_e \frac{1 - \frac{v_0}{v_e}\tanh\left(\frac{\alpha t}{v_0}\right)}{1 - \frac{v_e}{v_0}\tanh\left(\frac{\alpha t}{v_0}\right)} \quad \text{with} \quad \alpha := g_0 \frac{D}{L} \tag{10.8.3}$$

To see that v indeed decreases, we expand this equation into a series of $\alpha t \ll v_0$ for small time periods (exercise, Problem 10.6) and for $v_e < v_0$

$$v = v_e - \left(1 - \frac{v_e^2}{v_0^2}\right)\left[\alpha t + \frac{v_e}{v_0^2}\alpha^2 t^2 - \frac{\alpha^3 t^3}{v_0^2}\left(\frac{1}{3} - \frac{v_e^2}{v_0^2}\right) + \cdots \right] \tag{10.8.4}$$

We recognize that velocity decreases at an increasing rate. The corresponding deceleration is found by differentiating Eq. (10.8.4)

$$a = -\alpha \left(1 - \frac{v_e^2}{v_0^2}\right) \left[1 + \frac{2v_e}{v_0^2}\alpha t - \frac{1}{v_0^2}\alpha^2 t^2 \left(1 - 3\frac{v_e^2}{v_0^2}\right) + \cdots \right] \qquad (10.8.5)$$

The acceleration increases monotonically over time. For a winged body with $L/D = 1.3$, Eqs. (10.8.4) and (10.8.5) have an inaccuracy of about 10% after 15 min, so they are sufficiently accurate over the entire blackout phase.

10.8.4
Aerodynamic Flight Phase

At an altitude of about 50 km, the shuttle leaves the blackout phase with a velocity of only $v =$ Mach 11. Because of the further increasing atmospheric density, the Space Shuttle comes now into an aerodynamic state. Also the absolute value of the deceleration increases to $1.5\,g_0$. In order to maintain a constant $a = -1.5\,g_0$, the AOA of 40° is continuously and linearly reduced to 8° at $v =$ Mach 1 (see Fig. 10.17). With a decreasing AOA, drag is reduced whereas lift is increased, and thus the total D/L decreases. This reduction just compensates the increase of the term $1 - v^2/v_0^2$ in Eq. (10.8.2), so that we reach a constant deceleration. In this constant-deceleration mode the shuttle flies down to an altitude of about 25 km and Mach 2.5. Up to this point the shuttle has covered a distance of 8 000 km in 54 min.

The final phase is the so-called Terminal Area Energy Management (TAEM) phase, where the shuttle during the remaining 6 min undergoes flight maneuvers with changing AOA $= 4°\text{--}10°$ to align velocity and heading with the approach cone of the landing strip about 100 km away. The glide path angle ($\gamma = 20°$) in the landing phase is six times bigger than that of a commercial aircraft, and the touch down velocity ($v \approx 350\text{ km h}^{-1}$) is about twice as high.

Problems

Problem 10.1 LEO Deorbit
Prove Eq. (10.1.11).

Problem 10.2 Normalized Equations of Motion
From the equations of motion (10.2.1)–(10.2.3), derive the normalized equations of motion (10.2.11).

Problem 10.3 Reduced Equations of Motion
From the equations of motion (10.2.1)–(10.2.3), derive the reduced equations of motions (10.2.14) through the variable substitution $dt \to d\lambda$.

Problem 10.4 Low-Lift Reentry Trajectory
Prove the low-lift reentry trajectory, Eq. (10.5.3).

Hint: Derive the first-order differential equation of the trajectory equation from $\dot{h} = -v \sin \gamma$ and $\dot{x} = v \cos \gamma$.

Problem 10.5 High-Lift Reentry
We assume a high-lift reentry.

(a) Show that, for the reflection phase where gravitation is negligible and hence $H/R \to 0$, the exact solutions to the equations of motion

$$\frac{d\varepsilon}{d\lambda} = -\frac{\sin \gamma_e}{\sin \gamma}\varepsilon$$

$$\frac{d(\cos \gamma)}{d\lambda} = \frac{\sin \gamma_e}{2}\frac{L}{D}$$

read

$$\cos \gamma = \cos \gamma_e \left[1 + \frac{\tan \gamma_e}{2}\frac{L}{D}(\lambda - \lambda_e)\right]$$

and

$$\ln \frac{\varepsilon}{\varepsilon_e} = -2\frac{D}{L}\left[\arcsin\left(\cos \gamma_e + \frac{\sin \gamma_e}{2}\frac{L}{D}(\lambda - \lambda_e)\right) - \arcsin(\cos \gamma_e)\right]$$

(b) Using for $x \to 0$ the functional approximation

$$\arcsin(\cos \alpha + x \sin \alpha) - \arcsin(\cos \alpha)$$

$$= x + \frac{1}{2}\cot \alpha \cdot x^2 + \frac{1}{2}\left(\frac{1}{3} + \cot^2 \alpha\right)x^3 + \cdots$$

prove that the latter solution passes over into

$$\ln \frac{\varepsilon}{\varepsilon_e} = -(\lambda - \lambda_e) - \frac{1}{4}\frac{L}{D}\cot\gamma_e\,(\lambda - \lambda_e)^2$$

and hence is in accordance with Eq. (10.3.15) for $c \to 0$ and Eq. (10.5.2).

Problem 10.6 Deceleration in Blackout Phase
Prove Eq. (10.8.4) from Eq. (10.8.3) for $at \ll v_0$.

11
Three-Body Problem

11.1
Overview

Until now, we've looked at two point masses that were moving under their mutual gravitational influence. Formally speaking we were dealing with two bodies, each with six degrees of freedom (three position vector components and three velocity vector components). To describe their motion, in total 12 quantities had to be determined, specified by six coupled equations of motion of second order (see Eq. (7.1.14)) or equivalently 12 coupled equations of motion of first order:

$$\dot{v}_1 = +\frac{Gm_2}{|r_1 - r_2|^3}(r_1 - r_2), \quad v_1 = \dot{r}_1$$

$$\dot{v}_2 = -\frac{Gm_1}{|r_1 - r_2|^3}(r_1 - r_2), \quad v_2 = \dot{r}_2$$

By transforming the origin of the reference system into the center of mass of the two bodies (see Section 7.1.5), we were able to split the differential equations into two independent sets with three coupled equations of second degree each, namely $\ddot{r} = -\mu r/r^3$ and $\ddot{r}_{cm} = 0$. We succeeded in directly integrating them, thus finding unambiguous analytical solutions.

A world with just two bodies is too idealistic in most cases. The motion of the Moon, for example, which circles the Earth, and at the same time is subject to the influence of the Sun, cannot be described adequately by just a two-body system. For these three bodies and for the general case of n bodies, one has to go back to the $6n$ coupled differential equations of first order, analogous to the above, which describe the acceleration and velocity of each body under the gravitational forces of all the other bodies. The specific motion of the bodies is determined by the $6n$ quantities $(r_1, v_1), (r_2, v_2), \ldots, (r_n, v_n)$, which follow from integrating the differential equations. The motion and hence the $(r_1, v_1), (r_2, v_2), \ldots, (r_n, v_n)$ are however restricted due to the ear-

Astronautics. Ulrich Walter
Copyright © 2008 WILEY-VCH Verlag GmbH & Co. KGaA, Weinheim
ISBN: 978-3-527-40685-2

lier discussed conservation laws. Mathematically, the conservation laws are 10 constraint equations for these $6n$ quantities, namely:

- six (i.e. 2×3) equations for the conservation of momentum of the center of mass (corresponds to the non-accelerated motion of the center-of-mass vector or equivalently to the initial values v_0 and r_0 of the center of mass; see Eq. (7.1.18));
- three equations for the conservation of total angular momentum;
- one equation for total energy conservation.

Each constraint equation defines a conserved quantity – a so-called "integral of motion" (here "integral" means a quantity that is independent of the motion and is thus constant). So in essence there are $6n - 10$ degrees of freedom, which entails that $6n - 10$ quantities remain to be determined. Already in 1896 the Frenchman Poincaré showed that, for the general n-body problem (i.e., $n \geq 3$ bodies with arbitrary masses and arbitrary initial conditions), there cannot exist any further algebraic integrals of motion. So the general n-body problem is analytically not integrable, and thus cannot be solved analytically. Because in general the effective gravitational force on a body is no longer central, its path is non-periodic, and because energy conservation applies to the entirety of bodies, a single body might gain or loose energy, so that unbounded solutions may exist.

The n-body problem might not be solved analytically, but it is possible to solve it by other means. One possibility is to approximate the solution by convergent function series expansion. This mathematical method is rather complex, so we do not want to go into details here. We just mention that the expansion at the end of Section Section 7.4.3 is a function series expansion of the solution to the Keplerian equation (7.4.16). In addition, with today's computers it is quite simple to get point-by-point solutions with arbitrary accuracies by solving the differential equations numerically. Despite the superior numerical capabilities, which are exclusively applied for specific space missions, also the so-called patched conics!method, which we already got to know in Section 9.1, is regularly used to solve n-body problems by approximation. This method is used for interplanetary flights to gain preliminary insight into possible trajectories, which is indispensable to handle the complex calculation models of a detailed mission design.

There are two cases where the three-body problem has specific practical significance:

1. *Hill periodic system*. This is the quasi-Keplerian motion of two bodies (so-called *tight binaries*) within their sphere of influence and their simultaneous Keplerian motion around a central body far away from these bodies.

The motion of a moon around a planet which themselves move around a star is a good example. A Hill periodic system is one of very few stable solutions of the three-body problem. The stability of our planetary system rests on this fact. Here "stable" means that the orbits do not vary significantly under the influence of the small external perturbations due to other planets. Hill systems exist for arbitrary mass ratios of the three participating bodies. We will not go into details of Hill periodic systems.

2. *Restricted three-body problem.* The limiting case of a Hill periodic system is the so-called restricted three-body problem (R3BP). An example is the motion of a vanishingly small mass, usually a spacecraft, in the gravitational field of two celestial bodies (planet-Moon or Sun-planet), which however may not be spatially restricted to orbits around one or the other. The peculiarity is that the path of the extremely small mass is determined by the two primaries, but the inverse influence on the primaries can be neglected. This specific property will give us basic insights into the behavior of the R3BP.

11.2 Synchronous Orbits

Before we turn to the restricted three-body problem, let's have a look at some specific solutions of the general n-body problem. We already noticed that the general n-body problem might always exhibit unbounded non-periodic solutions. There are three special cases, however, where all bodies always display bounded periodic trajectories:

1. *Euler configuration* – three arbitrary masses moving synchronously in a linear configuration with fixed relative distances on Keplerian orbits (ellipses, parabolas, or hyperbolas).

2. *Lagrange configuration* – three arbitrary masses moving synchronously in an equilateral triangle configuration with fixed relative distances on Keplerian orbits (ellipses, parabolas, or hyperbolas).

3. *Isomass configurations* – three or more bodies with same masses moving synchronously on one closed, winding, axially symmetric orbit.

In all three cases all the masses move in a single plane. We shall now look at these three cases, starting with the last.

11.2.1
Isomass Configurations

In this special case all the masses are exactly the same. Only recently, in 2000, a solution was found where three bodies revolve around each other in the form of an 8 (see Figure 11.1). In contrast to many other similar, symmetrical, co-orbital orbits with more than three bodies (see Fig. 11.2 and Fig. 11.3), which are *not stable*, this 8-shaped orbit is *stable*. The so-called stability domain, that is the range of admissible deviations from the ideal path or mass, however is so small that one expects only one 8-shaped orbit in the universe or up to one per galaxy.

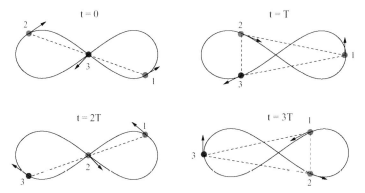

Figure 11.1 Stable co-orbital motion of three masses on an 8-shaped orbit at different time intervals.

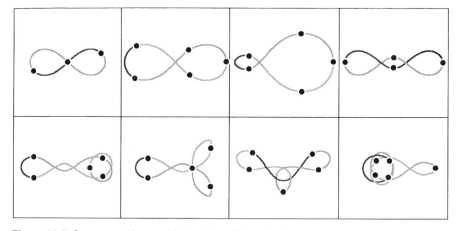

Figure 11.2 Some unstable co-orbital motions of three to five masses, showing the paths of the masses from their current position to their next position. See also color figure on page 465.

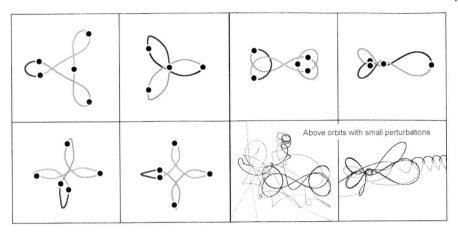

Figure 11.3 Further unstable co-orbital motions of four and five masses, showing the paths of the masses from their current position to their next position. See also color figure on page 465.

11.2.2
Euler Configuration

It can be shown that in the general three-body problem only two kinds of fixed configuration can exist: Euler configuration and Lagrange configuration. Already in 1765 the Swiss mathematician Leonard Euler showed that if three bodies with arbitrary masses take on certain positions on a straight line – called Eulerian points – their joint motion can be described as a rotation of this straight line where their mutual distances change such that the distance ratios, and hence the entire configuration, remain constant. Because the Euler configuration is of high relevance for the R3BP, we will take a closer look at its mathematical description.

Let m_1, m_2, m_3 be the three masses of arbitrary size, which may be positioned at r_i, $i = 1, 2, 3$, with regard to their common center of mass. Without loss of generality we label the mass located between the other two as m_2. It can be shown (see Guthmann (2000)) hat in the inertial reference system with origin at their common center of mass each rotating mass obeys one of the following interrelated Newton equations of motion

$$\ddot{r}_i + \frac{\mu_i}{r_i^3} r_i = 0 \quad \text{for} \quad i = 1, 2, 3 \qquad (11.2.1)$$

with

$$\frac{\mu_1}{G} = \frac{\beta^2 m_2}{(1-\beta)^2} + \frac{\beta^2 m_3}{(\alpha-\beta)^2}$$

$$\frac{\mu_2}{G} = -\frac{m_3}{(\alpha-1)^2} + \frac{m_1}{(1-\beta)^2} \qquad (11.2.2)$$

$$\frac{\mu_3}{G} = \frac{\alpha^2 m_2}{(\alpha-1)^2} + \frac{\alpha^2 m_1}{(\alpha-\beta)^2}$$

where $\alpha > 1$ and $\beta < 0$ describe the collinearity of the position vectors

$$r_3 = \alpha r_2, \quad \text{and} \quad r_1 = \beta r_2$$

Thus $\alpha > 1$ and $\beta < 0$ is just the implication that m_2 is the middle of the three masses. Therefore they obey the center-of-mass equation

$$\beta m_1 + m_2 + \alpha m_3 = 0$$

It can be shown (exercise, Problem 11.1, cf. Guthmann (2000)) that:

- α is the unequivocal *positive* root of $\mu_3 = \alpha^3 \mu_2$;
- β is the unequivocal *negative* root of $\mu_1 = -\beta^3 \mu_2$.

Coordinate System

Because the points are collinear and the distance ratios of the masses are fixed by the constants α and β, we choose the configuration line as our co-rotating coordinate x axis so that the masses are located at positions $r_i = (x_i, 0, 0)$ (see Fig. 11.4). Then, by the above definition, $x_3 = \alpha x_2$ and $x_1 = \beta x_2$. The Euler configuration can be characterized by the configuration parameter χ, which is defined in terms of the relative distances

$$x_{12} := x_2 - x_1$$

$$x_{13} := x_3 - x_1 = (1+\chi) x_{12} \qquad (11.2.3)$$

$$x_{23} := x_3 - x_2 = \chi x_{12}$$

$$\chi := \frac{x_{23}}{x_{12}} = \frac{x_3 - x_2}{x_2 - x_1} = \frac{\alpha - 1}{1 - \beta} > 0 \qquad (11.2.4)$$

With χ we relate the distances of the masses to the reference distance x_{12}, which may be chosen freely.

Rotation Dynamics

Because according to Eq. (11.2.1) each mass is subject to a central Newtonian force, the orbits must be conic sections. This follows from our considerations

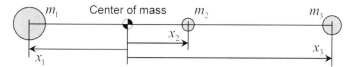

Figure 11.4 Collinear Euler configuration of three masses of arbitrary size.

in Sections 7.3 and 7.4. We now first show that the motion of the masses conserves their collinearity and relative configuration, if their relative distances initially obey the configuration parameter χ. To do so we prove that the angular velocity of each of the three orbits is the same at any time. From $\mu_3 = \alpha^3 \mu_2$ and because of $x_3 = \alpha x_2$ it follows directly that

$$\frac{\mu_2}{r_2^3} = \frac{\mu_2}{x_2^3} = \frac{\mu_3}{x_3^3} = \frac{\mu_3}{r_3^3}$$

According to Eq. (11.2.1) these relations are just the angular velocities of masses m_2 and m_3 due to $\ddot{r} = -\omega^2 r$. By the same token it can be proven that with $x_1 = \beta x_2$ and $\mu_1 = -\beta^3 \mu_2$ the following holds:

$$\frac{\mu_1}{r_1^3} = \frac{\mu_1}{-x_1^3} = \frac{\mu_2}{x_2^3} = \frac{\mu_2}{r_2^3}$$

which equals the angular velocity of masses m_1 and m_2. Therefore we have equal instantaneous angular velocities for all three orbits:

$$\frac{\mu_i}{r_i^3} = \omega^2(\theta) \quad \text{for} \quad i = 1, 2, 3 \tag{11.2.5}$$

> The Euler masses rotate on a straight line, with variable absolute distances, but constant relative distances between the masses – it's like a rotating rubber band.

As an example, Fig. 11.5 shows the dynamics of three collinear, rotating masses with $m_1 = 1/10$ Earth mass, $m_2 =$ Moon mass, and $m_3 = 1/2$ Moon mass. The relative distances on the rotating configuration line are obviously retained, and the individual masses move in ellipses with the same lines of apsides, and with the focus in the joint center of mass.

One may be surprised that the motion of the masses is indeed on a straight line, as this means a uniform orbital period for all masses, although their semi-major axes have different sizes. This seems to contradict Kepler's third law: $T = 2\pi \sqrt{a^3/\mu}$. But one has to consider that according to Eq. (11.2.2) every orbit has a different μ_i, because every mass moves in the gravitational field

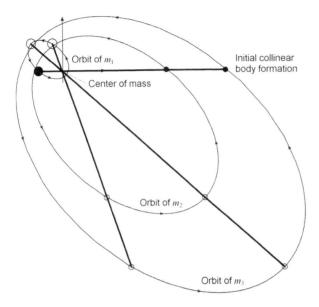

Figure 11.5 Dynamics of the collinear Euler configuration in the case of elliptical orbits, for $m_1 = 1/10$ Earth mass, $m_2 =$ Moon mass, and $m_3 = 1/2$ Moon mass.

of *two* other masses. Because the relative distances on the rotating line are fixed, the semi-major and the semi-minor axis of the three ellipses have the same ratio. From this and with Eq. (7.4.9) follows that the eccentricities are equal. Therefore and with Eq. (7.4.8a) we get at the periapsis $r_{i,per} = a_i(1-e)$. Inserting this into Eq. (11.2.5) yields

$$a_i = \frac{1}{1-e}\left(\frac{\mu_i}{\omega_{per}^2}\right)^{1/3} \propto \mu_i^{1/3} \tag{11.2.6}$$

Due to this dependence $T_i = 2\pi\sqrt{a_i^3/\mu_i} = const$ and thus independent of the orbit. The absolute values of a_i and hence the absolute value of the orbital period T are determined by the initial conditions.

Circular Orbits

If the initial conditions are such that the three bodies move on circular orbits, then even the absolute distances, and in particular x_{12}, are constant and the period for the circular orbit turns out (exercise, Problem 11.5) to be

$$\omega = \frac{2\pi}{T} = \sqrt{\frac{\mu}{x_{12}^3}\frac{m_1\chi^2 - m_3}{m_1 - m_3\chi}} = const \tag{11.2.7}$$

with $\mu = GM = G(m_1 + m_2 + m_3)$ The Euler configuration here is not only configuration-invariant, but also form-invariant. The fixed positions in the co-rotating reference system, the so-called *synodic system*, are called *stationary points*.

Determination of the Mass Positions

We now want to determine the positions of the masses on the line with only the masses given. Because the three masses are configuration-invariant on a straight line we do not need to solve the equation of motion, but it suffices to derive a conditional equation for χ determined only by the three masses. Equation (11.2.4) results in

$$\alpha - \beta = (1+\chi)(1-\beta) = \frac{1+\chi}{\chi}(\alpha - 1)$$

With this and Eq. (11.2.2) we can rewrite the conditional equation $\mu_3 = \alpha^3 \mu_2$ as

$$\alpha \left[-\frac{m_3}{(\alpha-1)^2} + \frac{m_1}{(1-\beta)^2} \right] = \frac{m_2}{(\alpha-1)^2} + \frac{m_1}{(\alpha-\beta)^2}$$

and

$$\alpha \left(\chi^2 m_1 - m_3 \right) = m_2 + m_1 \frac{\chi^2}{(1+\chi)^2}$$

From the center-of-mass equation $\beta m_1 + m_2 + \alpha m_3 = 0$ and the above follows that

$$1 - \beta = \frac{m_1 + m_2 + \alpha m_3}{m_1} = \frac{\alpha - 1}{\chi}$$

Hence

$$\alpha = \frac{m_1 + \chi(m_1 + m_2)}{m_1 - \chi m_3}$$

Inserting this into the above equation leads to

$$\frac{m_1 + \chi(m_1 + m_2)}{m_1 - \chi m_3} \left(\chi^2 m_1 - m_3 \right) = m_2 + m_1 \frac{\chi^2}{(1+\chi)^2} \tag{11.2.8}$$

After some trivial extensions and rearrangements, this equation can be transformed into **Lagrange's quintic equation**:

$$(m_1 + m_2)\chi^5 + (3m_1 + 2m_2)\chi^4 + (3m_1 + m_2)\chi^3 \\ - (m_2 + 3m_3)\chi^2 - (2m_2 + 3m_3)\chi - (m_2 + m_3) = 0 \tag{11.2.9}$$

For given masses m_1, m_2 and m_3, where m_2 is located in the middle, the only one positive root χ of Lagrange's quintic equation determines via the reference distance x_{12} all the other distances by means of Eqs. (11.2.3) and (11.2.4).

Note 1: *That there is only one positive root of Lagrange's quintic equation follows from Descartes' rule of signs, since the coefficients of the powers of χ change sign only once.*

Note 2: *Equation (11.2.9) holds for any type of orbit that obeys Eq. (11.2.1), be it bounded or unbounded.*

The synchronous motion of the Euler configuration only takes place if there are no perturbations. In the presence of even the tiniest perturbation of the configuration, the Euler configuration is always *unstable*, even in the R3BP limit. The masses then will run away from this configuration.

11.2.3
Lagrange Configuration

We are seeking other three-body configurations, with their bodies moving on Keplerian orbits around a common center of mass. If Keplerian, they have to obey the Newtonian equations of motion (11.2.1). It can be shown that, other than the Euler configuration, there can exist only one more such configuration, which was found in 1772 by the French mathematician Joseph Lagrange and named after him. The results will merely be summarized here without proof. A Lagrange configuration is a configuration of three bodies with arbitrary masses m_1, m_2, m_3, which show an equilateral triangular formation and obey one of the corresponding three Newton equations of motion:

$$\ddot{r}_i + \frac{\mu_i}{r_i^3} r_i = 0 \quad \text{for} \quad i = 1, 2, 3$$

with

$$\mu_1 = \frac{G}{M^2} \left(m_2^2 + m_3^2 + m_2 m_3 \right)^{3/2}$$

$$\mu_2 = \frac{G}{M^2} \left(m_1^2 + m_3^2 + m_1 m_3 \right)^{3/2} \quad (11.2.10)$$

$$\mu_3 = \frac{G}{M^2} \left(m_1^2 + m_2^2 + m_1 m_2 \right)^{3/2}$$

and

$$M = m_1 + m_2 + m_3$$

If the total energy of the system is negative, zero, or positive, this results in bounded ellipses (or circles) or unbounded parabolas or hyperbolas revolving around a common center of mass. The size of their semi-axis depends on the size of the individual mass. Figure 11.6 shows a bounded system with bodies of 1/10 Earth mass, Moon mass, and 1/2 Moon mass, from which we infer that the triangular Lagrange configuration, just like the Euler configuration, does not change its symmetry, but merely is stretched in space, i.e. its distances change, but their ratios remain constant. Obviously we can state this as follows:

> In the Lagrange configuration, the three bodies always form an equilateral triangle, which rotates and continuously changes its size – it is "breathing."

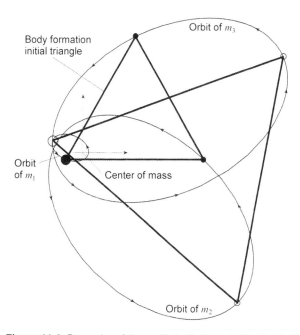

Figure 11.6 Dynamics of three elliptically bounded bodies in the coplanar Lagrange configuration.

Though the Lagrange configuration has a remarkable symmetry, it is generally *unstable*: it disintegrates after a certain period of time. It is only stable in the limit of a restricted three-body problem when one primary mass is significantly bigger than the other (see Eq. (11.4.11)).

Circular Orbits

If the initial conditions are such that the bodies move in circles, then, like the Euler configuration, the Lagrange configuration is not only configuration-invariant, but also form-invariant. It can be shown (exercise, Problem 11.4) that in this case the orbital period is given as

$$\omega = \frac{2\pi}{T} = \sqrt{\frac{\mu_i}{r_i^3}} = \sqrt{\frac{\mu}{r_{ij}^3}} \qquad (11.2.11)$$

with the distance between the bodies

$$r_{ij} := r_{12} = r_{23} = r_{13}$$

and

$$\mu = GM = G(m_1 + m_2 + m_3)$$

Initial Conditions

The initial conditions for each mass of viable Euler and Lagrange configurations are as follows:

1. The net resultant force on a mass is a radial vector through the system center of mass.

2. The velocity of a mass is proportional in magnitude to its distance to the center of mass.

3. The velocity vectors form equal angles with their corresponding radial position vectors

11.3
Restricted Three-Body Problem

The general Euler and Lagrange configurations are quite academic cases. From a practical point of view, only the special case of the *restricted three-body problem* (R3BP) plays a role. Here one of the three masses, m, is a spacecraft, and thus negligibly small compared with the other two, $m \ll m_1, m_2$, the so-called primary bodies. If not stated otherwise, we label the smaller of the two primaries m_2, that is $m_2 < m_1$. As an example the primaries might be Earth–Moon or Sun–Earth. Because in good approximation they move on circular orbits around a common center of mass, the R3BP is usually further restricted to circular orbits of the primaries m_1 and m_2 (*circular restricted three-body problem*, CR3BP), with angular velocity $\omega^2 = G(m_1 + m_2)/x_{12}^3$ according to Eq. (7.1.16). The following derivations however refer to the general R3BP, which includes the CR3BP.

Euler defined the R3BP in 1772, after he had already in 1765 discovered the collinear equilibrium points, the results of which also apply to the R3BP. Also in 1772 Lagrange discovered all five equilibrium points in the R3BP, including the two triangular points of the Lagrange configuration. Ever since then, the collinear equilibrium points of mass m in the R3BP have been called *Eulerian points*, L_1, L_2, L_3, and the triangular equilibrium points are called *Lagrangian points*, L_4 and L_5 (see Fig. 11.7 and 11.8). All five points together are the so-called *libration points*, but quite frequently and confusingly they are also simply called Lagrangian points. We now want to have a closer look at the characteristics of the libration points.

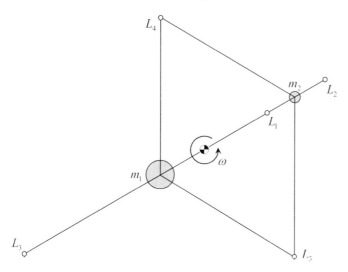

Figure 11.7 The Eulerian points L_1 to L_3 and Lagrangian points L_4 and L_5 in the restricted three-body problem.

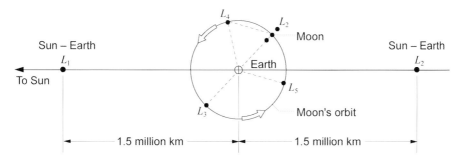

Figure 11.8 Libration points in the vicinity of the Earth.

Note: *In the literature the Eulerian points L_1, L_2, L_3 are not uniquely labeled. In this book L_1 is the point between the two primaries, L_2 the point beyond the minor primary, and L_3 is the point opposite to the minor primary. One often finds a reverse labeling of L_1 and L_2. The Lagrangian points are labeled consistently in the literature: L_4 is the leading and L_5 the trailing triangular point with respect to motion of the minor primary m_2.*

11.3.1
Eulerian Points

First of all, let's introduce more convenient distance variables: let $\Delta x(L_i)$ be the distance from the Eulerian point L_i to the center of the major primary body. Because x_{12} is the distance between the two primary bodies, the *relative distances* of the three Eulerian points to the center of the major primary body are according to Fig. 11.4 and Eq. (11.2.3)

$$\frac{\Delta x}{x_{12}}(L_3) = \frac{x_{23}}{x_{12}} = \chi =: \Delta_3 \tag{11.3.1a}$$

$$\frac{\Delta x}{x_{12}}(L_2) = \frac{x_{13}}{x_{12}} = 1 + \chi =: \Delta_2 \tag{11.3.1b}$$

$$\frac{\Delta x}{x_{12}}(L_1) = \frac{x_{12}}{x_{13}} = \frac{1}{1+\chi} =: \Delta_1 \tag{11.3.1c}$$

As our considerations in the last section are valid for general Euler and Lagrange configurations for any masses, we get the positions of the Eulerian points L_2 and L_3 from Eq. (11.2.8) for $m_3 = m \approx 0$

$$m_1 + \chi(m_1 + m_2) = \frac{m_2}{\chi^2} + \frac{m_1}{(1+\chi)^2} \tag{11.3.2}$$

If $m_2 > m_1$, then $m_3 = m$ is located at the L_3 point. In this case, we get

$$\mu_1 = \frac{m_1}{m_1 + m_2} = \frac{1}{1 + m_2/m_1} \quad \rightarrow \quad \frac{m_2}{m_1} = \frac{1-\mu_1}{\mu_1}$$

and from Eq. (11.3.2) with Eq. (11.3.1a) we get

$$\Delta_3 + \mu_1 = \frac{\mu_1}{(1+\Delta_3)^2} + \frac{1-\mu_1}{\Delta_3^2} \quad \text{L_3 point} \tag{11.3.3}$$

as the conditional equation for Δ_3. If $m_1 > m_2$, then $m_3 = m$ is located at the L_2 point. In this case, we get

$$\mu_2 = \frac{m_2}{m_1 + m_2} = \frac{1}{1 + m_1/m_2} \quad \rightarrow \quad \frac{m_1}{m_2} = \frac{1-\mu_2}{\mu_2}$$

and from Eq. (11.3.2) with Eq. (11.3.1b) we get

$$\Delta_2 - \mu_2 = \frac{\mu_2}{(1-\Delta_2)^2} + \frac{1-\mu_2}{\Delta_2^2} \quad L_2 \text{ point} \tag{11.3.4}$$

as the conditional equation for Δ_2. Finally, for $m_2 \to 0$ and $m_1 > m_3$, $m_2 = m$ is located at the L_1 point near m_3. Because of Eq. (11.2.8), we get

$$\chi^2 m_1 - m_3 = \frac{\chi^2}{(1+\chi)^3}(m_1 - \gamma m_3)$$

and

$$\mu_3 = \frac{m_3}{m_1+m_3} = \frac{1}{1+m_1/m_3} \quad \to \quad \frac{m_1}{m_3} = \frac{1-\mu_3}{\mu_3}$$

With this and Eq. (11.3.1c) we get

$$\Delta_1 - \mu_3 = -\frac{\mu_3}{(1-\Delta_1)^2} + \frac{1-\mu_3}{\Delta_1^2} \quad L_1 \text{ point} \tag{11.3.5}$$

as the conditional equation for Δ_1.

As usually the relevant $\mu_i \ll 1$, the solutions of Eqs. (11.3.3)–(11.3.5) can be determined (exercise, Problem 11.2) by series approximation:

$$\frac{\Delta x}{x_{12}}(L_1) = \Delta_1 = 1 - \lambda + \frac{1}{3}\lambda^2 + \frac{1}{9}\lambda^3 - \frac{58}{81}\lambda^4 + \cdots$$

$$\frac{\Delta x}{x_{12}}(L_2) = \Delta_2 = 1 + \lambda + \frac{1}{3}\lambda^2 - \frac{1}{9}\lambda^3 + \frac{50}{81}\lambda^4 + \cdots \quad \text{for } \lambda = \left(\frac{\mu}{3}\right)^{1/3} \tag{11.3.6}$$

$$\frac{\Delta x}{x_{12}}(L_3) = \Delta_3 = 1 - \frac{7}{12}\mu - \frac{1127}{20736}\mu^3 - \cdots$$

Note: *In Eq. (11.3.6) or each of the three relative distances of the Eulerian points μ is the ratio of the minor primary mass to the total mass of the primaries. This is why we have dropped its index.*

For the common Sun-planet or planet-Moon systems, Eq. (11.3.6) provides very good approximations. Just for the Earth–Moon system $\lambda = 0.159\,401$ is relatively big, leading to a correspondingly worse convergence. In Eq. (11.3.6), the Eulerian distances are given with regard to the distance between the two primary bodies. If both primary orbits are circular, then the Eulerian points also have fixed distances relative to the center of mass. If their orbits are elliptical, their mutual distance changes, and with it also the absolute distance to the Eulerian points.

> One has to bear in mind that from a physical point of view the positions of the Eulerian points result from the sum of the gravitational forces from both primaries, plus the centrifugal force of its rotation around the center of mass. This is why, for example, the L_1 point is not located where the gravitational forces of m_1 and m_2 just cancel each other out, but a little bit further in the direction of the larger of the two masses, to balance the centrifugal force by a somewhat larger gravitational force.

11.3.2
Lagrangian Points

The *two Lagrange points* are determined by the equilaterality of the triangle m_1, m_2, m. I.e.

$$\frac{\Delta x}{R}(L_4) = 0.5, \quad \frac{\Delta y}{R}(L_4) = +\frac{\sqrt{3}}{2}$$
$$\frac{\Delta x}{R}(L_5) = 0.5, \quad \frac{\Delta y}{R}(L_5) = -\frac{\sqrt{3}}{2}$$
(11.3.7)

where Δy measures the distance perpendicular to the configuration line of the two primaries.

Table 11.1 shows specific examples: the exact position of all libration points in the Earth–Moon system for $m_{earth}/m_{moon} = 81.3007$.

Table 11.1 Libration points in the Earth–Moon system. Here Δx is the distance of the libration point to the Earth in units of the distance x_{12} between the Earth and the Moon; Δy is the distance of the libration points perpendicular to the connecting line between the Earth and the Moon. C is the Jacobi constant (see Section 11.4.2) for $v = 0$.

Libration points	$\Delta x/x_{12}$	$\Delta y/x_{12}$	C
L_1	0.849 068	0	−1.6735
L_2	1.167 830	0	−1.6649
L_3	0.992 912	0	−1.5810
L_4	0.500 000	0.866 025	−1.5600
L_5	0.500 000	−0.866 025	−1.5600

11.4
Circular Restricted Three-Body Problem

11.4.1
Energy Conservation in the CR3BP

Already in the introduction of Section 11.3 we saw that for the general n-body problem (bodies with arbitrary mass and with arbitrary initial conditions) there exist no algebraic integrals of motion other than the classical conservation laws. In this case we would not have any information on the motion of the test body. The only possibility would be to solve the Newton equations numerically. In the following we consider the special case of a *circular restricted three-body problem* (CR3BP), with a zero mass (test mass) moving in between the primaries, which move on circles around their barycenter. In this special case there exists an additional integral of motion: the energy conservation of the test mass m in the reference system co-rotating with the primaries, the so-called *synodic system*. It reads

$$\frac{1}{2}mv^2 - \frac{1}{2}m\omega^2\left(x^2 + y^2\right) - \frac{Gmm_1}{\Delta r_1} - \frac{Gmm_2}{\Delta r_2} = E_{tot} \quad \text{energy conservation} \quad (11.4.1)$$

with

$$\Delta r_i = |\mathbf{r} - \mathbf{r}_i| = \sqrt{(x - x_i)^2 + y^2 + z^2}$$

In this synodic system the test body is at position $\mathbf{r} = (x, y, z)$ and has distances Δr_i to the two primary bodies located at positions $\mathbf{r}_i = (x_i, 0, 0)$, $i = 1, 2$, and v is the relative velocity of the test body. The synodic reference system has its origin in the center of mass of the primaries; according to Eq. (7.1.16) it rotates with angular velocity $\omega^2 = G(m_1 + m_2)/r_{12}^3$ along the z axis, $\boldsymbol{\omega} = (0, 0, \omega)$, where $r_{12} = x_2 - x_1$ is the distance between the primaries; its x axis is the connecting line between the two primaries, and the x-y plane is the configuration plane of the three bodies.

Conservation of energy holds for the test mass, because in the synodic system the primaries maintain fixed positions and the test mass just moves in their conservative gravitational potentials and the conservative rotational potential of the rotating synodic system. Note that in the inertial reference system the primaries move and exchange energy with the test mass via the gravitational interaction. Therefore, the energy of the test mass is not conserved in the inertial reference system where it constantly gains or loses energy. On this energy transfer property hinges the flyby maneuver.

The left side of Eq. (11.4.1) comprises four energy contributions for the test mass: its kinetic energy, its rotational energy, and the two gravitational energies with regard to the primaries m_1 and m_2.

Rotational Energy and Rotational Potential

It may be confusing that in the energy conservation Eq. (11.4.1) the rotational energy and with it the rotational potential $U_{rot} = E_{rot}/m = -\frac{1}{2}\omega^2(x^2+y^2) = -\frac{1}{2}\omega^2 r^2$ is negative, exactly opposite to what we would expect (see Eqs. (7.2.14) and (7.2.12)). The reason is the following. The centrifugal force of course always points outwards, i.e. $F_{rot} = m\omega^2 r$. According to Eq. (7.1.5) the corresponding rotational potential is determined by

$$U_{rot} = -\frac{1}{m}\int F_{rot} \cdot dr = -\int \omega^2 r \cdot dr$$

In an inertial (sidereal) reference system $h = \omega \cdot r^2$ (see Eq. (7.2.5)) is constant and therefore

$$U_{rot} = -\int \omega^2 r \cdot dr = -h^2 \int \frac{dr}{r^3} = \frac{1}{2}\frac{h^2}{r^2} = \frac{1}{2}\omega^2 r^2$$

However, in a co-rotating synodic reference system $\omega = const$ and therefore

$$U_{rot} = -\omega^2 \int r \cdot dr = -\frac{1}{2}\omega^2 r^2$$

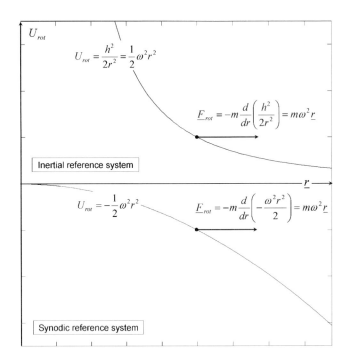

Figure 11.9 Centrifugal potential and centrifugal force in an inertial and a synodic reference system.

So, while in the inertial system the rotational energy $E_{rot} = mU_{rot} = 1/2mh^2/r^2$ is positive, it is negative in the synodic system: $E_{rot} = -1/2m\omega^2r^2$ (see Fig. 11.9). In both cases the energy decreases with increasing distance from the origin, in a way that the gradient always forms the same centrifugal force. Because the rotational potential (energy) is negative in the synodic system, one speaks of a potential field – comparable to a gravitational field, which is negative as well – which creates a fictitious force (centrifugal force). Here "fictitious" does not mean that the force is not real, but a real force that fictitiously acts from the outside.

11.4.2
Jacobi's Integral

Since the primaries circle each other at a constant distance r_{12}, we introduce for convenience the normalized coordinates

$$\xi := \frac{x}{r_{12}}, \quad \eta := \frac{y}{r_{12}}, \quad \zeta := \frac{z}{r_{12}}, \quad \rho = (\xi, \eta, \zeta). \tag{11.4.2}$$

With the dimensionless variables

$$v = \sqrt{\dot{\xi}^2 + \dot{\eta}^2} = \frac{v}{r_{12}\omega},$$

$$\mu_i := \frac{m_i}{m_1 + m_2}$$

and the relative distance

$$\Delta\rho_i = |\rho - \rho_i| = \sqrt{(\xi - \xi_i)^2 + \eta^2 + \zeta^2}$$

we can write the energy conservation Eq. (11.4.1) in the dimensionless form

$$\frac{1}{2}v^2 + U = \frac{E_{tot}}{mr_{12}^2\omega^2} =: C < 0 \quad \text{Jacobi's integral} \tag{11.4.3}$$

with

$$U := -\frac{1}{2}\left(\xi^2 + \eta^2\right) - \frac{\mu_1}{\Delta\rho_1} - \frac{\mu_2}{\Delta\rho_2} < 0. \quad \text{effective potential} \tag{11.4.4}$$

Equation (11.4.3) is called *Jacobi's integral of motion* with the non-dimensional constant *negative* energy C, which is called the Jacobi constant, and the *negative* effective potential energy U depicted in Fig. 11.10. For bounded orbits, which we will consider in the following, the total energy E_{tot} and with it also C are negative. Their absolute values are determined from the initial conditions.

Note 1: *In line with physical conventions, see Eq. (7.1.2), the effective potential (a.k.a. pseudo-potential due to the fictitious centrifugal force) U is defined negatively here. This is contrary to most relevant literature which defines it positively, $U > 0$, in which case $C := \frac{1}{2}v^2 - U$.*

Note 2: *In literature, the Jacobi constant often is alternatively defined as $C := U - \frac{1}{2}v^2 > 0$ or $C := 2U - v^2 > 0$, with $U > 0$.*

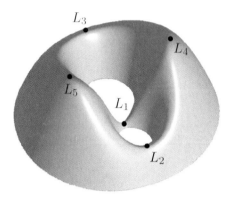

Figure 11.10 Effective potential U in the CR3BP.

11.4.3
Orbits in the Synodic System

How does a test mass move in the effective potential of the circular rotating primaries? The equation of motion is derived from Newton's second law of motion, Eq. (7.1.12), $m\ddot{\rho} = \sum_j F_j$ with all forces F_j acting on the test mass. For visual clarity we are seeking to describe the motion in the co-rotating synodic system. In the above section we have already derived the effective potential describing all static influences on the test mass in the synodic system, which according to Eq. (7.1.5) gives rise to an effective static force via $F_U = -m \cdot \omega^2 dU/d\rho$ including the centrifugal force $F_{rot} = m\omega^2\rho$. Since our test mass is moving in the non-inertial synodic system, we also have to take into account the Coriolis force $F_{Cor} = -2m(\omega \times \dot{\rho}) = -2m\omega(-\dot{\eta}, \dot{\xi}, 0)$ in the configuration plane, which results from and is perpendicular to any motion $\dot{\rho} = (\dot{\xi}, \dot{\eta}, \dot{\zeta})$ of the test mass. Its motion is therefore given by the equations of motion

$$\ddot{\xi} - 2\omega\dot{\eta} = -\omega^2 \frac{\partial U}{\partial \xi}, \quad \ddot{\eta} + 2\omega\dot{\xi} = -\omega^2 \frac{\partial U}{\partial \eta}, \quad \ddot{\zeta} = -\omega^2 \frac{\partial U}{\partial \zeta}$$

11.4 Circular Restricted Three-Body Problem

Defining the dimensionless time $\tau := \omega t$ and the differential operator $()' := d/d\tau$ yields the dimensionless forms

$$\xi'' - 2\eta'' = -\frac{\partial U}{\partial \xi}, \quad \eta'' + 2\xi' = -\frac{\partial U}{\partial \eta}, \quad \zeta'' = -\frac{\partial U}{\partial \zeta} \tag{11.4.5}$$

Finally, taking the partial derivatives of the effective potential yields

$$\xi'' - 2\eta' - \kappa^2 \xi = -\gamma_1^2 \tilde{\xi}_1 - \gamma_2^2 \tilde{\xi}_2$$
$$\eta'' + 2\xi' - \kappa^2 \eta = 0 \tag{11.4.6}$$
$$\zeta'' \quad\quad - \kappa^2 \zeta = 0$$

with

$$\kappa^2 = 1 + \gamma_1^2 + \gamma_2^2 > 0$$

and

$$\gamma_i^2 = \frac{\mu_i}{\left[(\xi - \tilde{\xi}_i)^2 + \eta^2 + \zeta^2\right]^{3/2}} > 0$$

In celestial mechanics the investigation of these equations of motion is a research field of its own.

Note: *The position of the libration points (see Eqs. (11.3.6) and (11.3.7)) can of course also be and usually is derived from the static equilibrium condition $\rho' = \rho'' = 0$ and thus according to Eq. (11.4.5) from $\partial U/\partial \xi = \partial U/\partial \eta = \partial U/\partial \zeta = 0$. We adopted here the derivation from the motion in the inertial system, because then it becomes clear that the libration points are just special cases of the general Euler and Lagrange configurations and therefore also bear their characteristics.*

The differential equations (11.4.6) are quite complex and can generally only be solved numerically. But because we only want to explore the general motion of the test body, we don't have to do that. We can use the Jacobi integral to determine the space that is accessible to the test mass. It ends where its velocity becomes zero. If we put $v = 0$ and with it $\nu = 0$ in Eq. (11.4.3), we get a curve which envelopes the space in which the body can possibly move. The test body cannot cross the envelope curve, it can only selectively touch it with velocity $v = 0$ at selected points. For a given C of the test body, the envelope curve is the line, the coordinates of which satisfy the equation $U(\xi, \eta) = C$ and is apparently the contour line of the effective potential in

Fig. 11.10. The envelope curve is also called *zero-velocity curve* or *Hill curve* after the astronomer Hill (1878), who studied it in detail.

Let's examine in detail the Hill curve and how it depends on the energy of the test mass. For a given negative C (total energy) and because kinetic energy is positive, the body can move only in those spatial areas where $U \leq C$. According to Eq. (11.4.4), this is the case whenever $\Delta\rho_1$ or $\Delta\rho_2$ is very small, i.e. when m is close to one of the bodies m_1 or m_2 (large negative gravitational energy), or when m is far a way from both (large negative rotational energy). These areas are indicated in white in Fig. 11.11 for the Earth m_1 and the Moon m_2. The inaccessible area $U > C$ in between is indicated in grey. If the energy of the test body and hence C is now gradually increased, the test body is able to access more and more space (Fig. 11.11). In Fig. 11.11b a trajectory from the Earth to the Moon via L_1 is possible for the first time. It shows that the flight to Moon via L_1 is energetically most favorable. If the energy of the test body is further increased, L_2 and L_3 also become accessible (Fig. 11.11c,d). The Lagrange points L_4 and L_5 are potentially the highest points in the Earth–Moon system, and are achieved in the end (Fig. 11.11e,f).

The Hill curves merely define the limits of motion of the test body. They don't tell us anything about how the test body moves within the permissible areas. We know that the most favorable path from the Earth to the Moon is via L_1, but we do not know what this path looks like. To find out, we have to solve the equations of motion with the initial conditions defining the energy of the test body. Figure 11.12 shows an example of an Earth–Moon L_1 transit with a ballistic capture at the Moon derived from solving the equation of motion.

11.4.4
Stability and Dynamics at Eulerian Points

From the general Euler configuration, we already know that quite generally the Eulerian points L_1, L_2, L_3 are *not stable*. This means that also in the CR3BP we have instability in the direction of the configuration line (blue arrows in Fig. 11.13). But perpendicular to this line there is a restoring force (in the motion plane as well as vertical to it, red arrows), which pushes the mass m back to the L_1 point. This is obvious, as the resultant from the gravitational forces and the centrifugal force points in the direction of L_1 (Fig. 11.14). So the potentials at the three points are saddle points. If you take a spacecraft to Eulerian points L_1 or L_2, and leave it alone without any orbit corrections, small initial deviations Δx_0 along the configuration line would exponentially increase according to

$$\Delta x = \Delta x_0 \cdot \exp\left(t/\tau\right)$$

11.4 Circular Restricted Three-Body Problem

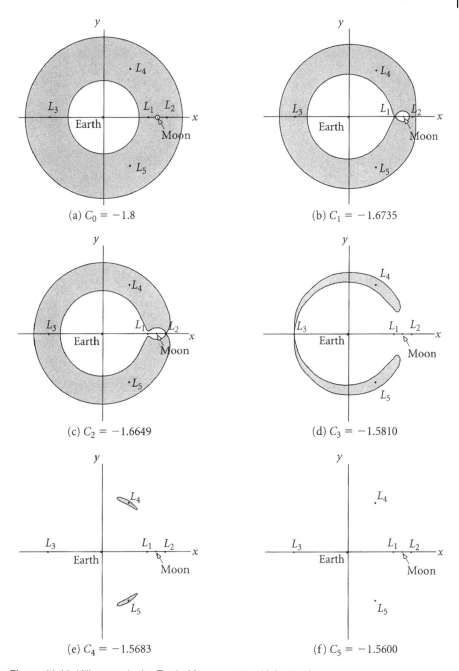

Figure 11.11 Hill curves in the Earth–Moon system with increasing Jacobi constant C.

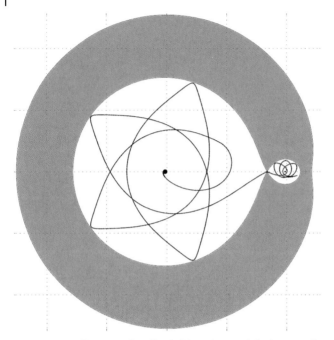

Figure 11.12 Example of an Earth–Moon L_1 transit in the synodic reference system.

with

$$\tau = \frac{T}{2\pi\sqrt{1+2\sqrt{7}}} = 0.063\,45 \cdot T \quad @ \ L_1 \text{ or } L_2 \tag{11.4.7}$$

where T is the orbiting time of the primary bodies. So, in the Earth–Moon system, the error would e-fold within $\tau \approx 1.8$ days. The Eulerian point L_3 turns out to be more stable. The following holds at this point:

$$\tau = \frac{T}{\pi}\sqrt{\frac{2m_1}{3m_2}} \quad @ \ L_3 \tag{11.4.8}$$

For the Earth–Moon system, this results in $\tau \approx 66$ days. So you only need little orbit corrections along the configuration line, which considerably decreases propellant consumption compared to any other fixed position in the Earth–Moon system. To remain at the point L_1 or L_2 only

$$\boxed{\Delta v \approx 10\ \mathrm{m\,s^{-1}} \quad \text{per year}}$$

is needed. For L_3 it's even less than that.

11.4 Circular Restricted Three-Body Problem

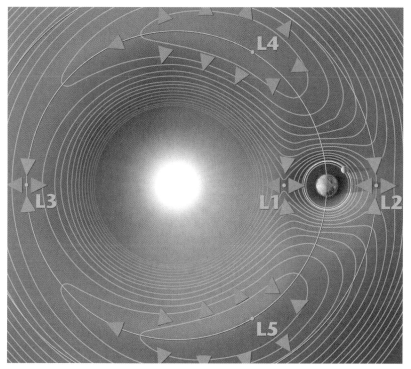

Figure 11.13 Equipotential lines of the effective potential U and the stabilizing (red) and destabilizing forces (blue) at the libration points resulting from it. See also color figure on page 466.

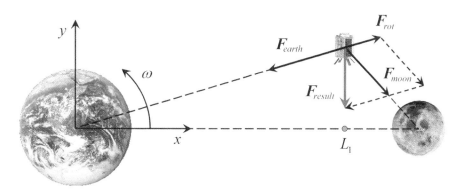

Figure 11.14 The restoring force (resultant arrow F_{result}) of a body near L_1 resulting from the gravitational forces (thick arrows F_{earth} and F_{moon}) and the centrifugal force (thick arrow F_{rot}).

Halo Orbits Around Eulerian Points

In many cases it is quite unfavorable to place a satellite exactly at an Eulerian point. Here are two examples:

- The first two probes to explore the Sun were ISEE-3 (NASA) and SOHO (ESA), both of which are still near the L_1 point in the Sun–Earth system. At or near the L_1 point the distance to Earth is constant, which is quite important for a continuous S-band communication. Being directly at the L_1 point though, the probes would be located in the line of view as seen from Earth. Therefore, solar radiation would jam the S-band communication.

- For a mission to the surface on the far side of the Moon, which is still unexplored, or if interference radio telescopes were to be placed there, which would be undisturbed by Earth's scattered radio signals, a relay satellite would be needed for communication with Earth. This would conveniently be placed at the L_2 point, so the antenna could have a fixed pointing. Unfortunately, L_2 lies in the radiation shadow of the Moon.

In both cases the satellite resides in an exclusion zone that renders impossible any communication with Earth. Solutions to these problems are periodic orbits, so-called halo orbits around the Eulerian points. They exist because the restoring forces act on any deviation from the Eulerian point perpendicular to the configuration line. Therefore one would expect circular, or more generally elliptical, orbits around an Eulerian point in the plane perpendicular to the configuration line. Only small correction maneuvers would be needed to keep the S/C in the perpendicular plane. If the orbital radii are chosen big enough, the S/Cs would be seen to orbit the Sun or the Moon, respectively, as viewed from the Earth, and this would be a smart solution to the problems.

However, no such perfect periodic motion in the perpendicular plane exists. This is because our reference system is the co-rotating synodic system, in which any motion $\dot{\rho}$ in the configuration plane immediately causes Coriolis forces $\mathbf{F}_{Cor} = -2m(\boldsymbol{\omega} \times \dot{\boldsymbol{\rho}})$ vertical to it in the configuration plane. To visualize the results of this, let's study the halo orbit of the SOHO satellite around the L_1 point in the Sun–Earth system (see Fig. 11.15). The projection of its periodic motion in the plane perpendicular to the configuration line (y-z plane) at L_1 has a component in the y direction that is perpendicular to the configuration line. Let's assume this sideway motion is initially in the direction of the revolving Earth – see upward arrow of SOHO orbit in Fig. 11.15a. Then this slightly increased orbit speed yields a slightly increased centrifugal force and hence a deflection in x direction, outward towards the Earth (This is just the Coriolis effect.) According to Eq. (7.4.3), $v = \sqrt{\mu/r}$, the increasing distance r to the Sun implies a reduction of the orbital speed until the satellite

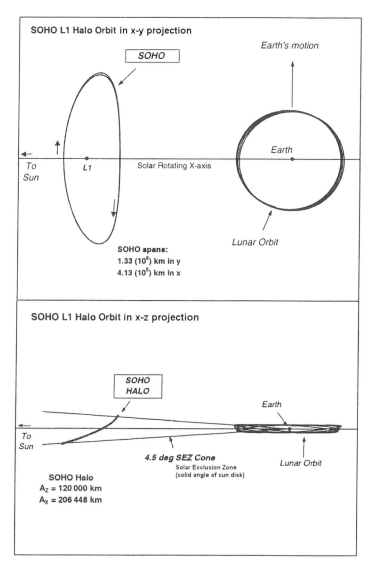

Figure 11.15 The halo orbit of SOHO around the L_1 point in the Sun–Earth system with amplitudes 206 500 km, 665 000 km, and 120 000 km in x, y, and z direction, respectively, and its projections onto the ecliptic x-y plane (above) and onto the x-z plane (below).

reverses its motion in y direction. Beyond the L_1 distance its centrifugal force is smaller than needed to balance the Sun's gravitational force. The satellite therefore begins to move towards the Sun, which in turn again increases the orbital speed. This brings the satellite back to its initial point.

In summary the body has described a nearly elliptical orbit in the *x-y* plane. In total we therefore have a three-dimensional halo orbit. The orbit in the *y-z* is an ellipse and via the Coriolis forces its *y* component induces in the *x-y* plane an orbit with an almost elliptical shape. For small deflection amplitudes the motions in the two perpendicular planes are not synchronous and so-called *Lissajous orbits* result. For larger amplitudes ($A_x \geq 200\,000$ km for SOHO) the increasing non-linearity of the effective potential leads to a synchronization of the two elliptical orbits, which results in a single elliptical orbit forming a certain angle with the configuration plane (see Fig. 11.15b), whereby the line of nodes is the *y* axis. This orbit is called *halo orbit*. If the probe orbits counterclockwise as seen from Earth, as with SOHO, these orbits are dubbed *class I orbits*. Clockwise orbits are called *class II orbits*, the only difference being that their slope in the *x-z* plane is flipped, i.e. due to the reversed Coriolis forces their upper (in *z* direction) turning point lies closer to the Sun, while that for class I orbits lies closer to the Earth.

Halo orbits around L_2 and L_3 points are very similar to those described here around the L_1 point. It can be shown that all these halo orbits are *dynamically stable*, so no propulsion effort is needed to keep a satellite in a halo orbit.

11.4.5
Stability and Dynamics at Lagrangian Points

We know from Section 11.2.3 that Lagrange configurations are statically unstable. The static instability in the CR3BP follows directly from Fig. 11.10: Since the Lagrange points are energetically maximal, any tiny deviation from a Lagrangian point lowers the potential energy of the S/C and therefore increases its kinetic energy – the potential gradient carries the S/C away from it (blue arrows in Fig. 11.13). So they are statically unstable. But surprisingly L_4 and L_5 are dynamically stable, though. Dynamic stability is achieved because, with increasing distance to a point, its velocity increases. Because of the rotation of the total system, the velocity causes a Coriolis force forcing the S/C back on an orbit around the Lagrange points. A mathematical analysis (see e.g. Murray and Dermott (1999) or Roy (2005)) shows that if

$$(\tilde{\xi}_0, \eta_0) = \left(0.5 - \mu_2, \pm\sqrt{3}/2\right)$$

are the coordinates of the two Lagrangian points in the synodic *barycentric* system, which unlike Eq. (11.3.7) are displaced by the amount $\mu_2 = m_2/(m_1 + m_2)$ on the configuration line, so the origin of the system is no longer at the center of m_1 but in the system's center of mass, then the equations

$$\tilde{\xi}(t) = \tilde{\xi}_0 + \tilde{\xi}_1 \sin(w_1 t + \varphi) + \tilde{\xi}_2 \sin(w_2 t + \varphi)$$
$$\eta(t) = \eta_0 + \eta_1 \cos(w_1 t + \phi) + \eta_2 \cos(w_2 t + \phi)$$
(11.4.9)

describe two modes of motion around a Lagrangian point, not too far away from it, where

$$\omega_{1,2} = \frac{\omega}{\sqrt{2}} \sqrt{1 \pm \sqrt{1 - 27(1-\mu_2)\mu_2}} \qquad (11.4.10)$$

are the angular frequencies of the two modes with $\omega = \sqrt{G(m_1 + m_2)/x_{12}^3} = 2\pi/T$ the angular frequency of the primaries (T = orbital period), with (ξ_i, η_i) their amplitudes and (φ, ϕ) their phases. These quantities are determined from the initial conditions and they are coupled via the angular frequencies and the curvature of the potential. Equation (11.4.9) describes periodic motions around the Lagrangian point if and only if both frequencies are real. This is the case if $1 - 27(1-\mu_2)\mu_2 \geq 0$, implying $\mu_2 \leq \left(27 - 3\sqrt{69}\right)/54 = 0.0385$, in turn implying

$$m_1 \geq \frac{25 + 3\sqrt{69}}{2} m_2 = 24.96\, m_2 \qquad (11.4.11)$$

This condition corresponds to a minimal curvature of the effective potential at the Lagrangian points, which is necessary to cause enough acceleration and hence speed and Coriolis force to curve the test mass on a bounded periodic orbit. In the Earth–Moon system, because $m_{earth} = 81.3007\, m_{moon}$, condition (11.4.11) is fulfilled. As all the Sun-planet and planet-Moon constellations in our solar system fulfill Eq. (11.4.11), all Lagrange points in our solar system are dynamically stable. Actually, more than 450 asteroids, so-called Trojans, have been found at the two Lagrange points of the Sun–Jupiter system, the first and most famous being Achilles, which moves in bounded orbits around these points. Also, in the Sun–Mars system there exists a Trojan named Eureka.

What do these bounded orbits look like? Because $\mu_2 \ll 1$, we can approximate Eq. (11.4.10) to

$$\omega_1 = \omega \sqrt{1 - \frac{27}{4}\mu_2} \approx \omega$$

$$\omega_2 = \omega \sqrt{\frac{27}{4}\mu_2}$$

According to Eq. (11.4.9) the test mass can move in two coupled modes around the Lagrangian point: one short-term mode with period $2\pi/\omega_1 \approx T$ and semi-major axis (ξ_1, η_1) and a long-term mode with period $2\pi/\omega_2 = T\sqrt{4/(27\mu_2)}$ and semi-major axis (ξ_2, η_2). The total movement can be considered as a short-term elliptical epicycle with semi-axis ratio $\xi_1/\eta_1 = \sqrt{3\mu_2}$, moving on a long-term ellipse with semi-axis ratio $\xi_2/\eta_2 = 1/2$ around the Lagrangian point (see Fig. 11.16). The composite motion is displayed in Fig. 11.17. The trajectory may (but doesn't have to) touch the Hill curve at some points with zero velocity, but cannot cross it. The smaller the excluded Hill zone around the Lagrangian point, the closer is the trajectory around the Lagrangian point.

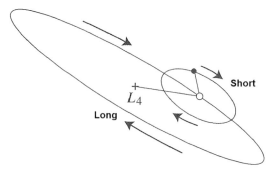

Figure 11.16 The epicyclic orbit of a body (full dot) around L_4, which may be considered as motion on a short-term ellipse, the center of which (open dot) in turn moves on a long-term ellipse around L_4.

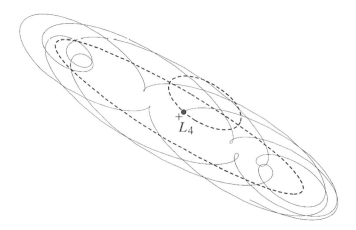

Figure 11.17 The epicyclic motion (full line) of a body with Jacobi constant $C = -1.563$ around the Lagrangian point L_4 in the Earth–Moon system, as a composite of the two basic modes (dashed lines) over 13 orbital periods.

In summary, bounded periodic orbits affected by Coriolis forces exist both around Eulerian and Lagrangian points. As in physics periodic movements around an equilibrium point is denoted as "libration," the term "libration points" just stems from the existence of such periodic orbits around these points.

Tadpole and Horseshoe Orbits

The epicyclic motions described above occur only if they do not deviate too far from the Lagrangian points. What does an orbit look like if its excursions become bigger? In this case approximate analytical solutions cannot be provided any more. But numerical solutions show that orbits with decreasing orbital energies and increasing initial tangential velocities become more and more elongated. Figure 11.18(a,b) depicts so-called *tadpole orbits*, which occur if a test mass with relatively small Jacobi constant starts at the Hill curve, i.e. quite far away from the Lagrangian point, with zero velocity. While the short-term mode of motion is preserved, the semi-major axis of the long-term ellipse stretches to a circular arc along the circular orbit of m_2. With further decreasing Jacobi constant and/or increasing initial tangential velocity, the tadpole orbits elongate and arc until both meet at the L_3 point and then merge. So-called *horseshoe orbits* have formed (Fig. 11.18c,d). For suitable initial tan-

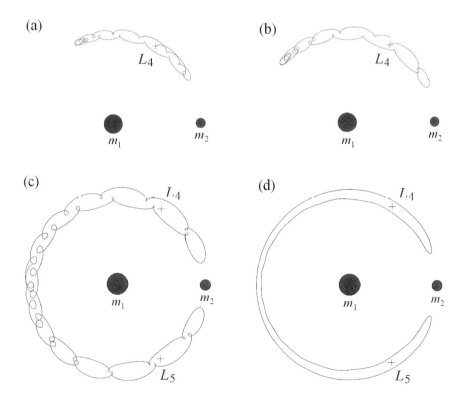

Figure 11.18 Librations in the Sun–Jupiter system with $m_2 = 0.001 m_1$ over 13 orbital periods. (a,b) Tadpole orbits with initial condition $v_0 = 0$, and with smaller Jacobi constant (b). (c,d) Horseshoe orbits with different initial tangential velocities.

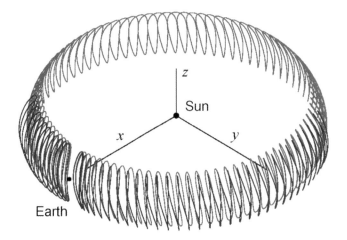

Figure 11.19 Three-dimensional representation of the horseshoe orbit of asteroid 2002 AA29 in the synodic Sun–Earth system. The looping is caused by a finite value of the z component of the initial velocity. The blue trajectory is the horseshoe orbit after reversal at the end points of the horseshoe. See also color figure on page 466.

gential velocities the amplitude of the short-term mode can be suppressed, forming a nearly smooth horseshoe orbit. If the initial condition is such that the tangential velocity has a component in z direction that is out of the configuration plane, then looping horseshoe orbits occur, as depicted in Fig. 11.19.

All Jupiter and Mars Trojans known so far move on tadpole orbits. But only recently the asteroids 3753 Cruithne and 2002 AA29 were found as two examples of horseshoe orbits in the Sun–Earth system. However, 3753 Cruithne moves on a horseshoe orbit with a high eccentricity, $e = 0.515$, and high inclination, $i = 19.81°$, to the ecliptic, which is why it is sometimes not counted as a horseshoe object. In addition, in 1980, Voyager 1 found the two equally massive asteroids Janus and Epimetheus with identical orbital radii around Saturn, which move on horseshoe orbits relative to each other. Trojans on tadpole orbits, asteroids on horseshoe orbits, and so-called quasi-satellites are called *co-orbital objects*, because they move in the same or nearly the same orbit of a celestial body around a central body.

With all these complex co-orbital objects it should be remembered that in the inertial barycentric system of the primaries the test mass always moves on elliptical orbits around the barycenter, which deviate just slightly from the minor primary. Only the transition into the synodic system results in the complex relative motions just discussed. In this way the horseshoe orbit from Fig. 11.19 corresponds to the full elliptical orbits in the inertial heliocentric system in Fig. 11.20.

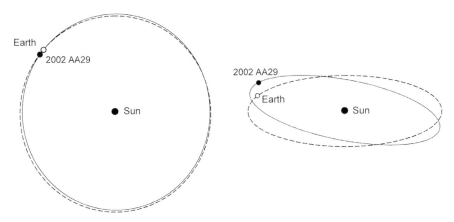

Figure 11.20 The orbit of asteroid 2002 AA29 (red) in the heliocentric system in the ecliptic (left) and viewed oblique (right).

Problems

Problem 11.1 Euler Configuration
Let Eqs. (11.2.1) and (11.2.2) be given. Prove that α is the unequivocal *positive* root of $\mu_3 = \alpha^3 \mu_2$.

Hint: cf. (Guthmann, 2000, pp. 242ff.).

Problem 11.2 Eulerian Points (Tedious but Good Practice)

(a) For the L_3 point the condition equation for Δ_3 is Eq. (11.3.3). By making the ansatz $\Delta_3 = 1 + a\mu + c\mu^3$ and assuming that the term of order $b\mu^2 = 0$, prove that

$$a = -\frac{7}{12} \qquad c = -\frac{12\,103}{13\,824}$$

(b) For L_1 and L_2 points make the ansatz $\Delta_i = 1 + a_i\lambda + b_i\lambda^2 + c_i\lambda^3$ with $\lambda = (\mu/3)^{1/3}$ and show that by inserting into the according condition equations (11.3.4) and (11.3.5) one gets $a_1 = -a_2 = -1$, $b_1 = b_2 = 1/3$, and $c_1 = -c_2 = 1/9$.

Problem 11.3 Jacobi Constant Approximations

By applying the above series expansion (Eq. (11.3.6)), show that for the Jacobi constant the following series expansions up to order $O(\mu)$ at the libration points hold:

$$C_{L1} = -\frac{3}{2} - \frac{3^{4/3}}{2}\mu^{2/3} + \frac{5}{3}\mu$$

$$C_{L2} = -\frac{3}{2} - \frac{3^{4/3}}{2}\mu^{2/3} + \frac{7}{3}\mu$$

$$C_{L3} = -\frac{3}{2} - \frac{\mu}{2}$$

$$C_{L4} = C_{L5} = -\frac{3}{2} + \frac{\mu}{2}$$

Hint: At the libration points $U = C$ due to $v = 0$.

Problem 11.4 Circularly Rotating Lagrange Configuration (Hard)

Show that for a circularly rotating Lagrange configuration $\Delta r = r_{12} = r_{23} = r_{13} = const$, Eq. (11.2.11)

$$\omega = \frac{2\pi}{T} = \sqrt{\frac{\mu_i}{r_i^3}} = \sqrt{\frac{\mu}{(\Delta r)^3}}$$

holds.

Hint: First show $\omega = \sqrt{\frac{\mu_i}{r_i^3}}$ and then $\frac{\mu_i}{r_i^3} = \frac{\mu}{(\Delta r)^3}$ by using the center-of-mass equation $m_1 r_1 + m_2 r_2 + m_3 r_3 = 0$.

Problem 11.5 Circularly Rotating Euler Configuration

Show that for a circularly rotating Euler configuration the orbital period is

$$T = 2\pi\sqrt{\frac{x_{12}^3}{\mu}\frac{m_1 - m_3\chi}{m_1\chi^2 - m_3}}$$

with

$$\mu = GM = G(m_1 + m_2 + m_3)$$

Hint: Starting from the center-of-mass equation $m_1 x_1 + m_2 x_2 + m_3 x_3 = 0$ show that

$$x_2 = x_{12}\frac{m_1 - m_3\chi}{M}$$

and finally because of $\omega^2 = \mu_i/x_i^3$ the wanted result.

12
Orbit Perturbations

12.1
Problem Setting

12.1.1
General Considerations

So far we have considered the motion of a body under a central Newtonian force, as given by a gravitational field of a point mass, which in turn led to Keplerian orbits as the solutions of the respective equations of motion. However, in reality there are many external forces acting on the body, which are neither point symmetric nor Newtonian. For instance, a massive central body usually is not quite homogeneous and isotropic, which in general gives rise to non-central and non-Newtonian forces. In addition, the gravitational forces of other celestial bodies, in particular neighboring planets, or interactions with the space environment will perturb the Keplerian orbit around the central body. In total there exist the following major disturbing forces:

- gravitational forces resulting from the non-spherical geometry and mass distribution of the central body;
- gravitational forces of other celestial bodies (like the Sun, Moon, planets);
- acceleration force resulting from the solar radiation pressure;
- acceleration force resulting from the drag of the remaining atmosphere.

Figure 12.1 provides a graphical representation of all the relevant perturbations acting on an Earth-orbiting spacecraft as a function of altitude. Obviously the Earth's anisotropy generates various perturbation terms $J_{n,m}$ of different strength, and atmospheric drag decreases rapidly with increasing altitude. In addition, Table 12.1 gives an overview of the essential external perturbations giving rise to accelerations of the S/C in a low Earth orbit (LEO) at 500 km altitude and in geostationary Earth orbit (GEO) for comparison.

Astronautics. Ulrich Walter
Copyright © 2008 WILEY-VCH Verlag GmbH & Co. KGaA, Weinheim
ISBN: 978-3-527-40685-2

12 Orbit Perturbations

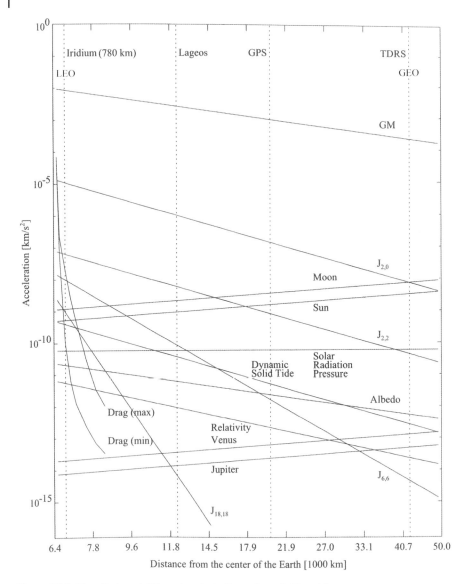

Figure 12.1 Magnitudes of different perturbations of a satellite orbit: GM = regular gravitational force of the Earth; $J_{n,m}$ = gravitational multipoles; Relativity = relativistic deviations; and the satellites Iridium, Lageos, GPS, TDRS at altitudes between LEO and GEO.

Judged from their magnitudes, the first four disturbing forces (except drag for GEO) have to be taken into account for real missions. The perturbations cause the orbits to be no longer Keplerian, but as long as the perturbations are small, which we will assume in the following, the orbit can be linearly

Table 12.1 External perturbational accelerations on a S/C with a given $A_\perp / m$, where $A_\perp$ is the effective surface area perpendicular to the impinging force and m the mass of the S/C.

Source of perturbation	Acceleration [m s^{-2}] in 500 km	Acceleration [m s^{-2}] in GEO
Drag (mean)	6×10^{-5} $A_\perp/m$	1.8×10^{-13} $A_\perp/m$
Solar pressure	4.67×10^{-5} $A_\perp/m$	4.67×10^{-5} $A_\perp/m$
Sun (mean)	5.6×10^{-7}	3.5×10^{-6}
Moon (mean)	1.2×10^{-6}	7.3×10^{-6}
Jupiter (maximum)	8.5×10^{-12}	5.2×10^{-11}

approximated by a sequence of Keplerian orbits, implying that the Keplerian elements $\vartheta = (a, e, i, \Omega, \omega, M - n(t - t_0))$ are changing with time according to

$$\vartheta(t) \approx \vartheta_0 + \dot{\vartheta} \cdot (t - t_0) \tag{12.1.1}$$

This reduces the problem to determining at a given point in time t_0 the so-called *osculating elements* $\vartheta_0 = (a_0, e_0, i_0, \Omega_0, \omega_0, M_0)$ by observation, and finding the time derivatives $\dot{\vartheta} = (\dot{a}, \dot{e}, \dot{i}, \dot{\Omega}, \dot{\omega}, \dot{M} - n)$ from theory. Finding the time derivatives for all major external perturbations in LEO and GEO as given in Table 12.1 is the objective of this chapter.

12.1.2
Gaussian Variational Equations

Any perturbing force $F_{perturb}$, or alternatively the acceleration $a_{perturb} = F_{perturb}/m$, acting on an orbiting body, can be decomposed into a radial component, a_r, a cross-radial component, a_θ, and into a component perpendicular to the orbital plane, i.e. along the angular momentum h, a_h, which perturbes the equation of motion according to

$$\ddot{r} + \frac{\mu}{r^3} r = a_{perturb} = (a_r, a_\theta, a_h)$$

Most generally $\dot{\vartheta}$ is derived by starting out from the functional dependency $r = r[\vartheta(t), t]$. Differentiation leads to $\dot{r} = v + \Phi$, with $v = \frac{\partial r}{\partial t}$ the velocity of the perturbed orbit and $\Phi = \sum_i \frac{\partial r}{\partial \vartheta_i} \dot{\vartheta}_i$. It can be shown (see e.g. Gurfil (2007)) that from differentiating again and substituting the result into the above perturbed equation of motion one can derive $\dot{\vartheta}$. Because this way of derivation is tedious and mathematically quite demanding (see e.g. Vallado (2001), or

Schaub and Junkins (2003)),) we will not attempt it here but defer to an exercise (see Problem 12.1) for an ab initio approach. From any of these approaches one finds the so-called **Gaussian variational equations (GVEs)**

$$\dot{a} = \frac{2a^2}{h} \left[e \sin\theta \cdot a_r + (1 + e \cos\theta) a_\theta \right]$$

$$\dot{e} = \frac{h}{\mu} \left[\sin\theta \cdot a_r + \left(\frac{e + \cos\theta}{1 + e \cos\theta} + \cos\theta \right) a_\theta \right]$$

$$\dot{i} = \frac{r \cos(\theta + \omega)}{h} a_h$$

$$\dot{\omega} = \frac{h}{e\mu} \left[-\cos\theta \cdot a_r + \frac{2 + e \cos\theta}{1 + e \cos\theta} \sin\theta \cdot a_\theta \right] - \frac{r \sin(\theta + \omega) \cos i}{h \sin i} \cdot a_h \qquad (12.1.2)$$

$$\dot{\Omega} = \frac{r \sin(\theta + \omega)}{h \sin i} a_h$$

$$\dot{M} = n - \frac{h\sqrt{1 - e^2}}{\mu} \left[\left(\frac{2}{1 + e \cos\theta} - \frac{\cos\theta}{e} \right) a_r + \frac{\sin\theta}{e} \left(\frac{2 + e \cos\theta}{1 + e \cos\theta} \right) a_\theta \right]$$

$$= n - \sqrt{1 - e^2} \left(\frac{2r}{h} a_r + \dot{\omega} + \dot{\Omega} \cos i \right)$$

Note: The elements $\dot{\vartheta}_i$ of the Gaussian variational equations become infinite or zero for $h \to 0$, i.e. for rectilinear orbits or orbits close to those. In addition $\dot{\omega}$ and $\dot{M}$ become infinite for $e \to 0$ and if a_r or a_θ accelerations occur, and $\dot{\omega}$ and $\dot{\Omega}$ become infinite for $i \to 0$ if an a_h acceleration occurs.

Remark: The vector function $\boldsymbol{\Phi}$ is called gauge function, because it constitutes three additional degrees of freedom, which can be chosen freely. As Lagrange already did, the so-called Lagrange constraints or osculation constraints, $\boldsymbol{\Phi} = 0$, are usually chosen, implying that the velocity vector of the perturbed orbit equals the one of the generating Keplerian orbit. This assumption however is fully arbitrary. Removing these constraints leads to the so-called gauge-generalized equations. Their evaluation is a very new and ongoing research and out of the scope of this book.

In the following sections we want to determine the perturbation accelerations listed above and calculate the corresponding variations of the Keplerian elements according to the Gaussian variational equations.

12.2 Gravitational Perturbations

12.2.1 Geoid

The gravitational field of the Earth is not absolutely isotropic, but it has slightly different values in different directions. The reason is the Earth's non-spherical shape and its density variations within. The geometric body representing the corresponding asymmetrical gravitational field in the different directions is called the *geoid* (see Fig. 12.2). Strictly speaking, the geoid is the body representing the equipotential surface of the gravitational potential at sea level, i.e. the surface to which the gravity vector g is always perpendicular. Graphically, the geoid is the equilibrium shape of the Earth, if its surface were fully covered with stationary water. In zero-order approximation the Earth's mass is distributed evenly and the geoid is a sphere. The largest deviation from the sphere is caused by the rotation of the Earth, which displaces the masses to the equator due to centrifugal forces. The Earth's radius around the equator is 6378.14 km and hence is some 21.4 km longer than through the poles. In first-order approximation the geoid therefore has the form of an oblate spheroid (see Fig. 12.3a). It never deviates more than 25 m from this spheroid.

Figure 12.2 Geoid of the Earth. See also color figure on page 467.

The Earth's interior is viscous with a high proportion of iron, magnesium, nickel, silicon, and oxygen. Temperature gradients between the hot interior of about 4500°C and the surface of the Earth cause convection flows on the

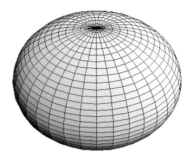 

Figure 12.3 (left) The Earth has a bulge of 21.4 km around the equator, deforming the geoid to a spheroid (excessively represented). (right) There are only small deviations from this spheroid (+20 m to −25 m), and they make up a shape like a potato (= geoid − spheroid).

one hand, and slightly asymmetrical density distributions on the other, which are reflected in the potato shape of the geoid, if you deduct the spheroid (see Figs. 12.3, right, and 12.9). Note that Fig. 12.2 depicts the real geoid, whereas Fig. 12.3 only shows the anisotropies obtained if you successively deduct the sphere and the spheroid. The deviations from the spheroid caused by the Earth's mass inhomogeneities are called *geoid undulations*.

As mountains mean a large accumulation of mass, the fine-structured geoid undulations clearly reflect the mountain ranges, which mainly account for the more complicated structure beyond the potato shape of the geoid. Because the finest details of the geoid reflect the mass proportions below the Earth's surface, one is able to determine from these structures, for instance, oil fields.

12.2.2
Gravitational Potential

The orbit of a spacecraft circling the Earth is influenced by these gravitational anisotropies, leading to slightly deformed Keplerian orbits. To be able to calculate the corresponding gravitational potential, we recall Eq. (7.1.3), according to which the potential of a point mass M at the origin (location $\mathbf{0}$) is given by

$$U(r) = -\frac{GM}{r} = -G\frac{M}{|\mathbf{r}-\mathbf{0}|}$$

Therefore, the potential of many masses M_i ($i = 1, \ldots, n$) at positions $\mathbf{r}_i$ is

$$U(\mathbf{r}) = -G\sum_i^n \frac{M_i}{|\mathbf{r}-\mathbf{r}_i|}$$

where the potential now is no longer isotropic in general. If we assume a celestial body with a continuous mass distribution described by the density distribution function $\rho(r)$, then we have to carry out the transition $M_i \to \rho(r) \cdot dV = \rho(r) \cdot d^3r$, whereby the sum becomes a volume integral

$$U(r) = -G \iiint_V \frac{\rho(r')}{|r - r'|} d^3r' \qquad (12.2.1)$$

This is the most general form to calculate the gravitational potential from a given mass distribution. Let's assume we know Earth's density distribution and we want to evaluate the above integral. To do so it can be shown that the expression $1/|r - r'|$ can be approximated by a series of spherical orthogonal functions, the so-called Legendre polynomials $P_n(x)$ of degree n,

$$\frac{1}{|r - r'|} = \frac{1}{r} \sum_{n=0}^{\infty} \left(\frac{r'}{r}\right)^n P_n(\cos\gamma)$$

where $\gamma = \angle(r, r')$. The familiar geographical latitude β and geographical longitude λ, i.e.

$$r = r(\cos\beta\cos\lambda, \cos\beta\sin\lambda, \sin\beta)$$

can be introduced by the *spherical harmonic addition theorem*

$$P_n(\cos\gamma) = P_n(\sin\beta) P_n(\sin\beta')$$
$$+ 2 \sum_{m=1}^{n} \frac{(n-m)!}{(n+m)!} P_n^m(\sin\beta) P_n^m(\sin\beta') \cos[m(\lambda - \lambda')]$$

where $P_n^m(x)$ are the so-called *(unnormalized) associated Legendre polynomials* of degree n and order m. If the volume integral with the dashed coordinates is now carried out over these functions weighted with the density $\rho(r')$, we obtain (see e.g. Kaplan (1976, p. 273)) expressions of the form $C_n^m P_n^m(\sin\beta)\cos(m\lambda)$ and $S_n^m P_n^m(\sin\beta)\sin(m\lambda)$, where the coefficients C_n^m and S_n^m represent the performed integrals. By this procedure Eq. (12.2.1) can be written in the standard form, adopted by the International Astronomical Union (IAU) in 1961, as

$$U(r, \beta, \lambda) = -\frac{\mu}{r}\left\{1 + \sum_{n=1}^{\infty}\sum_{m=0}^{n} \left(\frac{R_\oplus}{r}\right)^n \cdot P_n^m(\sin\beta)\left[C_n^m\cos(m\lambda) + S_n^m\sin(m\lambda)\right]\right\}$$

(12.2.2)

In this representation the origin of the reference frame is at the center of the Earth's mass and the z axis is the Earth's axis of rotation. Because the density distribution of the Earth is not known, the coefficients C_n^m and S_n^m are available only by measuring them with particular satellites. The most famous and most precise of them was the GRACE satellite mission. The values thereby obtained are given in Table 12.2.

Table 12.2 Multipole coefficients of the Earth's gravitational potential.

C_n^m	$m=0$	1	2	3	4
$n=0$	$+1.000\,000$				
1	0.00	0.00			
2	$-1.082\,627 \times 10^{-3}$	0.00	1.5745×10^{-6}		
3	2.5326×10^{-6}	-2.1926×10^{-6}	3.0899×10^{-7}	-1.0055×10^{-7}	
4	1.6196×10^{-6}	5.0880×10^{-7}	7.8418×10^{-8}	-5.9210×10^{-8}	-3.9841×10^{-9}

S_n^m	$m=0$	1	2	3	4
$n=0$	0.00				
1	0.00	0.00			
2	0.00	0.00	-9.0380×10^{-7}		
3	0.00	-2.6843×10^{-7}	-2.1144×10^{-7}	-1.9722×10^{-7}	
4	0.00	4.4914×10^{-7}	1.4818×10^{-7}	1.2008×10^{-8}	6.5257×10^{-9}

Note: The multipole coefficients $\overline{C}_{nm}, \overline{C}_{nm}$ as used in geodesy (see e.g. Earth Gravitational Model 1996 (EGM96), http://cddis.gsfc.nasa.gov/926/egm96/getit.html) are related to the more common coefficients C_n^m, S_n^m adopted also here through

$$\left\{ \begin{array}{c} C_n^m \\ S_n^m \end{array} \right\} = (-1)^m \sqrt{\frac{(2-\delta_{0m})(2n+1)(n-m)!}{(n+m)!}} \left\{ \begin{array}{c} \overline{C}_{nm} \\ \overline{S}_{nm} \end{array} \right\}$$

$$\text{with } \delta_{0m} = \left\{ \begin{array}{ll} 1 & @\ m=0 \\ 0 & @\ m>0 \end{array} \right.$$

Table 12.2 is incomplete. In fact, coefficients are known today of order up to $n = 360$. Some of them vanish, for the following reasons:

- because $\sin(m\lambda) = \sin(0) = 0$, the coefficients S_n^0 are undetermined and can be set to zero, i.e. $S_n^0 = 0$;

- if the center of mass is chosen to be at the origin of the reference frame, $C_1^0 = C_1^1 = S_1^1 = 0$;

- if the Earth's principal moment of inertia is chosen to be along the z axis, then $C_2^1 = S_2^1 = 0$.

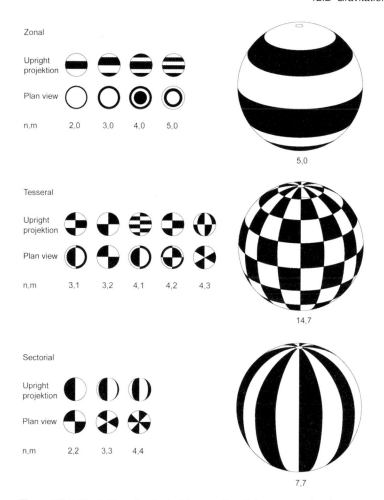

Figure 12.4 Illustration of spherical harmonics of degree n and order m: zonal harmonics (above), tesseral harmonics (middle), sectorial harmonics (bottom). A three-dimensional view of all spherical harmonics of degree 6 is provided in the color table on page 470.

Physically speaking, the successive terms of the sums in Eq. (12.2.2) correspond to a stepwise spherical approximation (multipole approximation) of the anisotropic gravitational potential, i.e. of the geoid. The terms $P_n^m(\sin\beta)\cos(m\lambda)$ and $P_n^m(\sin\beta)\sin(m\lambda)$ are so-called *spherical harmonics* (see Figure 12.4). They are the spherical distribution functions (multipoles) of order (n, m), and are called *zonal harmonics* for $(n, 0)$ because they describe just latitudinal variations, *sectorial harmonics* for (n, n) describing only longitudinal variations, and *tesseral harmonics* for $(n, m < n)$ describing mixed variations. The coefficients C_n^m and S_n^m determine the strength of these multipoles. The framed coefficient in Tab. 12.2 causes the predominant geoid sphere, the coefficient marked gray the oblate spheroid, and the bold coefficients the general shape of the **potato**. (See color figures of the potato on pages 468 and 469.)

Note: *In the earlier literature one may find the statement that the Earth is pear-shaped (symmetry axis = polar axis, pear stalk at south pole) beyond the spheroid. This discovery dates back to a publication of J. A. O'Keefe, A. Eckles, and R. Squires from 1959. They derived the pear shape from the analysis of long period perturbations of orbital eccentricities of the first US satellite Vanguard I. Because such secular perturbations stem only from zonal multipoles (see below), from this method only the C_n^0 coefficients (in particular the pear shaped C_3^0) are derivable and not the C_2^2 and C_3^1 coefficients, which extend the pear to a potato.*

Because the anisotropic terms of the gravitational potential are small, it is convenient to separate them from the spherical potential $-\mu/r$ and lump them into the perturbational term P

$$U(r, \beta, \lambda) = -\frac{\mu}{r} - P(r, \beta, \lambda) \tag{12.2.3}$$

with

$$P(r, \beta, \lambda) = \frac{\mu_\oplus}{R_\oplus} \sum_{n=1}^{\infty} \sum_{m=0}^{n} \left(\frac{R_\oplus}{r}\right)^{n+1} P_n^m(\sin \beta) \left[C_n^m \cos(m\lambda) + S_n^m \sin(m\lambda)\right]$$

12.3
Numerical Perturbation Calculation

12.3.1
Cowell's Method by Recurrence Iteration

According to Eqs. (7.1.5) and (7.1.12) we find with Eq. (12.2.3) the following equation of motion for a gravitationally perturbed orbit:

$$\ddot{r} = -\frac{\mu_\oplus}{r^3}\frac{dr}{dr} - \frac{dP}{dr} \tag{12.3.1}$$

For all practical cases perturbed orbits are calculated numerically by choosing an Earth-fixed Cartesian reference frame, in which the trajectory vector is $r = (x, y, z)$. With this it can be shown (see Montenbruck and Gill (2000)) that if the terms $(V_{n,m}(r), W_{n,m}(r))$; $n = 0, \ldots, n_{\max} + 1$; $m = 0, \ldots, n$ can be calculated by the the following iteration:

Recurrence Iteration

$$\alpha_x := \frac{xR_\oplus}{r^2}, \quad \alpha_y := \frac{yR_\oplus}{r^2}, \quad \alpha_z := \frac{zR_\oplus}{r^2}, \quad \alpha_R := \frac{R_\oplus^2}{r^2}$$

$$V_{0,0} = \frac{R_\oplus}{r}, \quad W_{0,0} = 0$$

Do $m = 0, n_{\max}$

  Do $n = m, n_{\max}$

  $$V_{n+1,m} = \alpha_z \frac{2n+1}{n-m+1} V_{n,m} - \alpha_R \frac{n+m}{n-m+1} V_{n-1,m}$$

  $$W_{n+1,m} = \alpha_z \frac{2n+1}{n-m+1} W_{n,m} - \alpha_R \frac{n+m}{n-m+1} W_{n-1,m} \quad @\; V_{m-1,m} = W_{m-1,m} = 0$$

  End Do

  $$V_{m+1,m+1} = (2m+1)\left(\alpha_x V_{m,m} - \alpha_y W_{m,m}\right)$$

  $$W_{m+1,m+1} = (2m+1)\left(\alpha_x W_{m,m} - \alpha_y V_{m,m}\right)$$

End Do

then Eq. (12.3.1) can be rewritten as follows:

$$\ddot{x} = -\frac{\mu_\oplus}{R_\oplus^2} \sum_{n=0}^{n_{\max}} \left[C_{n0} V_{n+1,1} + \frac{1}{2} \sum_{m=1}^{n} \left\{ C_{nm} V_{n+1,m+1} + S_{nm} W_{n+1,m+1} \right.\right.$$
$$\left.\left. - \frac{(n-m+2)!}{(n-m)!} \left(C_{nm} V_{n+1,m-1} + S_{nm} W_{n+1,m-1} \right) \right\} \right]$$

$$\ddot{y} = -\frac{\mu_\oplus}{R_\oplus^2} \sum_{n=0}^{n_{\max}} \left[C_{n0} W_{n+1,1} + \frac{1}{2} \sum_{m=1}^{n} \left\{ C_{nm} W_{n+1,m+1} - S_{nm} V_{n+1,m+1} \right.\right.$$
$$\left.\left. + \frac{(n-m+2)!}{(n-m)!} \left(C_{nm} W_{n+1,m-1} - S_{nm} V_{n+1,m-1} \right) \right\} \right]$$

$$\ddot{z} = -\frac{\mu_\oplus}{R_\oplus^2} \sum_{n=0}^{n_{\max}} \sum_{m=1}^{n} (n-m+1)\left(C_{nm} V_{n+1,m} + S_{nm} W_{n+1,m}\right) \tag{12.3.2}$$

with

$$\left\{ \begin{array}{c} C_{nm} \\ S_{nm} \end{array} \right\} = \sqrt{\frac{(2-\delta_{0m})(2n+1)(n-m)!}{(n+m)!}} \left\{ \begin{array}{c} \overline{C}_{nm} \\ \overline{S}_{nm} \end{array} \right\}$$

$$= (-1)^m \left\{ \begin{array}{c} C_n^m \\ S_n^m \end{array} \right\} \quad \text{with} \quad \delta_{0m} = \left\{ \begin{array}{ll} 1 & @\ m=0 \\ 0 & @\ m>0 \end{array} \right.$$

and $\overline{C}_{nm}, \overline{S}_{nm}$ as given at
ftp://cddis.gsfc.nasa.gov/pub/egm96/general_info/egm96_to360.ascii
ftp://cddis.gsfc.nasa.gov/pub/egm96/general_info/readme.egm96
and

$$\mu_\oplus = 3.986\,004\,415 \times 10^5 \text{ km}^3\text{s}^{-2}$$

$$R_\oplus = 6378.1363 \text{ km}$$

Solving Newton's differential equation using the direct method as given above is called **Cowell's method**. If one takes into account only the biggest perturbation, the spheroid, $n_{\max} = 2$ and $m = 0$, the explicit equations read:

$$\ddot{x} = -\frac{x\mu_\oplus}{r^3} \left[1 + C_2^0 \frac{3}{2} \frac{R_\oplus^2}{r^2} \left(5\frac{z^2}{r^2} - 1 \right) \right]$$

$$\ddot{y} = -\frac{y\mu_\oplus}{r^3} \left[1 + C_2^0 \frac{3}{2} \frac{R_\oplus^2}{r^2} \left(5\frac{z^2}{r^2} - 1 \right) \right] = \frac{y}{x} \ddot{x} \qquad (12.3.3)$$

$$\ddot{z} = -\frac{z\mu_\oplus}{r^3} \left[1 + C_2^0 \frac{3}{2} \frac{R_\oplus^2}{r^2} \left(5\frac{z^2}{r^2} - 3 \right) \right]$$

So, for calculating a trajectory, first the maximum order $n_{\max}$ of the perturbation is chosen depending on the accuracy of the initial data and of the orbit needed. Then the coupled differential equations (12.3.2) or (12.3.3) are solved, whereby one of the currently best solvers is the Runge–Kutta–Nyström algorithm RKN12(10)17M (see Brankin et al. (1989)), which can be found in the NAG Library under the name of D02LAF. The initial step size should be $h = 0.1$ for all Keplerian problems. Note that also other perturbations like neighboring planets can easily be taken into account by Eq. (12.3.2) by just adding the perturbational terms in the Cartesian coordinate form. Observe that the solution is given in an Earth-fixed reference frame. In order to find the result in an inertial (sideral) system an appropriate coordinate transformation has to be applied to the solution $r = (x, y, z)$.

12.3.2
Encke's method

If the perturbations are very small, as in our case of gravitational perturbation, they can be separated and calculated as such. The undisturbed orbit $r_{osc}(t)$ then is called *osculating orbit*, and the residual $\delta(t) := r - r_{osc}$. Figure 12.5 illustrates the defined values. It can be shown that for this residual the following equation of motion holds (see e.g. Schaub and Junkins (2003))

$$\ddot{\delta} = -\frac{\mu}{r_{osc}^3}\left(3q\frac{1+q+q^2/3}{1+(1+q)^{3/2}}r + \delta\right) - \frac{\partial P}{\partial r} \approx \frac{\mu}{r_{osc}^3}\left(3\frac{\delta r}{r^2}r - \delta\right) - \frac{\partial P}{\partial r} \quad (12.3.4)$$

with

$$q = \frac{\delta \cdot \delta - 2\delta r}{r^2}$$

This differential equation is also solved with a high-quality Runge–Kutta algorithm. This approach of separating the perturbation from the osculating orbit is called **Encke's method**. It is a very accurate method, as the numerical integration only treats the perturbation in question, and does not have to "drag along" the full Keplerian orbit. However, due to the increased numerical precision of today's CPUs it is more convenient to apply the above recurrence type of Cowell's method.

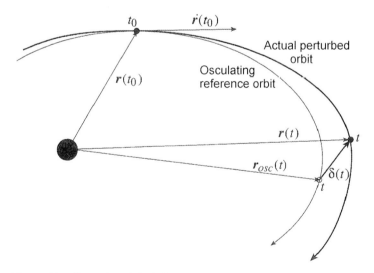

Figure 12.5 Illustration of Encke's method.

12.4
Analytical Perturbation Calculation

12.4.1
Lagrange's Planetary Equations

In order to understand the key effects of the gravitational anisotropy on an orbit, we shall now look for analytical perturbation solutions. If one derives the perturbational acceleration from the gravitative perturbation P according to $a_{perturb} = F_{perturb}/m = -dP/dr$ and decomposes it according to Section 12.1.2 into the components $a_{perturb} = (a_r, a_\theta, a_h)$ (see e.g. Roy (2005)) and inserts this into the Gaussian variational equations (12.1.2), then the time-derivatives of the Keplerian elements caused by the gravitational perturbations can be provided analytically (see e.g. Schaub and Junkins (2003, p. 508ff.)) by the so-called Lagrange's planetary equations (LPEs)

$$\dot{a} = \frac{2a}{h}\sqrt{1-e^2}\left[\frac{\partial P}{\partial M}\right]$$

$$\dot{e} = \frac{1-e^2}{he}\left[\sqrt{1-e^2}\frac{\partial P}{\partial M} - \frac{\partial P}{\partial \omega}\right]$$

$$\dot{\omega} = \frac{1}{h}\left[\frac{1-e^2}{e}\frac{\partial P}{\partial e} - \cot i \frac{\partial P}{\partial i}\right]$$

$$\dot{\Omega} = \frac{1}{h\sin i}\frac{\partial P}{\partial i}$$

$$\dot{i} = \frac{1}{h\sin i}\left[\cos i \frac{\partial P}{\partial \omega} - \frac{\partial P}{\partial \Omega}\right]$$

$$\dot{M} = n - \frac{\sqrt{1-e^2}}{h}\left[\frac{1-e^2}{e}\frac{\partial P}{\partial e} + 2a\frac{\partial P}{\partial a}\right]$$

Lagrange's planetary equations (12.4.1)

12.4.2
Gravitational Perturbations of First Order

Table 12.2 shows that the spheroid, represented by the gray-marked coefficient, constitutes by far the strongest perturbation of the gravitational potential. To determine its influence on the Keplerian orbits, we develop P in Eq. (12.2.3) only up to $n = 2, m = 0$, getting

$$P(r,\beta) = \frac{\mu_\oplus R_\oplus^2}{r^3}C_2^0 P_2^0(\sin\beta) = \frac{\mu_\oplus J R_\oplus^2}{3r^3}\left(1 - 3\sin^2\beta\right) \quad (12.4.2)$$

12.4 Analytical Perturbation Calculation

with

$$P_2^0 = \frac{1}{2}\left(3\sin^2\beta - 1\right)$$

$$J := -\frac{3}{2}J_2 := -\frac{3}{2}C_2^0 = 0.001\,623\,94\underline{5} \cdot 10^{-3}$$

$$J \approx \frac{f}{1+f} - \frac{\omega_\oplus^2 R_\oplus^3}{2\mu_\oplus} \approx f - \frac{\omega_\oplus^2 R_\oplus^3}{2\mu_\oplus}$$

We have introduced the quantities J and J_2 which are frequently used in the literature. They have the indicated relation to the flattening $f := \left(R_{equatorial} - R_{polar}\right)/R_{equatorial} \approx 1/298.2572$ of the Earth (exercise, Problem 12.2) and $R_\oplus = R_{equatorial}$.

Secular Perturbations

Any deviation of the gravitational potential $U(r)$, represented by the geoid, from $U = -\mu/r$ results in a local deviation from the Keplerian orbit. Thus a local increase δU brings about an enhancement of the gravitational force by δF and this in turn causes a tiny plunge down of the orbit. However, we are not interested in such local, temporary variations of the Keplerian orbit, but rather in those non-periodic, so-called *secular* variations, which accumulate during various rotations and thus modify the Keplerian elements. We therefore have to average U and with it P over an orbital time T, or $\theta = 2\pi$, respectively.

$$P_{sec} = \frac{1}{2\pi}\int_0^{2\pi} P \cdot d\theta$$

Carrying out this elementary but tedious integration (see e.g. Ruppe (1966), p. 172)), one gets for the gravitational perturbation in first order

$$P_{sec} = \frac{\mu J R_\oplus^2}{3a^3\left(1-e^2\right)^{3/2}}\left(1 - \frac{3}{2}\sin^2 i\right) \qquad (12.4.3)$$

We now replace $P \to P_{sec}$ in Eq. (12.4.1), carry out the partial derivatives, which can be done straightforwardly, and get the secular changes of the Keplerian elements

$$\dot{\Omega}_{sec} = -nK\cos i \begin{cases} > 0 & @\ i > 90° \\ = 0 & @\ i = 90° \\ < 0 & @\ i < 90° \end{cases} \quad \text{regression of nodes} \qquad (12.4.4)$$

$$\dot{\omega}_{sec} = nK\left(2 - \tfrac{5}{2}\sin^2 i\right) \begin{cases} > 0 & @\ i < 63.42° \\ = 0 & @\ i = \arcsin\tfrac{2}{\sqrt{5}} \\ < 0 & @\ i > 63.42° \end{cases} \quad \text{progression of line of apsides}$$

$$\dot{M}_{sec} - n = nK\sqrt{1-e^2}\left(1 - \tfrac{3}{2}\sin^2 i\right) \begin{cases} > 0 & @\ i < 54.74° \\ = 0 & @\ i = \arcsin\sqrt{\tfrac{2}{3}} \\ < 0 & @\ i > 54.74° \end{cases} \quad \text{progression of epoch}$$

$$\dot{a}_{sec} = \dot{e}_{sec} = \dot{i}_{sec} = 0$$

with

$$K = \frac{JR_\oplus^2}{a^2\left(1-e^2\right)^2} > 0 \quad \text{and} \quad n = \frac{2\pi}{T}$$

The flattening of the Earth therefore causes a shift of the nodes. This can be explained as follows. The flattening can be considered as a bulge around the equator of a spherical Earth (see Fig. 12.6), which attracts the body on its orbit. This causes a torque triggering the rotation of the orbital plane (see Fig. 12.7), just as with a spinning top.

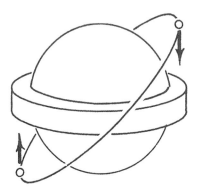

Figure 12.6 Flattening of the Earth interpreted as an equatorial bulge attracting the orbiting body.

The total effect of the equatorial bulge on the orbit can also be understood, if its gravitational pull is considered as continuous kick-burns of type $\delta v_{\perp\perp} \propto \sin(\theta + \omega)$ and type $\delta v_{\perp\odot} \propto \cos(\theta + \omega)$. According to Eq. (8.4.3) this causes $\delta e(-\theta) = -\delta e(\theta)$ and $\delta i(-\theta) = -\delta i(\theta)$, i.e. the effects cancel out over one period, and therefore $\langle\delta a\rangle_\theta = \langle\delta e\rangle_\theta = \langle\delta i\rangle_\theta = 0$. On the other hand $\delta\omega(-\theta) = \delta\omega(\theta)$ and $\delta\Omega(-\theta) = \delta\Omega(\theta)$, i.e. they sum up and hence $\langle\delta\Omega\rangle_\theta \neq 0$, implying

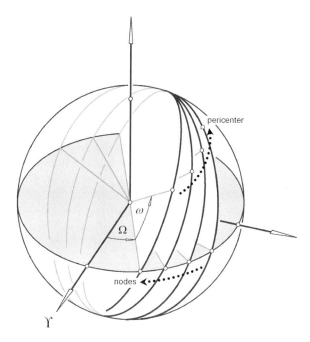

Figure 12.7 The joint orbital regression of nodes and progression of the line of apsides.

a shift of the node, and $\langle \delta\omega \rangle_\theta \neq 0$, a shift of the periapsis and thus also a shift of the epoch, δt_0. For inclinations $i < 63.4°$ the periapsis shifts along the line of motion (progression of the line of apsides). Therefore the orbital period increases, and accordingly the mean anomaly and the epoch. That the two bordering inclinations, at which the shift of the line of apsides and the shift of epoch change direction, do not coincide, as one would expect, is due to the minor contribution of the shift of nodes to the orbital period. In conclusion the following can be said:

> The flattening of the Earth – the by far biggest contribution to a secular variation of orbital elements – only changes the orientation of the elliptical orbit, but not its size and its shape.

12.4.3
Higher-Order Gravitational Perturbations – Triaxiality

A detailed analytical description of the influence of perturbations of a higher order can be found in Groves (1960), Campan et al. (1995), and Fortescue et al. (2003, p.133).

Triaxiality

According to the LPEs (12.4.1) and (12.2.3) the variations of orbital elements caused by the perturbation P of order n decline on average with $P/h \propto 1/hr^{n+1} \approx 1/a^{n+3/2}$ with the orbital radius. Because the radius of a geostationary orbit is larger by a factor 6.2 than those of LEO, the secular changes of the orbital elements caused by the biggest perturbation, i.e. the spheroid (Fig. 12.2a), are smaller by a factor of $6.2^{-3.5} \approx 0.17\ \%$ and therefore negligible in the short term. According to Tab. 12.2 this is even more so for non-spherical perturbations of higher order. Their coefficients (bold entries) are smaller by three orders of magnitude or more than the ones for the spheroid. Nonetheless, for a high-precision long-term GEO orbit $n = 2$ perturbations have still to be considered. In practice, this can only be done numerically (see Section 12.3.1).

Even though the secular impacts of gravitational anisotropies are smaller in GEO than in LEO the sensitivity on sectorial perturbations strongly increases in GEO, because the satellite circles around the Earth in synchrony with the rotation of the Earth and therefore the averaging effect of a fast orbit no longer applies. So the spacecraft integrates even tiny anisotropic perturbations over time. This is called *orbital resonance*. Because $i = 0$ for geostationary orbits the azimuthal-symmetric perturbations of zonal perturbations of the geoid vanish. The impacts of sectorial perturbations depend on the latitudinal position of the S/C: If the sectorial perturbations are symmetric with respect to the radial vector, they vanish too; if they are asymmetric they don't cancel and cause orbital changes, which we will investigate for the most important perturbation, the so-called *triaxiality*.

The term "triaxiality" refers to the triaxiality of the "potato" potential (see Fig. 12.3b), which includes zonal, sectorial, and tesseral terms (bold coefficients in Table 12.2). But, as discussed above, only terms with latitudinal modulations have an perturbing effect on the orbit, with the dominant coefficient J_{22} given by

$$J_{22} := \sqrt{C_{22}^2 + S_{22}^2} = 1.8155 \times 10^{-6}$$

The corresponding spherical harmonic term, which looks like a dumbbell, is shown in Fig. 12.8 and represents the most prominent part of the "potato" as depicted in Fig. 12.2b. It stretches along the 15°W–165°E direction.

Equation of Motion

If one takes also the weaker perturbational terms J_{31} and J_{33} into account, the induced longitudinal acceleration $\ddot{\lambda}$ of the satellite can be derived analytically from the LPEs (12.4.1) and described by the following equation of motion (see

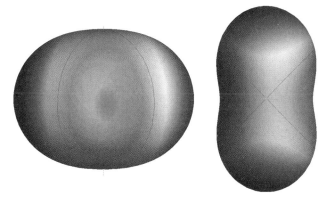

Figure 12.8 The sectorial perturbation term J_{22}: (left) observed from the terrestrial equatorial plane; (right) observed from the terrestrial pole. See also color figure on page 471.

Campan et al. (1995))

$$\ddot{\lambda} = \frac{1}{2}\omega_\lambda^2 \cdot \left[\underbrace{\sin 2(\lambda + 14.93°)}_{J_{22}} \underbrace{- 0.046 \cdot \sin(\lambda - 6.98°)}_{J_{31}} \right.$$

$$\left. + \underbrace{0.138 \cdot \sin 3(\lambda - 21.00°)}_{J_{33}} \right] \quad (12.4.5)$$

with

$$\omega_\lambda = \frac{6R_\oplus \sqrt{J_{22}}}{a_{GEO}} \omega_\oplus = 2.814 \text{ year}^{-1}$$

which is depicted in Fig. 12.9a. It has four positive roots at $\lambda = 75.1°E$, $162.1°E$, $104.9°W$, $11.4°W$ with negative slopes (indicating stable positions) at $\lambda_0 = 75.1°E$, $104.9°E$ and positive slopes (indicating metastable positions) at the other two positions. Because the perturbation terms J_{31} and J_{33} are small (see Fig. 12.9a), equation (12.4.5) can be approximated, particularly well around the stability points, by just the J_{22} term

$$\ddot{\lambda} = -\frac{1}{2}\omega_\lambda^2 \cdot \sin 2(\lambda - \lambda_0)$$

By simply considering only the dumbbell the acceleration and the stability conditions can be explained as follows. The masses at the ends ob the dumbbell cause a constant lateral gravitational pull on the body in GEO. Its impact on the body depends on its initial position. If the position is exactly on

12 Orbit Perturbations

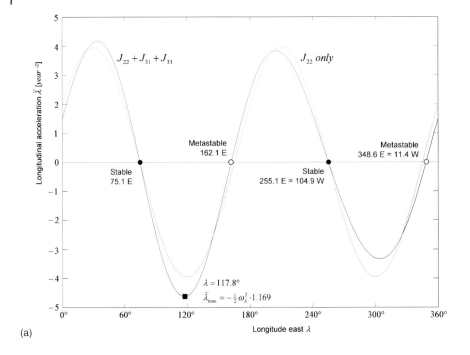

(a)

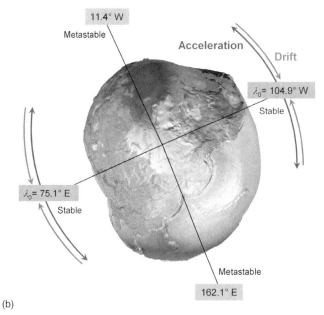

(b)

Figure 12.9 Dynamics of a body in geostationary orbit. Longitudinal acceleration as given by Eq. (12.4.5) (a), and dynamics at the abeam points of the triaxial "potato" (b). See also color figure on page 471.

the dumbbell's lateral axis, i.e. displaced by 90° from the longitudinal 15°W–165°E axis at 75°E or 105°W, then it is subject to the same gravitational pull from both ends of the dumbbell. So the effective gravitational pull is indifferent, and its position would be metastable. If it is displaced by only 45° with regard to the longitudinal axis, it is subject to a stronger force from the closer end of the dumbbell. Then the body is minimally accelerated tangentially on its orbit towards the closer end (inner red arrows in Fig. 12.9b). Let us denote this acceleration as $a_{||}$. From Problem 8.6 we see that $a_{||}$ causes an increase in the orbital radius of

$$\dot{r} = 2r\sqrt{\frac{r}{\mu}}a_{||} = \frac{2}{\omega}a_{||}$$

where $\omega = v/r = \sqrt{\mu/r^3}$ is the angular velocity of the S/C. We can neglect the induced ellipticity because the mean orbital radius is not dependent on it. The enlarged orbital radius causes a decrease of the angular velocity which equals a longitudinal deceleration $\ddot{\lambda}$ according to

$$\ddot{\lambda} = \dot{\omega} = -\frac{3}{2}\frac{\omega}{r}\dot{r} = -\frac{3}{r}a_{||} \quad (12.4.6)$$

So, both the larger orbital radius and the lower orbital velocity in the higher orbit lead to a decrease of the angular velocity and hence to a drift of the body (outer green arrows in Fig. 12.9b) in the opposite direction (minus sign), i.e. along the lateral axis back to the metastable positions, contrary to first expectations. So we see that the abeam positions at 75°E and 105°W, though being statically unstable, are actually dynamically stable. This situation is comparable with the dynamical stability of a S/C near the statically unstable L4 and L5 Lagrange points (see Section 11.4.5). If the body is positioned right above the ends of the dumbbell this would also be an indifferent position. But just a small deviation would cause a tangential acceleration back to the indifferent position, causing a drift even more away from the initial position according to Eq. (12.4.6). The positions above the ends of the longitudinal axis hence turn out to be metastable. The actual values of the stable and metastable positions, are slightly different from these, which is due to the J_{31} and J_{33} perturbations.

Dynamics at the Stable Positions

If the body is near a stable position λ_0, in the approximated equation of motion above the sine for small arguments can be approximated linearly to give

$$\ddot{\lambda} = -\omega_\lambda^2 \cdot (\lambda - \lambda_0)$$

the general solution of which is

$$\lambda = \lambda_0 + A\sin(\omega_\lambda t + \varphi)$$

This is a harmonic oscillation with amplitude A and phase φ around the stable positions with frequency $\nu = \omega_\lambda/2\pi = 0.447 \text{ year}^{-1}$, which equals one oscillation per 2.24 years.

Dynamics at Non-Stable Positions

Let's have a closer look at the time behavior when the body is far away from one of the two equilibrium positions. For small time periods the instantaneous position λ can be considered almost constant. So the right side of the differential equation (12.4.5) is a constant, and the equation can simply be integrated directly

$$\lambda = \lambda_i + \frac{1}{2}\ddot{\lambda} \cdot t^2 \tag{12.4.7}$$

So the body, initially in a resting position, moves away quadratically with time from its initial position λ_i. This is the so-called *east–west drift*.

To counteract the drift and keep the satellite within the standard longitudinal position box of $\Delta\lambda = 0.1°$ (in order not to collide with adjacent satellites), correction burns need to be fired at that side of the box to which the satellite tends to drift freely. A burn, so-called *east–west station-keeping*, causes the satellite to drift just to the other side of the box and then drift back. The recurrence time and the delta-v demand for one kickburn can be derived from Eq. (12.4.7) to be

$$\Delta t_{sk} = 2\sqrt{\frac{2 \cdot \Delta\lambda}{|\ddot{\lambda}|}} \geq 20.1 \text{ days}$$

and

$$\Delta v_{sk} = |a_\||| \cdot \Delta t_{sk} = \frac{2}{3} r \sqrt{2 \cdot \Delta\lambda \cdot |\ddot{\lambda}|} \leq 0.113 \, \frac{\text{m}}{\text{s}}$$

The limits are derived from the position $\lambda = 117.8°$ where the longitudinal acceleration has its maximum absolute value with $|\ddot{\lambda}|_{max} = \frac{1}{2}\omega_\lambda^2 \cdot 1.169$ (see Fig. 12.9a). In total for east–west station-keeping a yearly delta-v of

$$\boxed{\Delta v = \frac{1}{3} r |\ddot{\lambda}| \cdot 1 \text{ year} \leq 2.06 \text{ m/s per year}}$$

needs to be taken into account.

12.4.4
Lunisolar Perturbations in GEO

We recall from Table 12.2 that, apart from perturbations by the Earth's asymmetrical gravitational potential, also the Sun and the Moon perturb Earth orbits. Their effects are noticeable in particular in GEO because they are no

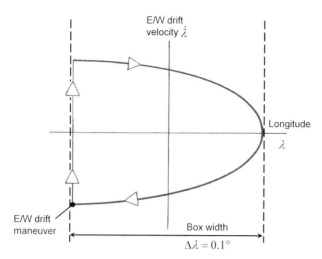

Figure 12.10 Satellite drift pattern inside an east–west station keeping box.

more concealed by the gravitational perturbations as in LEO. The analytical description of these lunisolar perturbations are very complicated and complex (see Campan et al. (1995) or Noton (1998, pp. 70 and App. A.1)). This is why we will treat them only qualitatively.

Because the terrestrial equatorial plane makes an inclination of 23.44° against the ecliptic and of $23.5 \pm 5.1°$ against the lunar plane, the gravitational force of the Sun and Moon can be decomposed into a component acting in the equatorial plane, i.e. in the GEO orbital plane, $F_\odot$, and one perpendicular to the equatorial plane $F_\perp$ (see Fig. 12.11). The effect of the constant action of $F_\odot$ is the same as that of the solar pressure of the Sun, which will be described in the next section. According to Eq. (8.4.3) such a constant force affects only the orbital elements e, a and ω, whereby due to $i = 0$, the change $\dot\omega$ is pointless, and $\dot a_{sec} = 0$ because a conservative (time independent) gravitational force is not able to secularly change the energy of a system and due to $\varepsilon = -\mu/2a$ also not a. This leaves $\dot e \neq 0$ to be investigated. Because in the geocentric equatorial reference frame (see Fig. 13.2), the Moon circles the Earth once in 27.55 days and the Sun circles Earth once a year, this average over all directions implies $\dot e_{sec} \propto e = 0$. Therefore $F_\odot$ does not affect a geostationary orbit secularly over one year. But there are non-secular periodic effects. Because the Sun revolves much slower around the Earth than the Moon, the periodic inclination amplitude of 0.02° induced by the Sun is much bigger than that of the Moon. Nevertheless, this is still irrelevant for practical purposes.

The component $F_\perp$ perpendicular to the orbital plane, on the other hand, is secular. Over one orbital period it is constant and acts like the equatorial bulge on the satellite. Therefore, and according to Eq. (8.4.3), it affects only

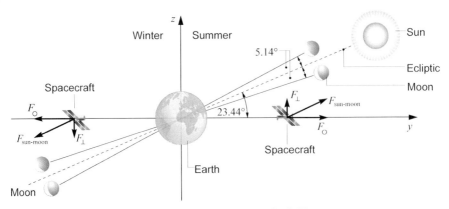

Figure 12.11 Lunisolar perturbation forces on a body in GEO. The resulting force component $F_\bigcirc$ in the orbital plane is orbit-periodic, i.e. non-secular, whereas the component $F_\perp$ perpendicular to the orbital plane is secular.

the orbital elements i, ω and Ω. But due to $e = i = 0$, the variations $\dot\omega$ and $\dot\Omega$ are pointless. Hence only the variation of the inclination remains to be considered, which for a disturbing planet p can be calculated in first-order perturbation calculation to be

$$\dot i_{sec} = \frac{3}{8}\frac{n_p^2 \mu_p}{n}\sin 2\langle i_p\rangle = \begin{cases} 0.539°/a & @\ p = \text{Moon} \\ 0.271°/a & @\ p = \text{Sun} \end{cases} \quad (12.4.8)$$

with $\langle i_p \rangle$ its mean inclination with regard to the equatorial plane; $n_p = 360°/T_p$, its mean motion; $n = 359°/$day the mean terrestrial rotation, and $\mu_p = m_p/(m_p + m_\oplus)$. If one allows also higher orders $k \cdot n_p$ of perturbation, then one obtains for the yearly rate of the inclination variation $\dot i_{sec} = 0.478°/a$ by lunar and $\dot i_{sec} \approx 0.319°/a$ by solar perturbations, whereby the latter varies from year to year. So in total we find

$$\boxed{\dot i_{sec} = 0.797° \text{ per year}}$$

Because from Earth the movement of a body in an inclined geostationary orbit looks like a vertical oscillation, it is called *north–south drift*. The so-called *north–south station-keeping effort* can be calculated from Eq. (8.2.4) to be $\Delta v = 2v \sin(\Delta i/2) = 2 \cdot 3.066 \cdot \sin 0.3985°$ km s^{-1}a^{-1} and therefore

$$\Delta v = 42.7 \text{ m/s per year}$$

This is more than one order of magnitude bigger than the east–west station-keeping and hence more decisive for the propulsion demand for orbit control.

This result confirms the rule of thumb that inclination changes in astronautics always imply high propulsion efforts because of the high orbital velocities in $\Delta v = 2v \sin(\Delta i/2)$.

In GEO there are no drift constraints in north–south direction, because these don't interfere with adjacent satellites. Depending on the drift requirements, but usually not later than at $i = 3°$, inclination corrections are carried out at the nodes (see Fig. 8.11) by twice the amount of accumulated inclination.

12.5 Solar Radiation Pressure

A spacecraft orbiting a planet at a distance of r from the Sun will be affected by solar radiation unless it happens to be in the shadow of the planet. A light particle (photon) does not possess mass, but according to quantum mechanics it still carries linear momentum h/λ (h is Planck's constant, λ is the wavelength of the light particle), which, depending on the reflectivity ρ of the surface of the S/C, transfers momentum $\rho h/\lambda$. If the surface is absorbing, then $\rho = 1$; if it is reflecting, $\rho = 2$, and if it is transparent, $\rho = 0$. The solar radiation thus produces a total radiation pressure $p_{S/C}$ on the S/C, which via the mass of the S/C creates the acceleration a given by

$$\frac{ma}{A_\perp} = p_{S/C} = \rho\, p_{sun}$$

where

$$p_{sun} = \frac{E}{c} = p_0 \left(\frac{r_\oplus}{r}\right)^2 \frac{\text{N}}{\text{m}^2} \quad \text{radiation pressure of the Sun} \quad (12.5.1)$$

with

$c = $ velocity of light

$E = (1372 \pm 45)\left(\dfrac{r_\oplus}{r}\right)^2 \dfrac{\text{W}}{\text{m}^2}$, intensity of radiation

$p_0 = (4.58 \pm 0.15) \times 10^{-6}\ \dfrac{\text{N}}{\text{m}^2}$ (seasonal)

$r_\oplus = 1.495\,9787 \times 10^8$ km, the mean radius of the Earth orbit

$A_\perp = $ the surface of the S/C projected onto the direction of radiation

This results in the following acceleration of the S/C:

$$a_{sun} = p_{sun} \cdot B_r \quad (12.5.2)$$

where we have defined a *ballistic radiation coefficient*

$$B_r = \rho \frac{A_\perp}{m} \qquad (12.5.3)$$

similar to the ballistic drag coefficient given in the next section (Eq. (12.6.2)).

12.5.1
Effect of Solar Radiation

Qualitative Considerations

To derive the effects of solar radiation on the orbital elements, let's assume a circular or a weakly elliptical orbit of the S/C at any altitude with $e \approx 0$ in addition to $i \approx 0$. The direction of radiation shall be in the orbital plane and perpendicular to the line of apsides, i.e. the Sun shines "laterally" onto the orbit; see Fig. 12.12. (As we will see in a moment, if we start out with a circular orbit, the solar pressure will cause an eccentricity with the line of apsides perpendicular to the radiation direction. So this assumption always holds.) Due to $e \approx 0, i \approx 0$ we can neglect $\dot\omega, \dot\Omega$ effects, and, since the radiation force does not have a component vertical to the orbital plane, Table 8.1 tells us that no inclination changes result. So we only have to focus on $\dot a, \dot e$ effects. Table 8.1 also tells us that forces perpendicular to the path in the orbital plane do not change a, yet Eqs. (8.4.8) and (8.4.9) tell us that such forces change e, but, because the effect of a force with constant direction is opposite on the two sides of the ellipse, they cancel out. So we only have to consider along-track forces at apoapsis and periapsis affecting $\dot a, \dot e$.

Let's assume that the orientation of the orbit is such that the radiation produces a minute thrust δv along-track at the periapsis, and the same thrust in the opposite direction at apoapsis. Then the induced change of the semi-major axis and the eccentricity according to Table 8.1 is

$$\frac{\delta a}{a} = \frac{2}{1+e}\frac{(-\delta v)}{v_h} \quad \text{and} \quad \delta e = -2\frac{(-\delta v)}{v_h} \quad \text{at the apoapsis}$$

$$\frac{\delta a}{a} = \frac{2}{1-e}\frac{(+\delta v)}{v_h} \quad \text{and} \quad \delta e = +2\frac{(+\delta v)}{v_h} \quad \text{at the periapsis}$$

with

$$v_h = \frac{\mu}{h} = \sqrt{\frac{\mu}{a(1-e^2)}} \approx \sqrt{\frac{\mu}{a}} = v_0$$

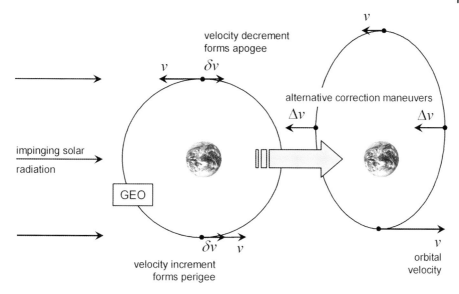

Figure 12.12 The solar pressure with acceleration effect δv deforms a geostationary orbit to an ellipse, which can be circularized by correction maneuvers Δv towards the Sun at positions $\theta = 90°, 270°$.

where the approximation holds for small eccentricities. So we get

$$\frac{\delta a}{a} \approx 2(e-1)\frac{\delta v}{v_0} \quad \text{and} \quad \delta e = 2\frac{\delta v}{v_0} \quad \text{at the apoapsis}$$

$$\frac{\delta a}{a} \approx 2(e+1)\frac{\delta v}{v_0} \quad \text{and} \quad \delta e = 2\frac{\delta v}{v_0} \quad \text{at the periapsis}$$

If we add up both impacts, we get the following changes:

$$\frac{\delta a}{a} = 4e\frac{\delta v}{v_0} \quad (12.5.4\text{a})$$

per revolution

$$\delta e = 4\frac{\delta v}{v_0} \quad (12.5.4\text{b})$$

The result $\delta e > 0$ positively feeds back our assumption that the solar radiation is parallel to the orbital velocity vector at periapsis. So starting out with a circular orbit, solar radiation will increase the orbital speed on one lateral side of the orbit and decrease it on the opposite side, thus inducing an eccentricity with a lateral line of apsides. With this constellation, the eccentricity will constantly increase with each revolution. So, we can say the following:

> For circular or nearly circular geostationary orbits the semi-major axis and hence the orbital period remains constant. However, an eccentricity laterally to the radiation direction develops, which constantly increases, independent of the size of eccentricity.

Quantitative Perturbation Calculation

We start out to determine the variation of the eccentricity quantitatively by refining the above considerations. Because the change of eccentricity comes from the along-track accelerations around the periapsis and apoapsis, we have to estimate the integral of their impact over one revolution around these points. To estimate this, we apply Eq. (12.5.4b) with $\delta v = a_{sun} \delta t$, where δt is the time of influence per revolution, which we estimate to be $\delta t \approx T/2$. As $T = 2\pi \sqrt{a^3/\mu}$ holds for an elliptical as well as a circular orbit, with $v_0 = \sqrt{\mu/a}$ we get

$$\delta e = 4\pi \cdot p_{sun} B_r \frac{a^2}{\mu} \quad \text{per revolution}$$

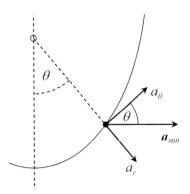

Figure 12.13 Acceleration components due to solar radiation.

This is a first rough estimation. In order to derive the orbit changes exactly, we need to determine the components of the solar force in the radial direction and vertically to it and then integrate their effects over one orbit according to the Gaussian variational equations (12.1.2). Because the line of apsides is lateral to the radiation, we find for the acceleration components (see Fig. 12.13) due to the solar pressure

$$a_{sun} = (a_r, a_\theta, a_h) = a_{sun} (\sin \theta, \cos \theta, 0)$$

and with Eq. (12.1.2) for the variations of the semi-major axis and eccentricity per revolution

$$\delta a = \frac{2a^2}{h} a_{sun} \int_0^T \left(e \sin^2 \theta + \cos \theta + e \cos^2 \theta \right) dt$$

$$\delta e = \frac{a}{h} a_{sun} \int_0^T \left(\sin^2 \theta + e \cos \theta + \cos^2 \theta - e \cos^3 \theta + \cos^2 \theta \right) dt$$

where we have linearized all terms because $e \ll 1$. Applying Eq. (7.3.16) we find

$$dt = \frac{h^3/\mu^2}{(1+e\cos\theta)^2} d\theta \approx \frac{h^3}{\mu^2}(1 - 2e\cos\theta) d\theta$$

from which with $h^2 = \mu a(1-e^2) \approx \mu a$ we obtain

$$\delta a = \frac{2a^3}{\mu} a_{sun} \int_0^{2\pi} \left(e + \cos\theta + e\cos^2\theta\right) d\theta$$

$$\delta e = \frac{a^2}{\mu} a_{sun} \int_0^{2\pi} \left(1 + e\cos\theta + \cos^2\theta - e\cos^3\theta - 2e\cos\theta - 2e\cos^3\theta\right) d\theta$$

$$= \frac{a^2}{\mu} a_{sun} \int_0^{2\pi} \left(1 - e\cos\theta + \cos^2\theta - 3e\cos^3\theta\right) d\theta$$

Because from symmetry considerations $\langle\cos\theta\rangle_\theta = \langle\cos^3\theta\rangle_\theta = 0$ and because

$$\int_0^{2\pi} \left(1 + \cos^2\theta\right) d\theta = \left(\theta + \frac{1}{2}\sin\theta\cos\theta + \frac{\theta}{2}\right)\Big|_0^{2\pi} = 3\pi$$

we finally derive with Eq. (12.5.2)

$$\delta a = 6\pi\, p_{sun} B_r \frac{a^3}{\mu} e + O\left(e^2\right) \approx 0$$

$$\delta e = 3\pi\, p_{sun} B_r \frac{a^2}{\mu} + O\left(e^2\right)$$

per revolution, @ $e \ll 1$ (12.5.5)

This agrees nicely with our qualitative and rough quantitative considerations above. Because δe increases quadratically with semi-major axis, this effect is 50 times bigger in GEO than in LEO. For an average reflecting surface $\rho = 1.5$ and a typical $A_\perp/m \approx 0.08$ m^2 kg^{-1} for communication satellites (large solar panels) in GEO ($r = r_\oplus$), we get

$$\delta e \approx 2.3 \times 10^{-5} \quad \text{per revolution in GEO} \tag{12.5.6}$$

We shall now concern ourselves with the other orbital elements. Because $a_h = 0$ and because the integrals over odd powers of circular functions vanish over one revolution, we find with Eq. (12.1.2)

$$\delta i = \delta\Omega = 0.$$

Equation (12.1.2) is not applicable to $\dot{\omega}$ and $\dot{M}$ since $i \to 0$ and $e \to 0$. Table 8.1, however, shows that forces $\delta v_\| \neq 0$, $\delta v_{\perp O} \neq 0$, which act externally on the body, always rotate the line of apsides, independently of the point at which they act. Therefore

$$\delta\omega \neq 0 \quad \text{and} \quad \delta M \neq 0$$

In summary: for $e \to 0$ all variations of orbital elements vanish except that the eccentricity increases and the line of apsides rotates. So, effectively only the eccentricity vector e changes.

12.5.2
Temporal Evolution of the Orbit

We want to determine the temporal evolution of e. To do so we need an equation of motion for it. We first introduce an appropriate reference frame with the x axis along the line of apsides and the y axis vertical to it. The above considerations dealt with e_x, for which we get the equation of motion

$$\frac{de_x}{dt} = \frac{\delta e}{T} = \frac{3\pi}{2\pi} \cdot p_{sun} B_r \sqrt{\frac{\mu\, a^2}{a^3\, \mu}} = \frac{3}{2} \cdot p_{sun} B_r \sqrt{\frac{a}{\mu}} = \frac{3 p_{sun} B_r}{2na} =: \kappa B_r$$

with

$$\kappa = \frac{3 p_{sun}}{2na} = (2.23 \pm 0.07) \times 10^{-9}\, \frac{\text{kg}}{\text{m}^2 \text{s}}$$

if the Sun's rays are vertically to the line of apsides of a GEO. In case they are not vertical, the angle between the solar radiation direction and the line of apsides is defined as the mean longitude of the Sun: $\lambda_* = n_* t + \lambda_0$ with $n_* = 0.9856°/\text{day}$ the mean motion of the Sun. With this we obtain for the Sun-angle-dependent variation of the eccentricity component along the line of apsides

$$\frac{de_x}{dt} = -\kappa B_r \sin(n_* t + \lambda_0) \tag{12.5.7}$$

It can be shown (see Campan et al. (1995)) that the line of apsides follows the change of the Sun's angle. We therefore obtain for the e_y component

$$\frac{de_y}{dt} = \kappa B_r \cos(n_* t + \lambda_0) \tag{12.5.8}$$

The solutions to the above equations of motion are easy to find:

$$e_x(t) = e_x(t_0) + \frac{\kappa B_r}{n_*} \cos(n_* t + \lambda_0)$$

$$e_y(t) = e_y(t_0) + \frac{\kappa B_r}{n_*} \sin(n_* t + \lambda_0) \tag{12.5.9}$$

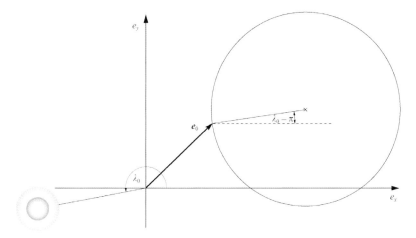

Figure 12.14 The circular motion of the tip of the GEO eccentricity vector with initial value e_0 on a circle within a year.

The tip of the eccentricity vector with initial value $e_0 = (e_x(t_0), e_y(t_0))$ therefore describes a circle with radius $\kappa B_r / n_*$ (see Fig. 12.14). This description is based on the assumption that the Sun moves on a circle in the equatorial plane of the Earth, which is not quite true. Its inclination (angle between equatorial plane and ecliptic) causes the circle to be actually a weak ellipse. In addition other perturbations (J_{20} term of the geoid, lunisolar perturbation) lead to rosette-type deviations from the circle including the effect that the initial and final points no longer coincide (see Fig. 12.15), implying a secular component ($\dot{e}_{sec} \propto e$, see Section 12.4.4)

12.5.3
Correction Maneuvers

The yearly drag-induced motion of the eccentricity vector e however is of no practical relevance, because in GEO the absolute amount of eccentricity has to be regularly reduced by correction maneuvers after about one month. To understand why, we have to know what the implications of a non-vanishing eccentricity are. Let's examine the periodic deviations of the orbital radius and the true anomaly caused by a body in an orbit with small eccentricity relative to a body in a circular orbit (so-called *guiding center*) with the same semi-major axis. This deviation is the apparent periodic horizontal motion of the position of the body in GEO as observed from the rotating Earth. To do this, we recall Eq. (7.3.17)

$$\frac{\mu^2}{h^3}(t - t_0) = \int_0^\theta \frac{d\theta'}{(1 + e \cdot \cos \theta')^2}$$

372 | *12 Orbit Perturbations*

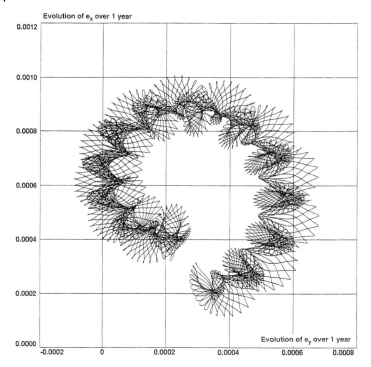

Figure 12.15 The yearly motion of the GEO eccentricity vector under the action of all orbital perturbations.

Because $e \ll 1$, $h^2 = \mu a \left(1 - e^2\right) \approx \mu a$ and $n = \sqrt{\mu/a^3}$, and by choosing $t_0 = 0$, we get

$$nt \approx \int_0^\theta \left(1 - 2e \cdot \cos \theta'\right) d\theta' = \theta - 2e \sin \theta$$

So, in zero order approximation we get for $\theta(t)$ a circular orbit with $\theta = nt$, and in first order approximation we have

$$\theta \approx nt + 2e \sin(nt) =: nt + \Delta\theta$$

i.e. compared with a circular orbit with $r = a$ the deviations are

$$\Delta\theta(t) = 2e \sin(nt)$$
$$\Delta r(t) = -ea \cos(nt)$$
(12.5.10)

The latter is obtained from the orbit equation (7.4.13a) with $E \approx M = nt$ (see also the function series expansions at the end of Section 7.4.3). Because

12.5 Solar Radiation Pressure

for small lateral deviations Δs the relation $\Delta s = a \cdot \Delta\theta$ holds, Eq. (12.5.10) describes an elliptical motion in the orbit plane (see Fig. 12.16) around the guiding center. Its semi-minor axis is ea in the radial direction and its semi-major axis is $2ea$ in the lateral direction, both of which increase with growing eccentricity. For the lateral oscillation range we find $\delta\theta = \Delta\theta_{max} - \Delta\theta_{min} = 4e$. As geostationary satellites are allowed to move only within their assigned box of standardized width $\Delta\theta = 0.1° = 1.745 \times 10^{-3}$ rad, we get the following limit for the slowly increasing eccentricity

$$e \leq \frac{1.745 \times 10^{-3}}{4} \tag{12.5.11}$$

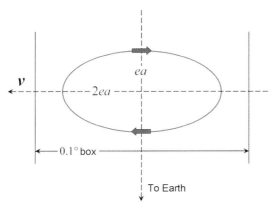

Figure 12.16 Apparent motion of a geostationary satellite with eccentricity e in the guiding center system, i.e. the motion as seen in an Earth-fixed reference frame.

When the eccentricity reaches this limit value, which occurs after 19 days for a typical $\delta e \approx 2.3 \times 10^{-5}$ per revolution, a correction maneuver has to be carried out to cut eccentricity back to zero. Since the correction maneuver should not change the semi-major axis and with it the orbiting time, this has to be done according to Eqs. (8.4.8) and (8.4.9) by a maneuver perpendicular to the direction of motion within the orbital plane of amount

$$|\Delta v_{\perp\odot}| = \frac{1.745 \times 10^{-3}}{4} \sqrt{\frac{\mu}{a}} = 1.34 \text{ m s}^{-1} \tag{12.5.12}$$

at $\theta = 90°$ in inward direction ($\Delta v_{\perp\odot} < 0$) or at $\theta = 270°$ with the same value in outward direction ($\Delta v_{\perp\odot} > 0$), i.e. in both cases into the direction of the Sun (see Fig. 12.12).

12.6
Drag

In low Earth orbits the atmospheric density cannot be neglected and therefore may exert a marked resistance on a circulating spacecraft. In this chapter we want to calculate how the orbit – specifically, how the Keplerian elements – will be affected by it.

12.6.1
General Considerations

Let's assume that the spacecraft experiences atmospheric drag, which depends on its specific shape with regard to the flight direction through the residual atmosphere. Let ρ be the atmospheric density, m the mass of the vehicle, $A_\perp$ its cross-sectional area with regard to the flight direction (wetted surface), and v_a its velocity relative to the atmosphere. To determine the atmospheric drag theoretically, we have to know the momentum transfer Δp of the atmospheric particles to a given part of the surface of the S/C. We usually don't have knowledge about how much momentum is transferred. It depends on the specific interaction (elastic or diffuse scattering) of the particles with the surface. From a qualitative point of view, we can only say that a certain proportion of the momentum p of the impacting particles is transferred, which somehow depends on the incidence angle θ on the surface as well as on surface property s, i.e.

$$\Delta p = -p \cdot f(\theta, s)$$

The function $f(\theta, s)$ is known theoretically only for some elementary surface shapes, so in general it is determined experimentally. Because of the continuity equation (1.2.9), the mass flow rate of air particles wetting the cross-sectional area $A_\perp$ is $\dot{m} = \rho v_a A_\perp$. So the drag force on $A_\perp$ is

$$D = \frac{d(\Delta p)}{dt} = -f(\theta, s) \frac{dp}{dt} = -f(\theta, s) \frac{d(mv_a)}{dt}$$
$$= -f(\theta, s) v_a \dot{m} = -f(\theta, s) \rho v_a^2 A_\perp$$

To calculate the drag force of the entire S/C, the drag forces $dD = -f(\theta, s) \rho v_a^2 \cdot dA_\perp$ of all partial areas $dA_\perp$ with different θ and/or surface properties have to be integrated over the total area

$$D = -\iint_{A_\perp} f(\theta, s) \rho v_a^2 \cdot dA_\perp = -\rho v_a^2 \iint_{A_\perp} f(\theta, s) \cdot dA_\perp$$

Hence

$$D = -\frac{1}{2} \rho v_a^2 C_D A_\perp \qquad (12.6.1)$$

with

$$C_D := \frac{2}{A_\perp} \iint_{A_\perp} f(\theta, s(\theta)) \cdot dA_\perp = 2\langle f \rangle_\theta \quad \text{drag coefficient}$$

We have defined the so-called *drag coefficient*, which is a dimensionless number and depends only on the shape of the body (not on its size) as well as on the detailed interaction of the impacting particles with the surface. For altitudes where molecule flow freely, i.e. where the average free path length $l = 7.6 \, [\text{g km}^{-2}]/\rho$ of a molecule is bigger than the dimension ϕ of the S/C, i.e. for

$$\rho < \frac{7.6 \, [\text{g km}^{-2}]}{\phi} \quad \Rightarrow \quad h > 150 \text{ km}$$

the following good approximate value for the drag coefficient can be given for spherical bodies or tumbling bodies with asymmetrical shape:

$$C_D = 2.2 \pm 0.2 \quad @ \; h > 150 \text{ km}$$

This implies a mean momentum transfer $\langle \Delta p \rangle_\theta = -pC_D/2 = -(1.1 \pm 0.1)\,p$ according to the derivation above. So, the impacting particles transfer a bit more than their total momentum to the surface, i.e. they are scattered on average by an angle of 96°, i.e. slightly backwards. The atmospheric deceleration can thus be described as a function of the drag D as

$$a_D = \frac{D}{m} = -\frac{C_D}{2} \frac{A_\perp}{m} \rho v_a^2 = -\frac{B}{2} \rho v_a^2 \tag{12.6.2}$$

with

$$B := C_D \frac{A_\perp}{m} \quad \text{ballistic coefficient}$$

The ballistic coefficient is a characteristic parameter of the S/C. The negative sign in Eq. (12.6.2) denotes that the S/C is decelerated by the drag.

Example
What will be the mean atmospheric drag of the International Space Station ($C_D \approx 2.3$, $m = 456 \, t$, $A_\perp \approx 2500 \text{ m}^2$, $h = 400 \text{ km}$) when assembly is complete?

Answer: The ballistic coefficient according to Eq. (12.6.2) is $B \approx 0.0126 \text{ m}^2 \text{ kg}^{-1}$. At an altitude of 400 km, $\langle \rho \rangle_t = 3.725 \text{ g km}^{-3}$. With this we calculate the deceleration to be $a_D = -1.4 \times 10^{-6} \text{m s}^{-2} \approx -0.1 \, \mu g$, where g is the Earth's gravitational acceleration.

Remark: *In the 1980s and 1990s Shuttle missions with Spacelab or Spacehab scientific laboratories used to take place at an altitude of 300 km. There the atmospheric density is already ten times larger as at ISS, and thus the drag acceleration $a_D \approx -1 \, \mu g$. This is where the expression "microgravity research" for scientific research in space comes from.*

Variation of Orbital Elements

We now want to determine the variation of the orbital elements due to atmospheric drag. We identify the drag force as an orbital perturbation which we split into radial, cross-radial and vertical components, and derive from Fig. 12.17 with the above results

$$\mathbf{a}_D = (a_r, a_\theta, a_h) = -\frac{B}{2} \rho(r) \cdot v_a^2 \, (\sin \gamma, \cos \gamma, 0)$$

with the flight path angle γ, which according to Eqs. (7.3.12) is related to the true anomaly by

$$\sin \gamma = \frac{e \sin \theta}{\sqrt{1 + 2e \cos \theta + e^2}}$$

$$\cos \gamma = \frac{1 + e \cos \theta}{\sqrt{1 + 2e \cos \theta + e^2}}$$

We insert this perturbation into the Gaussian variational equations (12.1.2)

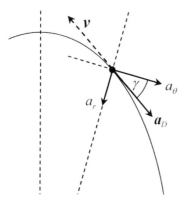

Figure 12.17 Decomposition of the atmospheric drag into radial and azimuthal components.

and find for the wanted variations of orbital elements

$$\dot{a} = -\frac{B\rho v_a^2}{n}\frac{1}{\xi(e,\theta)}$$

$$\dot{e} = -\frac{B\rho v_a^2}{na}\xi(e,\theta)(e+\cos\theta)$$

$$\dot{\omega} = -\frac{B\rho v_a^2}{nae}\xi(e,\theta)\sin\theta \qquad (12.6.3)$$

$$\dot{M} - n = \frac{B\rho v_a^2}{nae}\xi(e,\theta)\frac{\sin\theta\cdot(1+e\cos\theta+e^2)}{(1+e\cos\theta)}\sqrt{1-e^2}$$

$$\dot{i} = \dot{\Omega} = 0$$

with

$$\xi(e,\theta) = \frac{\sqrt{1-e^2}}{\sqrt{1+2e\cos\theta+e^2}} = \sqrt{\frac{\mu}{a}}\frac{\sqrt{1-e^2}}{v}$$

In the following we are only interested in the secular variations of the orbital elements. So we have to integrate over one orbit. From the above equations we establish $\dot{\omega}(-\theta) = -\dot{\omega}(\theta)$ and $\dot{M}(-\theta) = -\dot{M}(\theta)$, implying that each periodic variation is compensated by an equally negative variation on the other side of the orbit. We therefore find

$$\dot{\omega}_{sec} = \frac{1}{2\pi}\int_{-\pi}^{\pi}\dot{\omega}(r,\theta)\cdot d\theta = 0, \quad \dot{M}_{sec} - n = \frac{1}{2\pi}\int_{-\pi}^{\pi}\left[\dot{M}(r,\theta)-n\right]\cdot d\theta = 0$$

So we arrive at the following important result:

> Drag secularly affects only the semi-major axis and the eccentricity.

12.6.2
Elliptical Orbits

We now want to figure out how large these secular variations are. The secular variation is the integral over the total atmospheric drag of one orbit. Due to the exponential dependence of the atmospheric density, drag is by far the strongest around the periapsis, $\theta \approx 0$, for an elliptical orbit. In order to be able to carry out the integration, we need to know the θ dependences of all factors in Eq. (12.6.3). This is particularly true for the atmospheric density $\rho(\theta)$. To find it out, we expand the orbit equation $r = a(1-e^2)/(1+e\cos\theta)$ for small angles at the periapsis

$$r \approx r_{per}\left[1+\frac{e}{2(1+e)}\theta^2\right] = r_{per} + a\frac{1-e}{1+e}\frac{e}{2}\theta^2$$

With this expression $\rho(\theta)$ can be written as

$$\rho(\theta) = \rho_{per} \exp\left(-\frac{r - r_{per}}{H_{per}}\right) \approx \rho_{per} \exp\left(-\frac{ae}{2H_{per}} \frac{1-e}{1+e} \theta^2\right)$$

hence

$$\rho(\theta) = \rho_{per} \exp\left(-\frac{\theta^2}{2\sigma^2}\right) \qquad (12.6.4)$$

with

$$\sigma = \sqrt{\frac{H_{per}}{ea} \frac{1+e}{1-e}}$$

where $\rho_{per} = \rho(r_{per})$ is the atmospheric density at periapsis. By the same token we expand the other terms in Eq. (12.6.3) and find with Eqs.(7.4.10a) and (7.4.10b)

$$v_a^2 = \frac{\mu \left(1 + 2e\cos\theta + e^2\right)}{a(1 - e^2)} \approx \frac{\mu}{a} \frac{1+e}{1-e}\left[1 - \frac{e}{(1+e)^2}\theta^2\right]$$

and

$$\zeta(e, \theta) \approx \sqrt{\frac{1-e}{1+e}} \left[1 + \frac{e}{2(1+e)^2}\theta^2\right]$$

We insert these expressions into Eq. (12.6.3) and obtain

$$\dot{a} \approx -B\rho_{per} \sqrt{\mu a} \left(\frac{1+e}{1-e}\right)^{3/2} \exp\left(-\frac{\theta^2}{2\sigma^2}\right) \left[1 - \frac{3e}{2(1+e)^2}\theta^2\right]$$

$$\dot{e} \approx -B\rho_{per} \sqrt{\frac{\mu}{a}} \frac{(1+e)^{3/2}}{(1-e)^{1/2}} \exp\left(-\frac{\theta^2}{2\sigma^2}\right) \left[1 - \frac{1+2e}{2(1+e)^2}\theta^2\right]$$

We now carry out the secular integration by assuming that $\sigma < \theta \ll 1$, i.e. $e \gg H_{per}/a$. This condition ensures that the density decays within the θ-range where our expansion is valid. Note that for the Earth $0.001 \leq H_{per}/a < 0.01$, where the lower limit holds for $h < 120$ km and the upper limit for $h > 300$ km. We then get with

$$\frac{1}{2\pi}\int_{-\pi}^{\pi} \exp\left(-\frac{\theta^2}{2\sigma^2}\right) \cdot d\theta \approx \frac{\sigma}{\sqrt{2\pi}}, \quad \frac{1}{2\pi}\int_{-\pi}^{\pi} \exp\left(-\frac{\theta^2}{2\sigma^2}\right) \theta^2 \cdot d\theta \approx \frac{\sigma^3}{\sqrt{2\pi}}$$

finally the wanted secular equations of variation:

$$\dot{a}_{sec} = -B\rho_{per}\left(\frac{1+e}{1-e}\right)^2\sqrt{\frac{\mu H_{per}}{2\pi e}}\left[1 - \frac{3H_{per}}{2a(1-e^2)}\right]$$

$$\approx -B\rho_{per}\left(\frac{1+e}{1-e}\right)^2\sqrt{\frac{\mu H_{per}}{2\pi e}} = \dot{a}_{sec,\,circle}\frac{\sigma}{\sqrt{2\pi}} \ll \dot{a}_{sec,\,circle}$$

(12.6.5)

$$\dot{e}_{sec} = -B\rho_{per}\frac{(1+e)^2}{1-e}\frac{1}{a}\sqrt{\frac{\mu H_{per}}{2\pi e}}\left[1 - \frac{H_{per}}{2ea}\frac{1+2e}{1-e^2}\right]$$

$$\approx -B\rho_{per}\frac{(1+e)^2}{1-e}\frac{1}{a}\sqrt{\frac{\mu H_{per}}{2\pi e}} = (1-e)\frac{\dot{a}_{sec}}{a}$$

where $\dot{a}_{sec,\,circle}$ is the secular variation of a circle with $a_{circle} = r_{per}$. So, drag decreases the semi-major axis and the eccentricity of an elliptical orbit at about the same rate. The semi-major axis, however, does not decline as fast as that of a comparable circular orbit. The reason, of course, is the much smaller range around the periapsis within which the body is decelerated as compared to the circumference of the circle.

Decoupling the Differential Equations

These coupled differential equations can be solved by first separating the variables in the second equation (even though we will deal only with secular variations in the following, we drop the index *sec* for convenience)

$$\frac{de}{1-e} = \frac{da}{a}$$

with the solution (see Eq. (7.4.8a))

$$r_{per} = a(1-e) = a_0(1-e_0) = const \qquad (12.6.6)$$

The periapsis hence remains unaffected, while the apoapsis

$$\dot{r}_{apo} = \dot{a}(1+e) + a\dot{e} = \dot{a}(1+e) + \dot{a}(1-e) = 2\dot{a} < 0$$

decreases. This behavior can be seized quite easily if we consider the temporary drag at the periapsis as a deceleration kickburn. According to Eq. (8.4.13) such a kickburn lowers the apsis on the opposite side of the orbit. With that we arrive at the important result:

> Drag circularizes elliptical orbits by lowering the apoapsis, but maintaining the periapsis.

This property is often used after planetary or aerocapture at Mars or Venus to turn the highly elliptic initial orbit down to a circular target orbit without any propulsion effort.

By making use of Eq. (12.6.6) one finally obtains from Eq. (12.6.5) the decoupled differential equations

$$\dot{a}_{sec} = -B\rho_{per}\sqrt{\frac{\mu H_{per}}{2\pi}}\left(\frac{2a}{r_{per}}-1\right)\sqrt{\frac{1}{1-r_{per}/a}}$$

$$\dot{e}_{sec} = -\frac{B\rho_{per}}{r_{per}}\sqrt{\frac{\mu H_{per}}{2\pi}}\frac{(1+e)^2}{\sqrt{e}}$$

(12.6.7)

The results of this section are shown in Fig. 12.18.

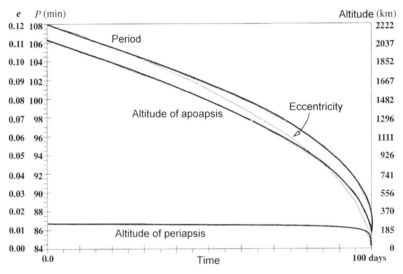

Figure 12.18 Drag-induced decay of orbit parameters with $e_0 = 0.12$ and $h_{per} = 200$ km. Here P is the orbital period and *alt* is the altitude of the perigee or apogee respectively.

12.6.3
Circularization

We are now seeking for solutions of the differential equations with the initial conditions $a_0 = a(t_0), e_0 = e(t_0)$. In the second equation we separate the variables

$$\frac{\sqrt{e}}{(1+e)^2}de = -\frac{B\rho_{per}}{r_{per}}\sqrt{\frac{\mu H_{per}}{2\pi}}dt$$

and find from relevant mathematical tables of integrals the following time dependence of the eccentricity

$$\arctan\sqrt{e} - \frac{\sqrt{e}}{1+e} = \arctan\sqrt{e_0} - \frac{\sqrt{e_0}}{1+e_0} - \frac{B\rho_{per}}{r_{per}}\sqrt{\frac{\mu H_{per}}{2\pi}}(t-t_0) \quad (12.6.8)$$

Since

$$\arctan\sqrt{e} - \frac{\sqrt{e}}{1+e} = e^{3/2}\left(\frac{2}{3} - \frac{4}{5}e + \frac{6}{7}e^2 - \cdots\right)$$

we can approximate

$$e^{3/2}(t) = e_0^{3/2} - \frac{3B\rho_{per}}{2r_{per}}\sqrt{\frac{\mu H_{per}}{2\pi}}(t-t_0) \quad @ \quad \frac{H_{per}}{a} \ll e < 0.20$$

Because we have $a(1-e) = r_{per} = const$, and with Eq. (7.4.9b) it follows that

$$e(t) = 1 - \frac{r_{per}}{a(t)} = \frac{r_{apo}(t) - r_{per}}{r_{apo}(t) + r_{per}}$$

from which we obtain by insertion the analytical orbit equation also for $a(t)$ and $r_{apo}(t)$.

How long would it take to circularize an elliptical orbit? This can be determined quite easily. The circularization time t_{cir} is the time to $e = 0$, i.e. $t_{cir} = t(e = 0)$. Inserting this condition into the above equations, we find with $t_0 = 0$

$$t_{cir} = \frac{r_{per}}{B\rho_{per}}\sqrt{\frac{2\pi}{\mu H_{per}}}\left(\arctan\sqrt{e_0} - \frac{\sqrt{e_0}}{1+e_0}\right) \quad \text{circularization time} \quad (12.6.9)$$

or

$$t_{cir} = \frac{2r_{per}e_0^{3/2}}{3B\rho_{per}}\sqrt{\frac{2\pi}{\mu H_{per}}} \quad @ \quad \frac{H_{per}}{a} \ll e < 0.20 \quad (12.6.10)$$

with e_0 the initial eccentricity, ρ_{per} the atmospheric density and H_{per} the scale height at periapsis. When tracing back the cause of the $e_0^{3/2}$ dependence we find its origin in the fact that, for increasing eccentricities, the stretch s within which the S/C dives into the dense portion of the atmosphere decreases with $s \approx 2\sigma \propto 1/\sqrt{e}$ (see Eq. (12.6.4)). The integral of this dependence then leads to the $e_0^{3/2}$ dependence.

12.6.4
Circular Orbits

When the ellipse is circularized down to $e < H(a)/a$, the body encounters a constant drag force upon circling the planet with radius a. To determine the

12 Orbit Perturbations

secular changes of the orbital elements, the equations (12.5.3) with $e = 0$ and hence $\xi(e, \theta) = 1$ have to be averaged over one orbit. Because of $\langle \cos \theta \rangle_\theta = 0$ and as in a circular orbit $v^2 = \mu/a$ (Eq. (7.4.3)) and $n^2 = \mu/a^3$ (Eq. (7.4.14)), we get $v_a^2 = n^2 a^2$ and thus

$$\dot{a}_{sec} = -\frac{B\rho v_a^2}{n} = -B\rho(a)\sqrt{\mu a} \quad @ \; e < \frac{H(a)}{a} \quad (12.6.11a)$$

$$\dot{e}_{sec} = 0 \quad (12.6.11b)$$

So we can make the following statement:

> Drag constantly decreases the radius of a circular orbit, without changing its eccentricity.

To describe the orbit decay quantitatively we need to solve the differential equation. Separating the variables in Eq. (12.6.11a) results in

$$dt = -\frac{1}{B\sqrt{\mu}}\frac{da}{\sqrt{a}\rho(a)}$$

from which by integration follows

$$t - t_0 = -\frac{1}{B\sqrt{\mu}}\int_{a_0}^{a}\frac{da}{\sqrt{a}\rho(a)} = \frac{1}{B\sqrt{\mu}}\int_{a}^{a_0}\frac{da}{\sqrt{a}\rho(a)} \quad (12.6.12)$$

where a_0, t_0 are the initial values of the orbit. To further evaluate the integral analytically $\rho(a)$ has to be expressed, according to Section 6.1.4, in a piecewise exponential form

$$\rho(a) = \rho(a_i)\exp\left(-\frac{a - a_i}{H_i}\right) = \rho_i \exp\left(-\frac{h - h_i}{H_i}\right) \quad @ \; h_i < h < h_{i+1} \quad (12.6.13)$$

where $h_i := a_i - R_\oplus$ is the base altitude and H_i the scale height for the i-th altitude interval as given in Table 6.2. Correspondingly we also achieve only a piecewise description of the orbit trajectory. For the initial part of the decaying orbit we therefore get from Eq. (12.6.12) with Eq. (12.6.13) and the substitution $a = x + R_\oplus$

$$t - t_0 = \frac{1}{B\rho_0\sqrt{\mu}}\int_{h}^{h_0}\frac{e^{(x-h_0)/H_0}}{\sqrt{x + R_\oplus}}dx \quad (12.6.14)$$

Because the biggest contributions to the integral come from the initial altitude, we can further approximate

$$t - t_0 = \frac{e^{-h_0/H_0}}{B\rho_0\sqrt{\mu}\sqrt{a_0}} \int_h^{h_0} e^{x/H_0} dx = \frac{H_0 e^{-h_0/H_0}}{B\rho_0\sqrt{\mu a_0}} \left[e^{x/H_0}\right]_h^{h_0}$$

from which follows

$$t - t_0 = \frac{H_0}{B\rho_0\sqrt{\mu a_0}} \left[1 - e^{-(h-h_0)/H_0}\right] \tag{12.6.15}$$

With this we derive with $h - h_0 = a - a_0$ the orbit equation

$$a(t) = a_0 + H_0 \ln\left[1 - \frac{B\rho(a_0)\sqrt{\mu a_0}}{H_0}(t - t_0)\right] \tag{12.6.16}$$

for the initial trajectory. We recall that a_0, t_0 are initial orbit values and H_0 the scale height of the initial altitude interval. As an example Figure 12.19 depicts the altitude decay $a(t)$ of the International Space Station.

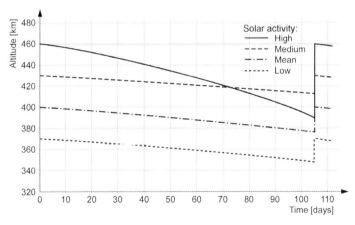

Figure 12.19 Simulated decay of the orbit altitude of the International Space Station at different solar activity phases and reboost after about 100 days.

12.6.5
Orbit Lifetime

When does a S/C without orbit maintenance burn up in the Earth's atmosphere? In the following, we want to determine this so-called orbit life t_L of a circular orbit from a given initial altitude. As we only want to estimate the orbit lifetime we apply Eq. (12.6.15) and use it with the initial values right down to $h = 0$, whereby we only slightly overestimate the orbit lifetime. With this approximation and setting $t_0 = 0$ we get

$$t_L = \frac{H_0}{B\rho_0\sqrt{\mu a_0}} \quad @\ e < H_0/a_0 \quad \text{orbit lifetime} \qquad (12.6.17)$$

with

- $a_0 = h_0 + R_\oplus$, initial radius of the circular orbit;
- $\rho_0 = \rho(h_0)$, atmospheric density at the initial altitude h_0 derived from Eq. (6.1.8);
- $H_0 = H(h_0)$, atmospheric scale height at the initial altitude, with the approximate value as given in Table 6.2 in Section 6.1.4.

A detailed analysis of the integral in Eq. (12.6.15) yields (exercise, Problem 12.5)

$$t_L = \frac{H_0}{B\rho_0\sqrt{\mu a_0}} \left[1 + \frac{H_0}{2R_\oplus} + \frac{3}{8}\left(\frac{h_0}{R_\oplus}\right)^2 - e^{-h_0/H_0} \right]$$

Because for $h_0 > 200$ km the additional terms in the square brackets are smaller than 1%, and for $h_0 < 200$ km orbit lifetime amounts to a few days anyhow, Eq. (12.6.17) is a good approximation for the orbit lifetime. Applying Eq. (12.6.13) the key result is:

> If the orbit lifetime of a given circular orbit is t_{L0}, then any change of the orbit radius by $h_0 \to h_0 + \Delta h$ results in an exponential change of the orbit lifetime according to
>
> $$t_L = t_{L_0} e^{\Delta h / H_0} \qquad (12.6.18)$$

Example

What is the mean orbit lifetime of the International Space Station at an altitude of 400 km at assembly complete?

Answer: For the ISS is $B = 0.0126$ m^2 kg^{-1}. For the altitude range $350\ \text{km} \leq h \leq 400$ km, $H_0 = 53.3$ km, $\rho_0 = 3.725$ g km^{-3}, and $\mu = 3.986 \times 10^5$ km^3 s^{-2}. With Eq. (12.6.17) this results in $t_L \approx 253$ days.

Remark: *To maintain the space station at an altitude of 400 km, it needs a reboost approximately every 100 days. The required propellant for this is about 7500 kg per year.*

If one evaluates the integral in Eq. (12.6.12) numerically with the atmospheric density as given by Eq. (6.1.8) one obtains the mean orbit lifetime as displayed in Fig. 12.20 for a spacecraft with various ballistic coefficients. $B = 0.005\,\mathrm{m^2 kg^{-1}}$ is a good average value.

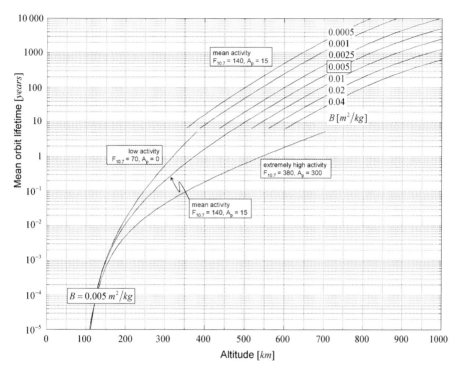

Figure 12.20 In the upper part the mean orbit lifetime as a function of altitude for various ballistic coefficients B is given. If the orbit life time of a S/C is less than six years it varies drastically by more than one order of magnitude due to the actual solar activity. In the lower part these variances are given for a mean $B = 0.005\,\mathrm{m^2 kg^{-1}}$.

Total Orbit Lifetime of Elliptical Orbits

If the initial orbit is elliptical, the total orbit lifetime is the circularization time plus the circular orbit lifetime. Which of the two is prevailing? Since, at the transition between the two phases, we have $r_0 = r_{per}$, we derive from Eqs. (12.6.10) and (12.6.17) the following ratio of the two contributions:

$$\frac{t_{cir}}{t_L} = \sqrt{2\pi} \left(\frac{r_{per}}{H_{per}}\right)^{3/2} \left(\arctan\sqrt{e_0} - \frac{\sqrt{e_0}}{1+e_0}\right) \tag{12.6.19}$$

$$\approx \sqrt{2\pi} \left(\frac{r_{per}}{H_{per}}\right)^{3/2} \frac{2}{3} e_0^{3/2}$$

12 Orbit Perturbations

From

$$e_0 \gg \frac{H_{per}}{a} \approx \frac{H_{per}}{r_{per}}$$

we get

$$\frac{t_{cir}}{t_L} \gg \frac{2}{3}\sqrt{2\pi} = 1.7 \qquad (12.6.20)$$

So, circularization time is much bigger than the circular orbit lifetime. This has the following practical consequence that for a planetary capture and a subsequent circularization to a circular target orbit, the periapsis has to be chosen lower than the target radius to more rapidly turn down the elliptical orbit. When the apoapsis attains the target orbit radius, a kick-burn at the apoapsis (see Eq. (8.4.13)) will increase the periapsis to the target radius. Though this maneuver requires some propulsion effort, it is much less demanding than to circularize the ellipse by propulsion only without making use of the atmospheric drag.

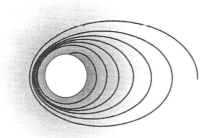

Figure 12.21 Circularization of an elliptical orbit and deorbit due to drag.

Problems

Problem 12.1 Gaussian Variational Equations

To prove the Gaussian Variational Equations (12.1.2) proceed as follows: First prove Eq. (8.4.3) and then apply the reverse transformation as in Problem 8.5. Therefore

1. Prove Eq. (8.4.3). Do this by first considering kick-burns only within the orbital plane and then out of the plane:

   (a) To derive the change of orbital elements for in-plane kick-burns first introduce the angle changes:

   $$d\gamma = \frac{dv_{\perp \odot}}{v}, \quad d\phi = \frac{dv_{\perp \perp}}{v} \quad \text{and} \quad d\eta := \frac{dv_{\|}}{v} = \frac{dv}{v}$$

   To simplify calculations also introduce the orbit number (see Section 14.4.1)

   $$k = \frac{rv^2}{\mu}$$

   With this rewrite the vis-viva Eq. (7.2.10) to

   $$r = (2-k)a \tag{a}$$

   From Section 7.3.2 follows

   $$v = \left(\frac{e\mu}{h}\sin\theta, \frac{h}{r}\right) = v(\sin\gamma, \cos\gamma)$$

   Therefore $h = rv\cos\gamma$. Show that from $h^2 = \mu a(1-e^2)$, Eq. (7.3.7), follows

   $$e^2 = 1 - k(2-k)\cos^2\gamma \tag{b}$$

   and

   $$\sin\theta = \frac{k}{2e}\sin 2\gamma \tag{c}$$

   Convince yourself by a drawing that if a kick-burn takes place at a certain position r in space, which remains constant during the kick-burn maneuver, then any change in θ corresponds to a negative change in ω, i.e. $d\omega = -d\theta$. With this and from (a), (b), (c) prove with the relations $k\cos^2\gamma = 1 + e\cos\theta$, $k\cos 2\gamma = 1 + 2e\cos\theta - e\cos E$, and $k\sin 2\gamma = 2e\sin\theta$

   $$da = \frac{a}{2-k}dk = \frac{2ak}{2-k}d\eta$$

$$de = 2(e + \cos\theta) \cdot d\eta + \frac{r}{a}\sin\theta \cdot d\gamma$$

$$e \cdot d\omega = 2\sin\theta \cdot d\eta - (e + \cos E)\, d\gamma$$

(b) To derive the changes of orbital elements for out-of-plane kick-burns consider the definitions made in the following figure

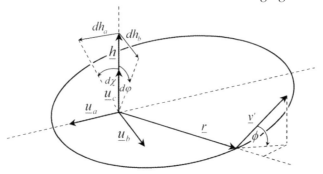

Show that from

$$dv = \frac{\partial v}{\partial \gamma} d\gamma + \frac{\partial v}{\partial \phi} d\phi + \frac{\partial v}{\partial \eta} d\eta$$

follows

$$dv = v \cdot \big[\cos(\theta - \gamma) \cdot d\gamma,\; \sin(\theta - \gamma) \cdot d\gamma,\; d\phi\big]$$
$$+ v \cdot \big[-\sin(\theta - \gamma),\; \cos(\theta - \gamma),\; 0\big] d\eta$$

and therefore

$$dh = d(\mathbf{r} \times \mathbf{v}) = \mathbf{r} \times d\mathbf{v}$$
$$= rv\big[\sin\theta \cdot d\phi,\; -\cos\theta \cdot d\phi,\; -\sin\gamma \cdot d\gamma + \cos\gamma \cdot d\eta\big]$$

Because on the other hand $dh = (dh_a, dh_b, dh_c) = (h \cdot d\chi,\; h \cdot d\varphi,\; dh)$ follows

$$d\chi = \frac{\sin\theta}{\cos\gamma} d\phi, \quad d\varphi = -\frac{\cos\theta}{\cos\gamma} d\phi, \quad \frac{dh}{h} = -\tan\gamma \cdot d\gamma + d\eta$$

Let $d\sigma_a, d\sigma_b, d\sigma_c$ be the positive deflections around the coordinate axes (u_a, u_b, u_c) for which holds (see e.g. (Kaplan, 1976, Eq.(1.28)))

$$\begin{pmatrix} d\Omega \\ di \\ d\omega \end{pmatrix} = \frac{1}{\sin i} \begin{bmatrix} \sin\omega & \cos\omega & 0 \\ \cos\omega \sin i & -\sin\omega \sin i & 0 \\ -\sin\omega \cos i & -\cos\omega \cos i & \sin i \end{bmatrix} \begin{pmatrix} d\sigma_a \\ d\sigma_b \\ d\sigma_c \end{pmatrix}$$

Show that

$$\begin{pmatrix} d\sigma_a \\ d\sigma_b \\ d\sigma_c \end{pmatrix} = \begin{pmatrix} -d\varphi \\ d\chi \\ -d\theta \end{pmatrix}$$

from which the desired change of orbital elements are derived.

2. Apply the reverse transformation as in Problem 8.5 to finally derive the Gaussian Variational Equations (12.1.2).

Problem 12.2 Earth's Flattening and J

Prove the relation (see Eq. (12.4.2))

$$J \approx \frac{f}{1+f} - \frac{\omega^2 R_\oplus^3}{2\mu} \approx f - \frac{\omega^2 R_\oplus^3}{2\mu}$$

between

$$J := -\frac{3}{2}J_2 := -\frac{3}{2}C_2^0 = 0.001\,623945 \times 10^{-3}$$

and Earth's flattening

$$f := \frac{R_{equatorial} - R_{polar}}{R_{equatorial}} \approx \frac{1}{298.253}$$

by showing that

$$J + \frac{\omega^2 R_\oplus^3}{2\mu} = \left(1 + \frac{\omega^2 R_\oplus^3}{2\mu}\right) \frac{f}{1+f} + O\left(f^2\right)$$

Hint: Use the fact that the free surface of the ellipsoid is a surface of equipotential and use Eq. (7.2.12).

Problem 12.3 Triaxial Motion

Show that the equation of motion (12.4.5) for a satellite at an unstable GEO position

$$\ddot\lambda = -\frac{1}{2}\omega_\lambda^2 \cdot \sin 2(\lambda - \lambda_0)$$

with initial condition $\lambda = \lambda_i \neq \lambda_0$ has the approximate solution

$$\lambda = \lambda_i - \frac{1}{4}\omega_\lambda^2 \sin[2(\lambda_i - \lambda_0)] \cdot t^2 + \frac{1}{96}\omega_\lambda^4 \sin[4(\lambda_i - \lambda_0)] \cdot t^4 - O(t^6)$$

Problems

Problem 12.4 Orbit Changes by Atmospheric Maneuvers

(a) After the Columbia accident on February 1, 2003, NASA administrator O'Keefe cancelled any Hubble repair mission, because from the Hubble telescope the Space Shuttle was supposed not to be able to reach the ISS as a safe haven. Given the orbit elements of the ISS (altitude $= 400$ km, $i = 51.63°$, $e \approx 0$) and the Hubble telescope (altitude $= 590$ km, $i = 28.5°$, $e \approx 0$) and the fact that the OMS engines of a Shuttle can only provide a delta-v of about $\Delta v = 200$ m s^{-1}, show that O'Keefe was right.

(b) We saw in Section 12.6.2 that it is possible to change the semi-major axis of an orbit by dragging through the atmosphere at the periapsis of an elliptical orbit. Suppose the vehicle also has lift. The vehicle then receives the delta-v changes $\delta v_\parallel$, $\delta v_\odot$ or $\delta v_\parallel$, $\delta v_{\perp\perp}$ depending on the orientation of the lift vector. Show by a similar procedure as in Section 12.6.2 that the delta-v change due to one fly-through the periapsis of an ellipse is given by

$$|\delta v_\parallel| \approx \frac{C_D \, \rho_{per} A_\perp}{2 \, m} (1+e) \sqrt{\frac{2\pi \mu H_{per}}{e}}$$

$$|\delta v_{\perp\odot}| = |\delta v_{\perp\perp}| \approx \frac{C_L \, \rho_{per} A_\perp}{2 \, m} (1+e) \sqrt{\frac{2\pi \mu H_{per}}{e}}$$

(c) Now consider a life threatening situation on-board the Shuttle during Hubble repair. Show that, though an inclination turn is in principle possible with an atmospheric maneuver, the following orbit maneuver would not be feasible: Lower one side of the Shuttle orbit by a small deorbit burn such that the now slightly elliptical orbit would touch the atmosphere at its periapsis. The Shuttle would receive a delta-v of $\delta v_\parallel$, $\delta v_{\perp\perp}$ thereby decelerating but also change the inclination $28.5° \rightarrow 51.63°$ due to its lift, $C_L/C_D = L/D \approx 1.3$. Finally the elliptical orbit would be raised to a circular ISS LEO and the Shuttle maneuvered to the ISS by some negligible rendezvous maneuvers.

Hint: Consult Table 8.1.

Problem 12.5 Orbit Life Time

Prove that detailed analysis of Eq. (12.6.15)) leads to the following orbit life time

$$t_L = \frac{H_0}{B \rho_0 \sqrt{\mu a_0}} \left[1 + \frac{H_0}{2 R_\oplus} + \frac{3}{8} \left(\frac{h_0}{R_\oplus} \right)^2 - e^{-h_0/H_0} \right]$$

Problem 12.6 King-Hele's Orbit Life Time

In his reputed book (King-Hele, 1987, p. 60ff) the author provides the following expression for the orbit life time of a satellite in a LEO

$$t_L \approx -\frac{3e_0 T_0}{4\dot{T}_0} \frac{I_0(z_0)}{I_1(z_0)} \left(1 + 2e_0 \frac{I_1(z_0)}{I_0(z_0)} - \frac{9e_0 z_0}{40} + \frac{H}{2a_0}\right) \quad @ \; z_0 = \frac{a_0 e_0}{H_0} < 3$$

with

$$I_0(z) = 1 + \frac{z^2}{4} + \frac{z^4}{64} + \frac{z^6}{2304} + \cdots \quad \text{Bessel function of first kind and order 0}$$

$$I_1(z) = \frac{z}{2}\left(1 - \frac{z^2}{8} + \frac{z^4}{192} - \frac{z^6}{9216} + \cdots\right) \quad \text{Bessel function of first kind and order 1}$$

Show that for an elliptical orbit from the above follows

$$t_L \approx \frac{H_0}{B\rho_0\sqrt{\mu a_0}} \left(1 + \frac{H_0}{2a_0}\right) \approx \frac{H_0}{B\rho_0\sqrt{\mu a_0}} \left(1 + \frac{H_0}{2R_\oplus}\right)$$

and therefore Eq. (12.6.17) holds.

13
Coordinate Systems

13.1
Space Coordinate Systems

To delineate the general trajectory of a body mathematically or graphically, one needs a space coordinate system. There is no restriction to the choice of the frame. Any frame will do. However, mathematical equations take on different forms in different systems. Inappropriate systems cause inefficient mathematical representations, which may even become untreatable, while being straightforward in a suitable system. So, choosing the right coordinate frame is of paramount importance. However, there is a class of frames that are special in their own: the inertial reference frames. Only in these systems do the physical laws and equations take on their established form. "Inertial" means "non-accelerated." A body in an inertial coordinate frame, which is not exposed to an external force (field), experiences a non-accelerated motion and therefore moves in a straight line. In a non-inertial coordinate frame, motion will not be in a straight line. In a rotating coordinate frame, a force-free body experiences a fictitious force, which fictitiously acts from the outside, because the body's trajectory deviates from the straight line. The cause of the deviation is the acceleration of the system itself – it rotates. This difference makes inertial systems so special – and the physics in particular especially easy.

Even though one commonly speaks of inertial systems in the plural, there is only one type. According to a postulate of the Austrian physicist Ernst Mach (1838–1916), inertia of a body emerges from the influence of all the masses in our universe on that body. If the inertial force is zero (non-accelerated body in a force-free state), the corresponding inertial reference frame must be fixed relative to the mean of these universal masses. The fixed stars are our reference to the universe, which is why inertial systems are also called "sidereal systems." Because at any practical scale the universe is infinite, this implies that any other reference frame, which relative to the inertial one is translationally and/or rotationally shifted, and/or moves with a constant velocity, experiences no acceleration as well and therefore is also an inertial reference

Astronautics. Ulrich Walter
Copyright © 2008 WILEY-VCH Verlag GmbH & Co. KGaA, Weinheim
ISBN: 978-3-527-40685-2

frame. Mathematically speaking, we have the following. If r and r' are vectors in the inertial reference frames F and F' respectively, then both vectors are interrelated through the transformation $r' = \boldsymbol{R} r + v \cdot t + r_0$, with time-constant rotation matrix $\boldsymbol{R}(t) = const$, constant relative velocity $v(t) = const$, and translation r_0 at $t = 0$. This is the most general form of a coordinate transformation, which is compatible with the requirement of a non-accelerated motion, that is $\ddot{r} = 0 \to \ddot{r}' = 0$. Note that rotating reference frames are excluded by this, because then $\boldsymbol{R}(t)$ and hence $\ddot{\boldsymbol{R}}$ would be time-dependent. In summary, we can state that:

> **Inertial (sideral) reference frames** are coordinate systems with a fixed relation to the entirety of the masses in the universe. They are related to each other through the transformation $r' = \boldsymbol{R} r + v \cdot t + r_0$ with $\boldsymbol{R}(t) = const$.

The international Celestial Reference Frame ICFR

What are the fixed points in our universe to which we can fix an inertial reference frame? They are stars so far away that their relative motion remains undetectable even with the best of our telescopes. The International Celestial Reference Frame (ICRF) is the internationally standardized inertial reference frame. It is based on the Fifth Fundamental Catalog (FK 5), which defines 608 (as of 1986) compact extragalactic (distance!) radio sources. Of these, 212 are adopted as fixed points to establish the ICRF coordinate axes with an accuracy of 0.02 milliarcseconds. The origin of this frame is the barycenter (center of mass) of our solar system. The base plane (vertical to the z axis) is the equatorial plane at January 1, 2000, 12.00 h (so-called J2000, today's reference (standard) epoch) and the x axis is the direction to the quasar 3C273. The y axis is determined from $y = z \times x$.

The Heliocentric Reference Frame

The ICFR is a perfect physical set-up, but almost useless in practice, because the measurement of extragalactic radio sources is not in an astronomer's daily work. But what can easily be measured from the observation of the Sun's movement is the ecliptic plane, in which the Earth moves around the Sun and which in good approximation is fixed in space. The vertical to this plane is the first axis (z axis) of a convenient inertial frame with the Sun at its origin. What is still needed is a distinguishable direction in the ecliptic. A convenient choice is its intersection with the terrestrial equatorial plane, which makes up a line of nodes (see Fig. 13.1). This line is chosen as the x axis, with the direction from the origin (Sun) to the ascending node (see Section 7.3.4) of the Earth's orbit as its positive part. As seen from the Earth, the Sun at vernal equinox

– around March 21 each year – points along this x axis to the so-called *First Point of Aries* ♈, which is an imaginary point in the celestial sphere.

Note: *Today the Sun at vernal equinox is positioned in the Sign of Pisces as seen from the Earth. 4000 years ago – at the dawn of western astronomy – the Sun at vernal equinox was positioned in the first star of the Sign of zodiac Aries as see from the Earth. This is why the direction at vernal equinox is still marked by Aries' symbol* ♈. *This direction is the same in which the Earth is positioned at autumnal equinox as seen from the origin (Sun).*

The final y axis is again determined from $y = z \times x$. This reference frame is called the *heliocentric ecliptical system*. Because Earth's axis precesses and nutates (see Section 15.3), the equatorial plane and thus the x axis are not fixed. So, the heliocentric ecliptical reference frame is not perfectly inertial. But because the precessional motion is very slow (one rotation within 27 500 years) and the nutational motion very small (9.2″ amplitude), and both are known very precisely, this is still a good reference frame for most practical purposes, if the variations are taken into account.

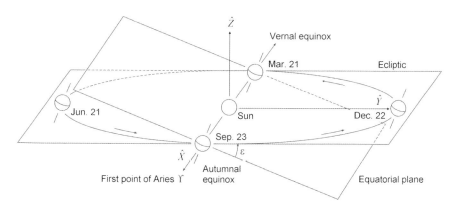

Figure 13.1 The heliocentric ecliptical reference system XYZ. The direction First Point of Aries ♈ is determined by the ascending node of the intersection between the terrestrial equatorial plane and the ecliptic, where the Sun is located at March 21 as seen from Earth.

This example nicely shows how reference frames in general are defined for practical purposes: Take a rotational movement (terrestrial rotation, orbit of the Earth around the Sun). It defines a rotation axis. Due to the conservation of angular momentum, this axis is inertial (as long as the body is not subject to perturbative forces). This is the z axis (polar axis). This axis defines an invariable plane (orbital plane, equatorial plane), in which a distinguished direction makes up the x axis. The y axis finally results from $y = z \times x$.

Terrestrial and Topocentric Reference Frames

The geocentric equatorial reference frame as shown in Fig. 13.2 is defined in exactly this way. Its x, y, z frame is denoted in celestial mechanics by I, J, K. Since the line of nodes made up by the ecliptic and the equatorial plane is common to both planes, the vernal equinox can be chosen for both geocentric equatorial and heliocentric ecliptical reference frames.

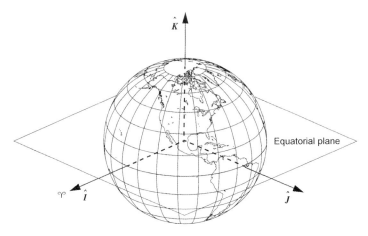

Figure 13.2 The inertial geocentric equatorial reference frame IJK.

The reference frames discussed so far are just a few of many possible reference frames, which can be classified according to the position of the point of origin, as well as the x-y plane, and the orientation of the x and y axes. Some of the frames and their corresponding terms are listed in Tab. 13.1.

Table 13.1 Some of the many possible frames of reference.

Position of the point of origin	Denomination
Center of the Sun	Heliocentric
Center of the Earth	Geocentric
Center of mass	Barycentric
Position of observation	Topocentric

Orientation of the x-y plane	Denomination
Ecliptic plane	Ecliptical
Equatorial plane	Equatorial

The three reference frames commonly employed for satellite applications are depicted in Fig. 13.2 to 13.4. The specific selection of a reference frame depends on whether one wants to study the orientation of the satellite's orbit (inertial IJK frame), its orbit shape (PQW frame), or other properties such

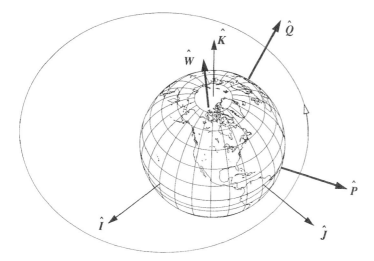

Figure 13.3 The perifocal reference frame PQW based on the satellite's orbit. The P-Q plane is the orbital plane and the P axis points to the perigee.

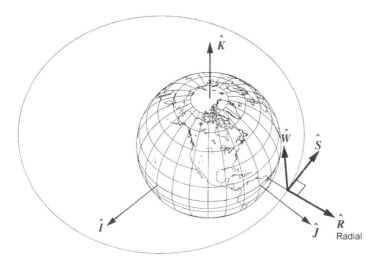

Figure 13.4 The topocentric reference frame RSW co-rotates with the satellite. The R-S plane is the orbital plane with R being the normalized vector to the position of the satellite.

as satellite attitude dynamics (RSW frame). For reference frames related to a satellite (PQW and RSW frames), its orbital plane naturally is the x-y plane, which is why it does not enter the denomination for these reference frames. A detailed account of these and any many other reference frames is given in Vallado (2001).

13 Coordinate Systems

If the vernal equinox is selected as orientation for the x axis in an equatorial reference frame (Fig. 13.5), the direction to an arbitrary celestial point is measured as the so-called *right ascension* α and *declination* δ. This is done by first projecting the observed direction onto the x-y plane of the reference frame. The right ascension α is then the angle between the vernal equinox and the direction to this projected point, while the declination δ is the angle from the direction to the projected point to the observed direction. Things change in an ecliptic system, where the terms *ecliptic longitude* λ and *latitude* ϕ are used. However, if one talks about longitude and latitude also in an equatorial system, longitude λ refers not to the vernal equinox, but to the 0° longitude of Greenwich. Latitude ϕ and declination δ are however always identical.

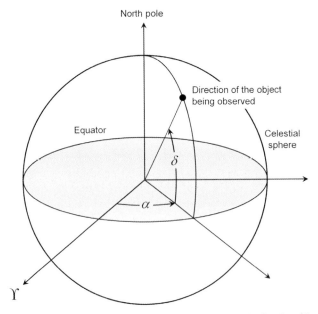

Figure 13.5 Definition of the right ascension α and declination δ in an equatorial reference frame.

For long-term celestial observations it is necessary to have a temporally fixed reference frame. The IJK frame however is not a fixed reference frame. As already mentioned, the gravitational perturbation of neighboring planets causes the Earth's axis to precess (see Section 15.3) and thus the vernal equinox shifts 1 degree in 72 years. So today, the vernal equinox is 56 degrees further in the direction of the sign of Pisces. This implies that for practical applications one has to define a specific reference frame by choosing an epoch (reference time). Presently the epoch January 1, 2000, 12.00h (J2000) is commonly used and named (astronomic) *standard epoch*. Astronomical data that refer to this standard epoch, are indexed with J2000; for instance α_{J2000} for the right ascension.

Transformation between Reference Frames

Frequently the problem arises to transform between two reference frames. The most general transformation between two vectors r and r' in the reference frames F and F' is $r' = (R + T) r$, where T is an arbitrarily time-dependent translation matrix and the rotation matrix may be arbitrarily time-dependent as well. For inertial frames we had $R(t) = const$, $Tr = v \cdot t + r_0$. To find the coordinate transformation equations between two given reference frames is an art in its own, which however is pure mathematics. Because the derivation of these transformations doesn't add to the understanding of astrodynamics, we merely refer to the work of Vallado (2001, Chapter 3), who treats coordinate transformations extensively.

13.2 Time Coordinates

In the following, we want to furnish the one-dimensional continuum of time with a coordinate, in order to determine a point in time unequivocally. To do this, we have to:

1. make time measurable, and
2. to determine a point of origin.

Measuring Time

The simplest way to measure time is with steady, periodic events, as periods are countable. Measuring time is thus reduced to counting periods of uniform length. The most obvious period for humans has always been a day. If you divide the day into 24 smaller periods, and these into 60 even smaller sub-periods, etc., then you get the well-known units of time: hour, minute, and second. The problem of this type of time-keeping starts with the question: What exactly is a day? If it is the time period between two successive culminations of the Sun, then we are talking about the *solar day* with duration of 24 h. If it is the time period between two successive culminations of a star, then we are talking about a *sidereal day* with duration of only 24 h 56 min 4 s. The small, but important, difference comes about because of the rotation angle of the Earth around the Sun during one day (see Fig. 13.6).

However, this ambiguity is only the beginning of the problem to find the right measure for time. Because of the irregularity of the orbital motion of the Earth (slight eccentricity) and the inclination of the ecliptic, a solar day does not always have the same duration. The time difference between the longest day (November 4) and the shortest day (February 12) of a year is almost 31 minutes. In 1925, there was an attempt to level out this flaw by defining a *mean*

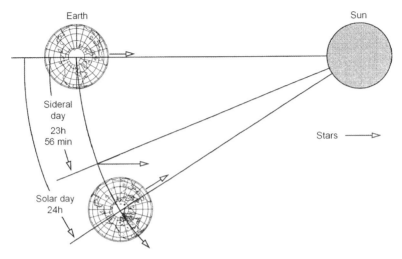

Figure 13.6 Difference between a solar day and a sidereal day.

solar day, the average of all the days the Earth needs to circle around the Sun once. The time determined by the mean solar day is called *Universal Time (UT)* or *Greenwich Mean Time (GMT)*, and it refers to the Greenwich meridian. But even UT is not constant. Because of the constant deceleration of the terrestrial rotation by means of the tidal impact of the Moon, the mean length of a day is reduced by about 2.1 ms per century. This flaw as well as other irregularities is considered in the time coordinate *UT1* derived from *UT*.

Due to these unsteadiness of the Earth's rotation, in 1972 it was agreed to use a new time standard, which is not based on astronomical observations, but on the extremely steady oscillation of atoms, the so-called *International Atomic Time (TAI* = Temps Atomique International). According to TAI, one second (1 s) is the time elapsed during 9 192 631 770 periods of the radiation of Cs-133 atoms corresponding to the transition between the two hyperfine structural levels of their ground state. To get this time in line with UT1, the *Coordinated Universal Time (UTC*, also called Zulu time) was introduced. UTC is a measure of time that is based on TAI. To be in accordance with the season, represented by UT1, deviations from UT1 are considered by introducing leap-seconds on December 31 and, if necessary, on June 30, if UTC and UT1 differ more than ± 0.9 s. Today UTC is the standard civil time and broadcasted by radio stations worldwide.

Referencing Time

As time can be measured now very exactly, the only open issue is to define an arbitrary point of origin in time. Once this point has been defined, any event in time (in astronomy, the moment (event) of a special position of a celestial body is also called an epoch) before or after that can be determined by counting the days, hours, minutes, etc. from the origin. The counted number is the date for the event. The time distance between two dates is a time interval. A list of successive dates is called a calendar. All calendars have one thing in common: a date is counted in units of days. Their only difference is that they have different points of origin, as they can be selected arbitrarily. The worldwide distributed civil calendar today is the Gregorian calendar, with its point of origin at the birth of Christ on January 1 of the year 1. The day before that was December 31, year –1 (December 31, 1 BC). So there is no year 0. The Gregorian calendar is based on a length of the year of 365.2425 days. In order to keep the vernal equinox on March 21, every four, 100, and 400 years a leap day has to be introduced.

The Julian Date

In astronomy, on the other hand, the so-called *Julian Date* JD, is used as the calendar – not to be confused with the ancient Roman Julian calendar. Its point of origin is 12 h noon on January 1, 4713 BC, which marks the beginning of Julian day 0. A Julian date is the decimal number of the time interval between this origin and the given point in time in unit of days. So a Julian date does not have weeks, months and years and therefore doesn't have the consistency problems that arise in civil calendars due to leap-days, leap-months and various calendar reforms. A date in the Gregorian calendar given in years (yr), months (mo), days (d), hours (h), minutes (min), and seconds (s) for the years $1901 \leq yr \leq 2099$ can be converted into JD as follows:

$$JD = 367 \cdot yr - INT\left\{\frac{7\left[yr + INT\left(\frac{mo+9}{12}\right)\right]}{4}\right\} + INT\left(\frac{275 \cdot mo}{9}\right)$$

$$+ d + 1721013.5 + \frac{\frac{s/60 + min}{60} + h}{24}$$

where the function $INT(x)$ truncates the real number x to the next lower integer number. Conversion algorithms between a Gregorian date and a Julian date can be found in literature as for instance in Montenbruck and Gill (2000, Appendix A.2).

Remark: *The odd origin 4713 BC was chosen, because in the early days several calendars had to be combined into the Gregorian calendar, where by the way of calculation January 1, 4713 BC turned out to be the lowest common denominator for the origin. Then 12.00 h noon was chosen in order not to have a day leap by using the ephemerides, because the ancient astronomers observed the sky at night.*

Other Astronomical Calendars

As nowadays the number of days of a Julian date is pretty high, two versions of the Julian date are commonly used today:

1. The modified Julian calendar, called *Modified Julian Date* (MJD), is defined as

   $$MJD = JD - 2400000.5$$

   The additional half-day was introduced so that the beginning of a day (midnight) is in line with UTC.

2. For numerical calculations in astronomy, *Julian centuries T* with origin J2000 = 2 451 545.0 = January 1, 2000, 12.00 h (*standard epoch*, see Section 13.1) are widely used:

   $$T = \frac{JD - 2\,451\,545.0}{36\,525}$$

A more detailed account on time coordinates and their conversions can be found in Vallado (2001).

14
Orbit Determination

Up to now we have conveniently used orbital elements to describe the orbit of a satellite in a two-body system. As already seen in Section 7.3.4, the so-called state vector (r, v) and the Keplerian elements $(a, e, i, \omega, \Omega, \theta)$ are two equivalent sets of the six orbital elements. The conversion between these two sets is a common problem, which we will examine more closely in Sections 14.4. But how are the orbital elements of an actual orbit determined? This chapter is to give an answer to this question.

To be able to specifically determine the six orbital elements for a given spacecraft, we need to observe at least six suitable components of r and/or v of the S/C. As observables always include errors of observation, they also bring about errors of the derived orbital elements. So, to improve the accuracy of the derived orbital elements, far more than six observations are usually taken. Usually the same orbit parameters are determined several times successively. Apart from observational errors, orbit perturbations occur due to gravitational asymmetries, solar winds, drag, etc. To determine the exact orbit under all these constraints, we need specific methods, which will be explained in the following.

14.1
Orbit Measurements

14.1.1
Radar tracking

Orbit tracking is usually done with a ground-based parabolic radar antenna (see Fig. 14.1), which is directed towards the satellite in question. A signal is sent to the satellite and either a passively reflected signal is detected (Fig. 14.2a) or a transponder on-board the satellite returns the signal with a well-known response time (Fig. 14.2b). Either way, radar tracking provides us with the following data:

- the *azimuth* and *elevation* (pointing angles) of the receiving antenna in the topocentric system of the ground station by tracking the direction of the maximum of the received signal (an antenna with diameter of 15 m has an angular resolution of typically 0.1°);

- the *distance* (two-way slant range) from the ground station to the S/C by measuring the runtime of the returned signal with accuracy between 1 m and 20 m;
- the *radial velocity* (range rate) with regard to the ground station by measuring the Doppler shift of the returned signal with an accuracy of 0.1–1 $mm\,s^{-1}$.

Figure 14.1 The 10 m parabolic antenna of the Space Science Laboratories at Berkeley, California.

Even though orbit determination without Doppler shift is basically possible, nowadays virtually every satellite has a transponder, making orbit determination much faster and easier. Orbit determination without Doppler shift is used only if satellites don't respond, for small satellites that don't have a transponder, and for space monitoring (space debris).

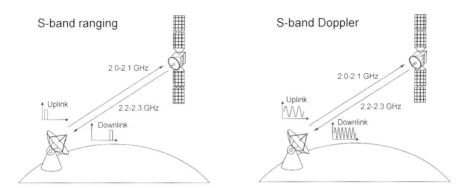

Figure 14.2 Current S-band methods using signal roundtrip time to determine the slant range (left side) and using Doppler shift to determine the two-way range rate (right side) of a satellite.

14.1.2
Other Tracking Systems

Tracking from ground stations has the disadvantage that the satellite can be observed only during the short time of passage, so it is relatively inaccurate. NASA solved this problem by taking orbital data with its *Tracking and Data Relay Satellites* (TDRS) in GEO. Currently nine satellites are in use centered around White Sands Ground Terminal in New Mexico and separated from each other by 130° longitude. However only two are active at the same time. Thus they cover 85–100% of all LEO satellites.

Today, also optical tracking systems are used, such as the satellite laser ranging of the US Natural Environment Research Council, NERC (see Fig. 14.3), or imaging systems such as the US Ground-based Electro-Optical Deep Space Surveillance Telescope (GEODSS). Imaging systems reach an angle resolution of typically one arcsecond, and thus are far more accurate than radars. Laser ranging systems also have a better angular resolution and have the additional advantage that they don't require a satellite transponder. On the other hand they don't work when it's cloudy. They determine the range from the runtime of a signal reflected from the satellite's surface quite accurately to just some centimeters.

All ground-based systems suffer from the refraction (changes of the ray's path due to varying atmospheric density) of signals by the atmosphere (atmospheric refraction and ionospheric refraction), which moreover is time-dependent. If the orbits are to be tracked accurately, these effects have to be

Figure 14.3 Satellite laser ranging at NERC.

accounted for, which is a quite complex task. So, the complexity of tracking a satellite increases with the required accuracy. This general rule seems to have been turned upside down lately by the advent of space-qualified GPS receivers on-board a spacecraft. These receivers are not only relatively cheap, but they allow position and velocity determination on-board the S/C in real time. The advantage here is that orbit determination and possible required orbit corrections can be determined on-board. Orbit determination efforts are thus transferred from the ground station to the spacecraft, which considerably reduces mission control efforts on the ground.

14.2
Method of Orbit Determination

The accuracy of position and velocity tracking nowadays is so high that it doesn't suffice to just determine the orbital elements with the measured orbit parameters, but one also has to consider variations of the orbital elements due to the perturbations of the gravitational potential or external forces. This is done by numerically solving the equations of motion that include these perturbations. The solution then provides the time-dependent state vector $(r(t), v(t))$, representing a comprehensive description of the orbit. The solution of the orbit tracking problem therefore is as follows:

1. Carry out a sufficient number of measurements of orbit-specific parameters.

2. Use these data to determine with a least-squares method (for batch operation on the ground) or Kalman filtering method (for on-board sequential real-time processing) the position and velocity of the S/C (so-called *orbit estimation*). This generates the so-called *measurement model* $(r, v)_m$ of the state vector.

3. Use this measurement model as initial values to solve with numerical methods the equations of motion (see Section 12.3.1), the precision of which is chosen corresponding to the accuracy of the measurement model. The solution propagates the orbital state into the future and is called the *theoretical trajectory model* of the state vector $(r, v)_t$.

4. Compare the predictions of the theoretical trajectory model with the measurement model updated by further measurements. To minimize the deviations of the two models – the so-called *residuals* – vary still unknown model parameters (such as drag coefficient or remaining atmospheric density). By this procedure the still unknown model parameters are determined.

5. When propagation limits are reached, noticeable by increasingly unresolved residuals, a new iteration starts. The measurement model is updated by measurements (so-called *differential correction*). It is then used as updated initial values for the solver of the equations of motion, whereby the theoretical prediction of the orbit is improved.

This procedure clearly shows not only that this orbit determination method does accurately determine the trajectory, but also that by adjusting the parameters of unknown perturbations one can determine their characteristics. This is exactly the way by which the coefficients of the terrestrial potential (see Section 12.2.1) were determined in past years with missions such as CHAMP or GRACE by high-precision measurements of their orbits.

In the following we will just explain some basic methods of the second step, because they are of general interest. A detailed explanation of how all these five steps are implemented in practice would be far beyond the scope of this book. For details, the interested reader should consult the books of Montenbruck and Gill (2000), Tapley et al. (2004), and Vallado (2001).

14.3
Orbit Estimation

How are the orbital elements determined from the measurements of components of r and/or v (so-called *orbit estimation*)?

In practice, the already mentioned three types of data (pointing angles, slant range, and range rate) as provided by orbital tracking are recorded at least twice for different times. From these one obtains with the so-called *Site-Track algorithm* (see Vallado (2001)) or the *Homotopy Continuation Method* (see Montenbruck and Gill (2000) a relatively robust measurement model of the state vector $(r, v)_m$, or of the orbital elements, which can be converted to each other (see the following Section 14.4).

These algorithms are based on fundamental methods, which were already developed by Gauss and Laplace (for at least three measurements of pointing angles) and Lambert (for at least two measurements of azimuth, elevation, and distance) to determine planetary orbits (for details see Vallado (2001)). Two of them will be outlined in the following.

14.3.1
Simple Orbit Estimation

For a preliminary estimation of the state vector of a satellite in a low Earth orbit, the following method is suitable. It is based on the position vectors r_0, r_1 measured at the time interval Δt in the course of a passage. Let $r_0 = r(0)$ be the position vector at the first observation time $t_0 = 0$. Any time later it can be expressed as a Taylor series

$$r = r_0 + \dot{r}_0 t + \frac{1}{2}\ddot{r}_0 t^2 + \frac{1}{6}\dddot{r}_0 t^3 \tag{14.3.1}$$

From this it follows by differentiation that

$$\ddot{r} = \ddot{r}_0 + \dddot{r}_0 t \tag{14.3.2}$$

We now apply Newton's equation of motion $\ddot{r} = -\gamma r$ with $\gamma := \mu/r^3$

$$-\gamma r = \ddot{r}_0 + \dddot{r}_0 t$$

from which with Eq. (14.3.2) and with the conditions $r(t_0 = 0) = r_0$ and $r(\Delta t) = r_1$ it follows that

$$-\gamma_0 r_0 = \ddot{r}_0$$

$$-\gamma_1 r_1 = \ddot{r}_0 + \dddot{r}_0 \cdot \Delta t$$

Solving these equations for $\ddot{r}_0$ and $\dddot{r}_0$ and inserting the results into Eq. (14.3.1) for $r(\Delta t) = r_1$ one obtains for $\dot{r}_0 = v_0$

$$v_0 = \frac{r_1 - r_0}{\Delta t} + \frac{\mu}{6}\left(\frac{2r_0}{r_0^3} + \frac{r_1}{r_1^3}\right) \cdot \Delta t \qquad (14.3.3)$$

Thus having determined the state vector (r_0, v_0) at time t_0, the orbital elements can be deduced as shown in Section 14.4.1.

14.3.2
Lambert's Method

We exemplify the derivation of orbital elements from observational data by Lambert's problem. Lambert's problem is as follows: given two orbit radii r_1, r_2, the change of true anomaly $\Delta\theta$, and the time of passage Δt between them. What are the orbital elements a, e? We only want to provide a summary and a limiting case of the solution; for details see e.g. Prussing and Conway (1993).

Lambert's Equation

Provided the measurements $(r_1, r_2, \Delta\theta, \Delta t)$ are given, we define the auxiliary variables

$$d := \sqrt{r_1^2 + r_2^2 - 2r_1 r_2 \cos(\Delta\theta)} \qquad \text{(orbit secant)}$$

$$\cos\alpha := 1 - \frac{r_1 + r_2 + d}{2a} \qquad (14.3.4)$$

$$\cos\beta := 1 - \frac{r_1 + r_2 - d}{2a}$$

by which the following conditional equation can be given (see e.g. Prussing and Conway (1993))

$$\Delta t = \sqrt{\frac{a^3}{\mu}}\left[(\alpha - \sin\alpha) - (\beta - \sin\beta)\right] \quad \text{Lambert's equation} \qquad (14.3.5)$$

Unfortunately, Lambert's equation provides the solution of the semi-major axis a at a given $\Delta\theta$ and Δt only very indirectly. In literature there are different semi-analytical methods to solve Lambert's equation for a. From a historical point of view, the Lambert–Gauss solution is quite interesting, which was used by Gauss in 1809 to determine the orbit of the asteroid Ceres, by which he rediscovered it one year after its last observation. All of them are however quite complicated (see Vallado (2001)) and are numerically unstable or don't converge quickly. In practice, Lambert's problem is solved today with a Newton iteration or – in the case of difficult hyperbolic orbits – with the bisection method.

Eccentricity Determination

When a is derived, e still has to be determined. To do this, we recall the orbit equation (7.3.5), which we solve for $\cos\theta$

$$\cos\theta_i = \frac{1}{e}\left[(1-e^2)\frac{a}{r_i} - 1\right]$$

$$\sin\theta_i = \frac{a\sqrt{1-e^2}}{er_i}\sqrt{e^2 - \left(1 - \frac{r_i}{a}\right)^2}$$

from which we finally get

$$\cos(\Delta\theta) = \cos\theta_1 \cdot \cos\theta_2 + \sin\theta_1 \cdot \sin\theta_2$$

As $r_1, r_2, \Delta\theta, a$ are known, we indirectly obtain, e.g. with Newton's method, the eccentricity e.

Because of the numerical instability for $a \to \infty$, it is preferable though not to determine e directly, but first the semi-latus rectum p. Because (exercise, Problem 14.1)

$$e\cos\theta_i = \frac{p}{r_i} - 1$$

$$e\sin\theta_i = \sqrt{\frac{p}{r_i}}\sqrt{2 - \frac{p}{r_i} - \frac{r_i}{a}}$$

we get

$$\left(\frac{p}{r_1} - 1\right)\left(\frac{p}{r_2} - 1\right) + \frac{p}{\sqrt{r_1 r_2}}\sqrt{\left(2 - \frac{p}{r_1} - \frac{r_1}{a}\right)\left(2 - \frac{p}{r_2} - \frac{r_2}{a}\right)}$$

$$-\cos(\Delta\theta)\left(1 - \frac{p}{a}\right) = 0 \quad (14.3.6)$$

The root of this equation, which can be found from Newton's method, is the solution for p, from which

$$e = \sqrt{1 - \frac{p}{a}} \quad (14.3.7)$$

can be determined. Because Eq. (14.3.6) is numerically simpler than the initial condition equation, one should always apply this latter procedure.

Highly Elongated Ellipses

Lambert's equation is valid for all Keplerian orbits. For the limiting case of highly elongated ellipses, $a \to \infty$, which happens for the important case of Earth-crossing comets, Lambert's equation however is numerically unstable. But it can be shown (exercise, Problem 14.2a) that in this limiting case the semi-major axis can be calculated in a first approximation as

$$a = \frac{3}{40} \cdot \frac{\Delta_+^{5/2} - \Delta_-^{5/2}}{6\Delta t \sqrt{\mu} - \left(\Delta_+^{3/2} - \Delta_-^{3/2}\right)} \quad @\ a \to \infty \qquad (14.3.8)$$

with

$$\Delta_\pm = r_1 + r_2 \pm d$$

To determine the eccentricity $e \approx 1$, we recognize from Eq. (14.3.7) that p/a is a first-order correction to the limiting case $e = 1$. To derive a first-order approximation for e it therefore suffices to determine p to zero order, i.e. for $a = \infty$. The equation determining p therefore follows from Eq. (14.3.6) as

$$\left(\frac{p}{r_1} - 1\right)\left(\frac{p}{r_2} - 1\right) + \frac{p}{\sqrt{r_1 r_2}}\sqrt{\left(2 - \frac{p}{r_1}\right)\left(2 - \frac{p}{r_1}\right)} - \cos(\Delta\theta) = 0 \quad (14.3.9)$$

It is straightforward but takes some effort to show (exercise, Problem 14.1b) that it has the two positive solutions

$$p = \frac{r_1 r_2}{d^2}\left[1 - \cos(\Delta\theta)\right]\left[r_1 + r_2 \pm 2\sqrt{r_1 r_2}\cos(\Delta\theta/2)\right] > 0 \quad @\ a = \infty \quad (14.3.10)$$

from which follows

$$e = \sqrt{1 - p/a}$$

Parabolic Orbit

The transition to a parabolic orbit $a = \infty$ is trivially found from zeroing the denominator in Eq. (14.3.8)

$$\Delta t = \frac{\Delta_+^{3/2} - \Delta_-^{3/2}}{6\sqrt{\mu}} \quad @\ a = \infty \qquad (14.3.11)$$

This just means that, if the measured values $r_1, r_2, \Delta\theta, \Delta t$ obey Eq. (14.3.9), then $a = \infty$. The semi-latus rectum is given by Eq. (14.3.10). We recall that the semi-latus rectum is the sole orbital element for a parabola, which exhaustively describes its shape.

14.4
Conversion of Orbital Elements

A frequent problem encountered in astrodynamics, in particular with orbit determination, is the conversion of the state vector (r, v) into Keplerian elements and vice versa. In the following two sections we want to have a look at these two problems.

14.4.1
Transformation $r, v \to a, e, i, \omega, \Omega, \theta$

First, we want to convert a state vector (r, v) to Keplerian elements. Let the geocentric-equatorial reference frame (u_I, u_J, u_K) have the following orientation (see Fig. 13.2): the unit vector u_I points to the vernal equinox, u_K to the north pole, and $u_J = u_K \times u_I$. Then, first perform

$$r = \sqrt{rr} \quad \text{and} \quad v = \sqrt{vv}$$

From the definition of the quantity

$$k := \frac{rv^2}{\mu} \quad \text{orbit number} \tag{14.4.1}$$

which we call orbit number, we can read off immediately the type of orbit:

$0 < k < 2$	ellipse	$(0 < e < 1)$,	$k = 1$ circle	$(e = 0)$
$k = 2$	parabola	$(e = 1)$,		
$k > 2$	hyperbola	$(e > 1)$		

Orbit number

The orbit number k is defined as the ratio of twice the specific kinetic energy $\varepsilon_{kin} = 1/2 \cdot v^2$ and the absolute value of the specific potential energy $\varepsilon_{pot} = \mu/r$.

$$k := \frac{rv^2}{\mu} = 1 + e \cos E = \frac{1 + 2e \cos \theta + e^2}{1 + e \cos \theta}$$

The orbit number k is specific for each type of orbit, as follows.

- Ellipse ($0 < e < 1$): because

$$-1 \leq \cos E \leq +1$$
$$\Rightarrow \quad 1 - e \leq k \leq 1 + e$$
$$\Rightarrow \quad 0 < k < 2 \quad \text{for} \quad 0 < e < 1$$

circle $k = 1$

- Parabola ($e = 1$): here

$$k = \frac{1 + 2\cos\theta + 1}{1 + \cos\theta} = \frac{2(1 + \cos\theta)}{1 + \cos\theta}$$
$$\Rightarrow \quad k = 2$$

- Hyperbola ($e > 1$): at the points of closest approach (periapsis) and furthest distance (infinity, see Eq. (7.4.21)) we find

$$k = 1 + e \quad @ \; \theta = 0°$$
$$k = \frac{e^2 - 1}{0} = \infty \quad @ \; \cos\theta = -\frac{1}{e}$$
$$\Rightarrow \quad k > 2$$

So, depending on whether k is smaller, larger, or equal to 2 we have an ellipse, hyperbola, or parabola.

The semi-major axis is determined from the vis-viva Eq. (7.2.10) as to

$$a = \frac{\mu r}{2\mu - rv^2} = \frac{r}{2 - k} \tag{14.4.2}$$

So, the semi-major axis and thus the orbital period $T = 2\pi\sqrt{a^3/\mu}$ can be determined merely from the absolute values of the position vector and orbital velocity. According to Eq. (7.3.3), the eccentricity vector of the orbit is

$$\mu e = v \times h - \mu u_r = v \times (r \times v) - \mu u_r = \left(v^2 - \frac{\mu}{r}\right) r - (rv) v$$

The latter is due to $v \times (r \times v) = (vv) r - (rv) v$. We then get

$$e = \sqrt{ee} = \sqrt{1 - \frac{1}{\mu^2}\left(\frac{2\mu}{r} - v^2\right)(r \times v)^2} \tag{14.4.3}$$

Now, establish the unit vectors

$u_r = r/r$

$u_v = v/v$

$u_e = e/e$ \qquad unit vector to periapsis

$u_h = u_r \times u_v$ \qquad unit vector of angular momentum

$u_k = u_K \times u_h$ \qquad unit vector to the ascending node

These unit vectors determine all angular elements

$$\Omega = \begin{cases} \arccos(u_I u_k) & @\ u_J \cdot u_k \geq 0 \\ 2\pi - \arccos(u_I u_k) & @\ u_J \cdot u_k < 0 \end{cases} \tag{14.4.4}$$

$$\omega = \begin{cases} \arccos(u_k u_e) & @\ u_K \cdot u_e \geq 0 \\ 2\pi - \arccos(u_k u_e) & @\ u_K \cdot u_e < 0 \end{cases} \tag{14.4.5}$$

$$\theta = \begin{cases} \arccos(u_r u_e) & @\ u_r \cdot u_v \geq 0 \\ 2\pi - \arccos(u_r u_e) & @\ u_r \cdot u_v < 0 \end{cases} \tag{14.4.6}$$

$$i = \arccos(u_K u_h) \tag{14.4.7}$$

To determine the orbit time, we derive the following relation from Eq. (7.4.13d)

$$\alpha := \tan\frac{E}{2} = \tan\frac{\theta}{2}\sqrt{\frac{1-e}{1+e}}$$

from which follows

$$E = \begin{cases} 2\arctan\alpha & @\ 0 \leq \theta < \pi \\ 2\pi + 2\arctan\alpha & @\ \pi \leq \theta < 2\pi \end{cases} \tag{14.4.8}$$

This together with Eq. (7.4.14) inserted into the Keplerian equation (7.4.16) furnishes the time after passing through the periapsis $E_0\,(t_0 = 0) = 0$

$$t = \sqrt{\frac{a^3}{\mu}} \cdot (E - e\sin E) \tag{14.4.9}$$

14.4.2
Transformation $a, e, i, \omega, \Omega, \theta \rightarrow r, v$

Let's assume the Keplerian elements $a, e, i, \omega, \Omega, \theta$ are given. In the co-rotating reference frame (u_r, u_θ, u_z), as shown in Fig. 7.6, the radius vector can be written with Eq. (7.3.5) as

$$r_{r\theta z} = \left[\frac{a(1-e^2)}{1+e\cos\theta}, 0, 0\right] \tag{14.4.10}$$

From Eq. (7.3.9) also the following is valid:

$$v_{r\theta z} = \frac{\mu}{h}(e\sin\theta, \; 1+e\cos\theta, \; 0) \tag{14.4.11}$$

To determine the required state vector in the inertial (sideral) geocentric equatorial system, the co-rotating orbital system (u_r, u_θ, u_z) has to be transformed via the elements (i, ω, Ω) into the geocentric-equatorial system (u_I, u_J, u_K)

$$\begin{aligned} r_{IJK} &= \mathbf{R} \cdot r_{r\theta z} \\ v_{IJK} &= \mathbf{R} \cdot v_{r\theta z} \end{aligned} \tag{14.4.12}$$

with the rotation matrix $\mathbf{R}$ (see e.g. Vallado (2001) or Chobotov (2002))

$$\begin{pmatrix} \cos\Omega\cos\omega - \sin\Omega\sin\omega\cos i & -\cos\Omega\sin\omega - \sin\Omega\cos\omega\cos i & \sin\Omega\sin i \\ \sin\Omega\cos\omega + \cos\Omega\sin\omega\cos i & -\sin\Omega\sin\omega + \cos\Omega\cos\omega\cos i & -\cos\Omega\sin i \\ \sin\omega\sin i & \cos\omega\sin i & \cos i \end{pmatrix}$$

Note that the elements θ and ω for circular orbits and Ω for equatorial orbits are undefined. In these cases the corresponding elements can be set to zero.

14.5
State Vector Propagation

A general and frequent problem, in particular in orbit determination (see step 3 in Section 14.2), which we will investigate here, is the following: Let (r, v) be the state vector at any point in time. What is the state vector (r', v') after time Δt? This determination $(r, v) \rightarrow (r', v')$ is called *state vector propagation*. A procedure to propagate is called a *propagator*. According to Section 14.4 an obvious and also possible approach to propagate would be as follows: (i) convert the initial state vector into Keplerian elements; (ii) determine the corresponding point in time t since the passing through the periapsis according to the Keplerian equation; (iii) use the Keplerian method (Sections 7.4.3–7.4.4) to determine $\theta' = \theta(t + \Delta t)$ at the new point in time; (iv) finally, and according to Section 14.4.2, reconvert the modified Keplerian elements to the wanted state vector. This procedure would be quite long-winded, but with today's computers we would get the result in the twinkling of an eye.

14.5.1
Propagation $r, v, \gamma \to r', v', \gamma'$

In fact, the above procedure is overdone, because the state vector doesn't imply six degrees of freedom (its six components), but only three, namely r, v, and $\gamma = 90° - \angle(r, v)$ (flight path angle, see Section 7.3.2). The other three degrees (corresponding to i, ω, Ω) remain unchanged. In this section we want to demonstrate that there exists a simple algorithm to propagate these minimal orbit state information: $r, v, \gamma \to r', v', \gamma'$. To do so, we first show that we can determine from r, v, γ the orbital elements a, e, θ. Thus we trace the problem back to the Keplerian problem, where we can calculate $\theta' = \theta(t + \Delta t)$ numerically. Then we simply reconvert $a, e, \theta' \to r', v', \gamma'$. The entire procedure thus looks as follows:

$$r, v, \gamma \to a, e, \theta \xrightarrow{\Delta t} a, e, \theta' \to r', v', \gamma'$$

First Step $r, v, \gamma \to a, e, \theta$

Just like in Section 14.4, we first determine the orbit number

$$k = \frac{rv^2}{\mu} \tag{14.5.1}$$

As we know, it describes the type of orbital curve:

$0 < k < 2$	ellipse	$(0 < e < 1)$,	$k = 1$ circle	$(e = 0)$
$k = 2$	parabola	$(e = 1)$,		
$k > 2$	hyperbola	$(e > 1)$		

With Eq. (14.4.2) we directly get the semi-major axis

$$a = \frac{r}{2 - k} \tag{14.5.2}$$

We use Eq. (7.3.6) to determine the eccentricity

$$1 - e^2 = \frac{h^2}{\mu a} \tag{14.5.3}$$

From Fig. 7.4 we derive

$$r\dot\theta = v \cos \gamma = \frac{h}{r} \tag{14.5.4}$$

the latter due to $h = r \times v = rv \sin(90° - \gamma) = rv \cos \gamma$. By applying Eqs. (14.5.1) and (14.5.2) we obtain from Eq. (14.5.3)

$$e = \sqrt{1 - k(2-k)\cos^2 \gamma} \tag{14.5.5}$$

We now use Eq. (7.3.8) to determine the true anomaly:

$$\dot{r} = \frac{e\mu}{h} \sin \theta = v \sin \gamma$$

where the latter is from Eq. (7.2.3) with Eq. (7.3.11). If we use Eqs. (14.5.4) and (14.5.1) in just a few steps we get

$$\sin \theta = \frac{1}{2}\frac{k}{e} \sin 2\gamma \tag{14.5.6}$$

Second Step $a, e, \theta \xrightarrow{\Delta t} a, e, \theta'$

To find $\theta' = \theta(t + \Delta t)$, let's have a look at the case of an elliptical orbit $0 < k < 2$. Because of Eq. (7.4.13d)

$$\alpha := \tan\frac{E}{2} = \tan\frac{\theta}{2}\sqrt{\frac{1-e}{1+e}} = \frac{k}{2}\frac{\sin 2\gamma}{1 - k\cos^2 \gamma}\sqrt{\frac{1-e}{1+e}}$$

from which it follows that

$$E = \begin{cases} 2\arctan \alpha & @ \ 0 \leq \theta < \pi \\ 2\pi + 2\arctan \alpha & @ \ \pi \leq \theta < 2\pi \end{cases} \tag{14.5.7}$$

Note: *Up to here one, of course, could determine a, e, θ, E also from Eqs. (14.4.2), (14.4.3), (14.4.6), and (14.4.8).*

Using the Keplerian equation (7.4.16) we can determine the corresponding time since the last passage through the periapsis

$$t = \sqrt{\frac{a^3}{\mu}} (E - e \sin E) \tag{14.5.8}$$

To determine the new state vector, we perform the Newton iteration (cf. Eq. (7.4.17))

$$E'_{i+1} = E'_i - \frac{E'_i - e\sin E'_i - n \cdot (t + \Delta t)}{1 - e \cos E'_i}$$

with the new time $t + \Delta t$ and with initial value

$$E'_1 = n(t + \Delta t) + 0.85\, e\, \text{sgn}\left[\sin n(t + \Delta t)\right]$$

to obtain the new eccentric anomaly E', and with Eq. (7.4.13a)–(7.4.13d) the required new true anomaly

$$\cos\theta' = \frac{\cos E' - e}{1 - e\cos E'}$$

Third Step $a, e, \theta' \rightarrow r', v', \gamma'$

The new state vector is then determined according to Eq. (14.4.10) and Eq. (7.3.10) as

$$r' = \frac{a(1 - e^2)}{1 + e\cos\theta'}$$

$$v' = \frac{\mu}{h}\sqrt{1 + 2e\cos\theta' + e^2} \tag{14.5.9}$$

$$\cos\gamma' = \frac{h}{r'v'} = \frac{1 + e\cos\theta'}{\sqrt{1 + 2e\cos\theta' + e^2}}$$

You can of course use the same approach for a hyperbolic orbit $k > 2$. For the conversion $\theta \leftrightarrow F$ you merely have to use Eq. (7.4.26), and Eq. (7.4.29) as the Keplerian method.

14.5.2
Propagation $r, v \rightarrow r', v'$

This relatively simple and clear method can, with a little addition, in principle also be used to propagate the full state vector, i.e.

$$r, v \rightarrow r, v, \gamma \rightarrow a, e, \theta \xrightarrow{\Delta t} a, e, \theta' \rightarrow r', v', \gamma' \rightarrow r', v'$$

To do that, first determine $r = \sqrt{rr}$, $v = \sqrt{vv}$, $\cos\gamma = rv/(rv)$, and the normalized angular momentum vector $\hat{h} = \hat{r} \times \hat{v} = (r \times v)/(rv)$. With these, perform the procedure as given in Eqs. (14.5.1) to Eq. (14.5.9). Now the reference frame with the orthonormal axes $(\hat{r}, \hat{h} \times \hat{r}, \hat{h})$ is rotated by the angle $\Delta\theta = \theta' - \theta$ along the angular momentum axis $\hat{h}$ to get the rotated reference system $(\hat{r}', \hat{h} \times \hat{r}', \hat{h})$. The new state vector then is given as

$$r' = r'\hat{r}' \tag{14.5.10}$$

$$v' = v'\left[\sin\gamma' \cdot \hat{r}' + \cos\gamma' \cdot \left(\hat{h} \times \hat{r}'\right)\right] \tag{14.5.11}$$

14.5.3
Universal Propagator

These are the solutions within the scope of this book. They may not seem to be very elegant. There are more direct and elegant propagation methods: for instance the *Lagrange–Gibbs f and g solution method* (see Schaub and Junkins (2003) or Chobotov (2002)), but most importantly the so-called *universal approach*. The latter is based on the so-called *Sundman transformation* and the *Stumpff functions*, which however go beyond the scope of this book. In the literature, a good overview is given by Chobotov (2002, Section 4.3), and there is a practicable algorithm for this propagator in Chobotov (2002, Section 4.5); there are many examples and a MATLAB code for this propagator in Curtis (2005, Section 3.7).

Problems

Problem 14.1 Preparatory Derivation for Eccentricity Determination

As mentioned in Section 14.3.2 show that, from

$$\sin \theta_i = \frac{a\sqrt{1-e^2}}{er_i}\sqrt{e^2 - \left(1 - \frac{r_i}{a}\right)^2}$$

with $p = a\left(1 - e^2\right)$, it follows that

$$e \sin \theta_i = \sqrt{\frac{p}{r_i}}\sqrt{2 - \frac{p}{r_i} - \frac{r_i}{a}}$$

Problem 14.2 Lambert's Problem for $a \to \infty$

(a) Prove that, from Lambert's equation (14.3.5) for $a \to \infty$, it follows that (see Eq. (14.3.8))

$$a = \frac{3}{40} \cdot \frac{\Delta_+^{5/2} - \Delta_-^{5/2}}{6\Delta t \sqrt{\mu} - \left(\Delta_+^{3/2} - \Delta_-^{3/2}\right)}$$

with

$$\Delta_\pm = r_1 + r_2 \pm \sqrt{r_1^2 + r_2^2 - 2r_1 r_2 \cos(\Delta\theta)}$$

(b) Prove that the two positive expressions for the semi-latus rectum p for the parabolic case given by Eq. (14.3.10) follow from straightforward algebraic manipulations of Eq. (14.3.9).

15
Rigid Body Dynamics

The general motion of a perfectly rigid body is the superposition of translation and rotation. We talked about translation under the influence of a gravitational field and possible external perturbations in the previous chapters. Now we want to have a look at the rotational behavior of a body.

A satellite in Earth orbit has certain tasks to perform, which usually impose requirements on its attitude and rotational behavior. The attitude control of an Earth observation satellite, for instance, has to ensure that in the course of its orbit the optical sensor always points to nadir or maybe somewhat obliquely. At the same time the satellite, say with a pushbroom sensor, is not allowed to have any rotational motions, because this would lead to distorted images. So it is important that rotational motions can be controlled. But there are questions to be asked: What do we have to control? What is the attitudinal and rotational behavior of a satellite in space?

15.1
Fundamental Physics of Rotation

A satellite can well be considered as a rigid body. Here "rigid" means that any in-flight change of distances between the components (mass particles) of the satellite does not have an impact on rotary dynamics. If, for instance, the satellite vibrates or performs other cyclic motions, e.g. thermal expansion, at such an amplitude that it alters its inertial matrix (see below) significantly, the satellite is not truly rigid. The vibrational dynamics of a satellite is a case of special importance, which we do not want to consider here.

15.1.1
Physical Basics

Principles of Rotation

The rotation of a rigid body forces all mass particles to circle around a *common line through the center of mass*, the rotation axis, with the *same angular velocity* ω. This circling motion can be described as the change of the orientation angle

15 Rigid Body Dynamics

φ of the rigid body in the time t. The instantaneous angular velocity is thus: $\omega(t) := \dot{\varphi}(t) = d\varphi/dt$. In general, the body spins around an axis, but also the spin axis may turn around one or even two axes perpendicular to it. Hence the angular velocity vector is made up of three components: $\omega = (\omega_x, \omega_y, \omega_z)$. For the mathematical description of such a rotation, it is very convenient to place the origin of the reference frame on the rotation line at the center of mass. Such a frame is called a *topocentric frame*. We still leave open whether the coordinate axes co-rotate with the body, or whether it's an inertial system, or else.

Angular Momentum and Moments of Inertia

Just as there is the linear momentum p in translational physics, we have the angular momentum L in rotational physics. According to Fig. 15.1 the angular momentum of a mass particle i with mass m_i is defined as

$$L_i := m_i r_i \times v_i \tag{15.1.1}$$

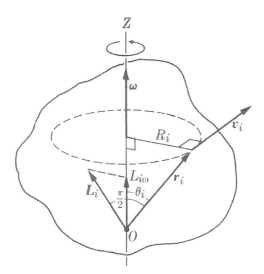

Figure 15.1 Angular momentum of a rotating rigid body.

As $v = \omega \times r$ holds quite generally for a rotation, we get the following for the total angular momentum of the body:

$$L = \sum_{\text{all } i} m_i r_i \times (\omega \times r_i) \tag{15.1.2}$$

where the sum comprises all particles of the rotating body. Observe that ω does *not* have an index i, as all the particles circle around a common axis with the same angular velocity.

Introductory Considerations on the Moment of Inertia

At first we are interested only in the component of the angular momentum $L_{i\omega}$ of a particle along the rotation axis. Let $\theta_i = \angle(\omega, r_i)$ with the origin O lying on the rotation axis (see Fig. 15.1). The absolute value of $v_i = \omega \times r_i$ then is

$$v_i = \omega \cdot r_i \sin \theta_i =: \omega \cdot R_{i\omega}$$

where $R_{i\omega}$ is the distance to the rotation axis. The z component of L_i with regard to ω is

$$L_{i\omega} = m_i r_i v_i \cdot \cos(\pi/2 - \theta_i)$$
$$= m_i (r_i \sin \theta_i)(\omega R_{i\omega}) = m_i R_{i\omega}^2 \omega$$

So the component L_ω of the total angular momentum along the rotation axis is

$$L_\omega = \sum_{\text{al } i} L_{i\omega} = \omega \cdot \sum_{\text{all } i} m_i R_{i\omega}^2$$

We now define the *moment of inertia* with regard to the rotation axis ω

$$I_\omega := \sum_{\text{all } i} m_i R_{i\omega}^2$$

resulting simply in $L_\omega = \omega \cdot I_\omega$.

Inertia Tensor

After these preliminary remarks to illustrate the concept of the moment of inertia with regard to one rotation axis, we want to derive a general expression for the moment of inertia with regard to any rotation axis. If one decomposes the expression $r \times (\omega \times r) = \omega(rr) - r(r\omega)$ in Eq. (15.1.2) into its axial components and performs the transition to infinitesimally small particles $\sum m_i \to \int dm$, one gets

$$L = \begin{pmatrix} \int (y^2 + z^2)\, dm & -\int xy\, dm & -\int xz\, dm \\ -\int yx\, dm & \int (x^2 + z^2)\, dm & -\int yz\, dm \\ -\int zx\, dm & -\int zy\, dm & \int (x^2 + y^2)\, dm \end{pmatrix} \begin{pmatrix} \omega_x \\ \omega_y \\ \omega_z \end{pmatrix}$$

We now define the matrix in the equation as

$$I := \begin{pmatrix} I_{xx} & I_{xy} & I_{xz} \\ I_{yx} & I_{yy} & I_{yz} \\ I_{zx} & I_{zy} & I_{zz} \end{pmatrix} \quad \text{inertia tensor} \tag{15.1.3a}$$

with

$$I_{kl} = \int_{total\ body} \left(r^2 \delta_{kl} - x_k x_l\right) dm \quad \text{moments of inertia} \tag{15.1.3b}$$

and where δ_{kl} is the *Kronecker symbol* (the Kronecker symbol is defined as: $\delta_{kl} = 1$ if $k = l$, and $\delta_{kl} = 0$ if $k \neq l$.). This results in

$$\mathbf{L} = \mathbf{I}\boldsymbol{\omega} \tag{15.1.4}$$

Obviously the inertia tensor is real and symmetric ($I_{kl} = I_{lk}$) with nine components. Because of the symmetry, only six of them are independent. The exact form of the inertia tensor depends on the selected reference frame. Mathematics holds that as $\mathbf{I}$ is symmetric, it is always possible to find a system of principal axes, where the tensor takes on a diagonal form:

$$\mathbf{I}_P = \begin{pmatrix} I_x & 0 & 0 \\ 0 & I_y & 0 \\ 0 & 0 & I_z \end{pmatrix} \tag{15.1.5}$$

Note: *Because the choice of the reference frame in body dynamics is important, we indicate the chosen frame as an index. If the index is omitted then no choice is necessary.*

The Cartesian axes of such a reference system are called *principal axes (of inertia)*, and the corresponding diagonal components I_x, I_y, I_z are called the *principal moments of inertia*. In general the three moments of inertia of the principal axis are different. We call the principal axis with the biggest moment of inertia the *major principal axis* and the one with the smallest moment of inertia the *minor principal axis*. The principal axes are to the largest extent axes of body symmetry, as illustrated in Fig. 15.2 for some simple bodies.

Remark: *For a general, non-diagonal inertia tensor $\mathbf{I}$, the principal moments of inertia are the eigenvalues of $\mathbf{I}$, and the principal axes of inertia are its eigenvectors.*

According to Eqs. (15.1.4) and (15.1.5) we can write the total angular momentum in this principal axes system as

$$\mathbf{L}_P = \left(I_x \omega_x, I_y \omega_y, I_z \omega_z\right)_P \tag{15.1.6}$$

Note that when no force is impacting from outside, the conservation of the angular momentum *vector* is only valid in an inertial system. If we choose

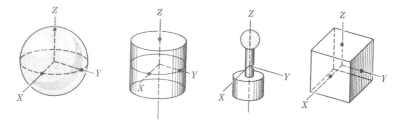

Figure 15.2 Principal axes of a few symmetric bodies.

a reference frame co-rotating with the body, the angular momentum vector rotates as well. Then just its amount

$$L^2 = L_P^2 = I_x^2 \omega_x^2 + I_y^2 \omega_y^2 + I_z^2 \omega_z^2 = const \qquad (15.1.7)$$

remains constant, because rotation does not change the length of a vector.

Why Rotation is So Odd

The angular momentum equation $L = I\omega$ has the same importance in rotational dynamics as the momentum equation $p = mv$ in linear dynamics. Because of momentum and angular momentum conservation, both are the hub of mechanics. This brings us for the first time to the important equivalence between rotational and translational dynamics. As we will see at the end of this section, this equivalence is even more profound. The equivalence is a 1:1 equivalence: $L \leftrightarrow p$, $\omega \leftrightarrow v$, $I \leftrightarrow m$. So far, however, there are two flaws in this equivalence:

1. Mass is a scalar, whereas inertia is a tensor.

2. Mass m is independent of coordinates, but unfortunately this is not true for the moment of inertia I; the definition in Eq. (15.1.3b) clearly exhibits the dependence of the components I_{kl} on the coordinates x, y, z.

The first flaw complicates the mathematical treatment of rotational physics quite a lot. But that's the nature of rotation, and we cannot get around it. We can live with this flaw, because it is manageable, even though the matrix algebra may be tedious.

The second flaw is a heftier problem. Let's see what it implies. Writing $L = I\omega$ for each component reads

$$L_i = I_{ii}\omega_i + I_{ij}\omega_j + I_{ik}\omega_k, \quad i = 1, 2, 3$$

In an inertial reference frame – which is the most natural choice of a coordinate system (see Section 13.1), which is why it sometimes is called *physical system* –

each of the $L_i = const$. But this restriction is not grave. It leaves the body the freedom to continuously change the way it rotates:

$$I_{ii}\omega_i(t) + I_{ij}\omega_j(t) + I_{ik}\omega_k(t) = const, \quad i = 1, 2, 3$$

So, an unsymmetrical body not only has no marked rotation axis – it just rotates around any given axis – but the rotation rates even change in time. This is why the motion of an unsymmetrical body looks so odd. Even worse, if the ω_i change, the coordinates x, y, z of the parts of the body change as well, and thus all moments of inertia $I_{ij}(t)$. This time dependence would practically render it impossible to calculate dynamic rotations, and thus is a real problem. But we are free to choose the reference frame we want. So, for specific mathematical calculations in rigid body dynamics, we always use a reference frame where the moments of inertia are constant with regard to time: a so-called *body system*. A body system is a special case of a topocentric system with its coordinate axes fixed relative to the geometry of the body. It co-rotates with the body.

We gain this advantage at the expense of intuitive clarity. This is because our understanding of the world around us is based on observations as an external inertial observer. Although mathematics in a body system is much easier, this doesn't help our intuition. As a rotating observer, we in general barely have a feeling for the pointing direction of the angular momentum vector and angular velocity vector (though theoretically we could determine it by evaluating the felt centrifugal forces), but we don't have the slightest idea how much the vectors are drifting, and why in which direction. The changing rotational rates would only make us quite dizzy due to the forces induced into the vestibular system.

This is why one considers the circumstances of a body dynamics first in an illustrative reference frame. Then one jumps to a body system in which the problem can be solved mathematically (but with little understanding why things behave as they are). The solution is finally transformed back into the earlier system to visualize how the body behaves and grasp an idea of why this is so. Because in the following we want to solve dynamical problems, we don't get around these transformations (see Eqs. (15.1.14) and (15.1.15)) and the temporary loss of intuitive understanding. In contrast to the current literature, however, we will limit the presentation of transformation equations to the absolute minimum, as they don't help us to understand rotational dynamics.

Coordinate Systems

In order to distinguish the coordinate systems from now on, we will denote or index body system coordinates by x, y, z, and those of non-body systems by

1, 2, 3. Physical parameters, presented in a body system, will have the index B (body axis) for a general body system. The principal axes system is a special body system with index P (principal axis). Inertial (physical) systems usually don't get an index, only when we want to distinguish them explicitly from others, and in this case they get the index I. So now we have

$$I \equiv I_I = (I_{ij} \neq const)$$
$$I_B = (I_{kl} = const)$$
$$\omega_B = (\omega_x, \omega_y, \omega_z)$$
$$\omega_{else} = (\omega_1, \omega_2, \omega_3)$$

Remark: *For matrix calculation a vector **a** has to be written as a column. From a formal point of view, a vector written as a line would be the transpose, i.e. $a = (a_x, a_y, a_z)^T$. For the sake of simplicity, we will do without this detail, and we'll just write $a = (a_x, a_y, a_z)$ hereafter.*

The transformation of a constant vector equation between reference frames is straightforward. Let C be the transformation matrix (rotation matrix with the three Euler angles) between two reference frames $I \to B$, i.e. $a_B = C a_I$. Then from $L = I\omega$ just follows that $L_B = CI\omega = CIC^{-1}C\omega = I_B\omega_B$. The transformation of a time-dependent vector equation is more complex and will be treated below.

Rotational Kinetic Energy

For a rotating body the rotational energy can be derived as

$$E_{rot} = \frac{1}{2}\sum_i m_i \dot{v}_i^2 = \frac{1}{2}\sum_i m_i (\omega \times r_i) \cdot (\omega \times r_i) = \frac{1}{2}\sum_i m_i \omega \cdot [r_i \times (\omega \times r_i)]$$

$$= \frac{1}{2}\omega \cdot \sum_i m_i r_i \times (\omega \times r_i)$$

With Eqs. (15.1.2) and (15.1.4) we thus get

$$E_{rot} = \frac{1}{2}\omega L = \frac{1}{2}\omega(I\omega) \quad \text{rotational kinetic energy} \tag{15.1.8}$$

Because of Eq. (15.1.6), the following is valid in the principal axes system (cf. Eq. (7.2.14))

$$E_{rot} = \frac{1}{2}\left(I_x\omega_x^2 + I_y\omega_y^2 + I_z\omega_z^2\right) \tag{15.1.9}$$

Because rotational energy is a conserved scalar quantity independent on the choice of reference frame, we can drop the reference frame index.

15.1.2
Equations of Rotational Motion

To be able to calculate the motions of a rotating body specifically, we are now looking for equations of motion that correspond to the Newtonian equations of motion in linear dynamics. Newton's law $F = dp/dt$, which by $p = mv$ leads to the linear equation of motion $F = m\dot{v}$, and the equivalence $p \leftrightarrow L$ inspires us to determine dL/dt. We use Eq. (15.1.1) to obtain

$$\frac{dL}{dt} = \sum_i m_i \left(\frac{dr_i}{dt} \times v_i + r_i \times \frac{dv_i}{dt} \right) = \sum_i \left[m_i \left(v_i \times v_i \right) + r_i \times \frac{d(m_i v_i)}{dt} \right]$$

First we have $v_i \times v_i = 0$. Also, as $m_i v_i = p_i$ is an instantaneous tangential momentum and $d(m_i v_i)/dt$ corresponds, according to Newton, to an external force on the particle i:

$$\frac{d(m_i v_i)}{dt} = \frac{dp_i}{dt} = F_i$$

And because

$$\sum_i r_i \times F_i = T \qquad (15.1.10)$$

is an external torque T, we finally get

$$\frac{d}{dt} L_I = T_I \qquad (15.1.11)$$

As the conservation of each angular momentum component is only valid in an inertial system, this is the equation of motion for the angular momentum of a body in an inertial reference frame (indicated by the index I). This equation is equivalent to Newton's law $dp/dt = F$ in linear dynamics.

Equations (15.1.10) and (15.1.11) imply the following. External forces F_i, that affect every point of the body make up a total torque T on the body, which causes a motion of the angular momentum, and via $L = I\omega$ a change of the rotation of the body. If there are no external forces, the angular momentum L is constant in the inertial system. External forces in astrodynamics are the gravitational effects of other celestial bodies such as the Sun and the Moon, the oblate spheroid form J_{20} and the atmospheric drag of the Earth, and the radiation pressure of the Sun, but also internal torques caused by engine propulsion or internal magnetic fields interacting with the magnetic field of the Earth.

In order to evaluate the motion equation $\dot{L}_I = T_I$ further, we would have to differentiate $L = I\omega$. Because in an inertial system the inertia tensor is not constant, we switch as announced to the body system for mathematical simplicity. To do so, we need to know how the time derivation of a vector is transformed between an inertial system and a rotating system, i.e. $I \rightarrow B$. Let a be an arbitrary vector, which transforms between the two systems as $a_B = C a_I$, where C is the rotation matrix with the three Euler angles of the three rotation steps. Then we find from the chain rule for the differentiation of a product

$$\dot{a}_B = \frac{d}{dt}(C a_I) = \dot{C} a_I + C \dot{a}_I = -\omega_B \times a_B + C \dot{a}_I$$

where we made use of the identity $\dot{C} a_I = -\omega_B \times (C a_I) = -\omega_B \times a_B$ (see e.g. Pisacane and Moore (1994, p. 266)). We now apply this transformation to $a \equiv L$. Thereby we consider that from $L = I\omega$ with $I_B = const$ we simply get

$$\dot{L}_B = \dot{I}_B \omega_B + I_B \dot{\omega}_B = I_B \dot{\omega}_B$$

This is the crucial simple equation, why we switched to a body system. Applying the Euler rotation to $\dot{L}_I = T_I$ yields

$$C \dot{L}_I = C T_I = T_B$$

With this we finally get

$$I_B \dot{\omega}_B = T_B - \omega_B \times (I_B \omega_B) \quad \text{Euler's equation (vectorial)} \quad (15.1.12)$$

for an arbitrary body system. If we choose the body system to be the principal axes system x, y, z, in which $I_P = diag(I_x, I_y, I_z)$ is diagonal, then this equation can be expressed for each component as

$$I_x \dot{\omega}_x = T_x + (I_y - I_z) \omega_y \omega_z$$
$$I_y \dot{\omega}_y = T_y + (I_z - I_x) \omega_z \omega_x \quad \text{Euler's equations} \quad (15.1.13)$$
$$I_z \dot{\omega}_z = T_z + (I_x - I_y) \omega_x \omega_y$$

where we dropped index P for the sake of simplicity. These are the equations of rotational motion for $\omega_x, \omega_y, \omega_z$ in the principal axes system x, y, z with constant I_x, I_y, I_z.

15.1.3
Reference Frames

The principal axis frame of a body is a given (except for the x-y axis pointing ambiguity for a fully rotational symmetric z axis). There is only the choice left of which axis to name x, y, and z. This we will do with respect to the external reference frame, into which we have to transform back the angular rates $\omega_x, \omega_y, \omega_z$, derived from the equations of solution, to make the motion representation particularly vivid. Figure 15.3 shows an external reference frame that is very suitable to describe a satellite in a planetary orbit, and it is frequently used to study attitude dynamics in such orbits. Its 3 axis points towards nadir and the 2 axis is normal to the orbital plane but opposite to the orbital momentum vector, so that in a circular orbit the 1 axis points in the flight direction: $u_1 = u_2 \times u_3$. In astronautics this frame is commonly denoted as "local vertical, local horizontal" (meaning: the local 3 axis is pointing vertically to nadir, and the local 1–2 plane is the local horizon), or *LVLH frame* for short. This frame emerges from the *RSW* reference frame of Fig. 13.4 by turning the 3 axis by 180° around the 1 axis.

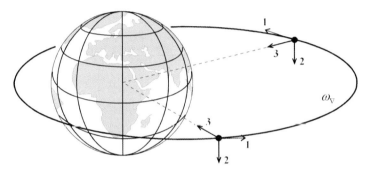

Figure 15.3 The common *LVLH* frame (1, 2, 3 reference frame) of a satellite circling the Earth. It emerges from the *RSW* reference frame of Fig. 13.4 by turning the 3 axis by 180° around the 1 axis.

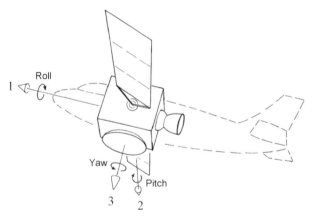

Figure 15.4 Satellite rotations denoted roll, pitch, and yaw in the *LVLH* frame.

15.1 Fundamental Physics of Rotation

We now overlay the principal axes frame with the *LVLH* frame by matching the axes according to $x, y, z \leftrightarrow 1, 2, 3$. By this procedure the principal axes x, y, z are determined such that the x axis points in the flight direction, and the z axis points "down." With this choice the denomination of the rotation rates are usually derived from aviation as *roll* (1), *pitch* (2), and *yaw* (3) (see Fig. 15.4).

Coordinate Transformations

We now allow only small deviations in roll (ϕ), pitch (θ), and yaw (ψ) between the axes, which we measure with respect to the *LVLH* frame. From Fig. 15.5 we recognize that any body rotation around the axes $1, 2, 3$ is given by

$$\omega_{\delta,P} = \omega_{\delta,LVLH} = (\dot{\phi}, \dot{\theta}, \dot{\psi})_{LVLH}$$

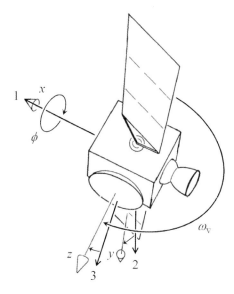

Figure 15.5 Overlay of the 1, 2, 3 system *LVLH* and x, y, z principal axes system with only small angular deviation ϕ between the axes, plus orbital revolution of the y axis around the 2 axis with angular velocity ω_∇. The ψ turn around the 3, z axes is similar.

In addition, we assume that the satellite orbit is circular. So, as shown in Fig. 15.5 the satellite turns with constant angular velocity ω_∇ around the 2 axis. This rotation in conjunction with small yaw and roll angles is written as

$$\omega_{\nabla,P} = \omega_\nabla (\sin\psi \cdot u_1 + u_2 - \sin\phi \cdot u_3) \approx \omega_\nabla (\psi, 1, -\phi)_{LVLH}$$

The rotation rates of the body expressed in the *LVLH* system therefore are

$$\omega_P = \omega_{\delta,P} + \omega_{\nabla,P} = (\dot{\phi} + \omega_\nabla \psi, \dot{\theta} + \omega_\nabla, \dot{\psi} - \omega_\nabla \phi)_{LVLH} \tag{15.1.14}$$

from which follows

$$\dot{\omega}_P = (\ddot{\phi} + \omega_\nabla \dot{\psi}, \ddot{\theta}, \ddot{\psi} - \omega_\nabla \dot{\phi})_{LVLH} \tag{15.1.15}$$

These are the wanted kinematic transformation equations for the Euler angles (so-called *Euler's rate equations*) between the two selected reference frames. They will be needed to analyze the solutions $\omega_x, \omega_y, \omega_z$ of the satellite motion equations in the *LVLH* reference frame.

15.1.4
Translation vs. Rotation

So far, we have been deriving equations for different physical parameters that are characteristic for rotational mechanics. If you compare these with those of translational mechanics (see Table 15.1), you will see that the already mentioned equivalence $L \leftrightarrow p$, $\omega \leftrightarrow v$, $I \leftrightarrow m$ holds without exception for the basic physical quantities. We have therefore found a fundamental symmetry between these two fields of physics.

Table 15.1 Equivalences between physical laws of translation and rotation.

	Translation		Rotation
Momentum	$p = mv$	Angular momentum	$L = I\omega$
Force	$F = dp/dt$	Torque	$T = dL/dt$
Kinetic energy	$1/2\,vp = 1/2\,mv^2$	Rotational energy	$E = 1/2\,\omega L = 1/2\,\omega I \omega$
Power	$P = Fv$	Power	$P = T\omega$

But be careful about applying this equivalence. From the equation of translational motion $F = m\dot{v}$ does not follow the rotational form $T = I\dot{\omega}$. Rather we have from $L = I\omega$ the derivation $T = \dot{L} = \dot{I}\omega + I\dot{\omega}$. Written in linear dynamics this reads $F = \dot{m}v + m\dot{v}$, and only because $\dot{m} = 0$ do we find $F = m\dot{v}$. But $\dot{I}$ is too complicated in rotational physics. This is why the equation $T = I\dot{\omega} + \dot{I}\omega$, though being valid, is useless. Only by switching to a body system we find Euler's equation $T_B = I_B \dot{\omega}_B + \omega_B \times (I_B \omega_B)$, which by $B = P$ is a treatable equation of motion.

Rotational Power

We still have to complete the table with the equation for the rotational power. Under the influence of an external torque, the rotational energy changes together with the angular momentum according to Eq. (15.1.8). The power P_{rot}

that the external torque has to bring up to change the rotational energy is defined as $P_{rot} = dE_{rot}/dt$. From Eq. (15.1.8) we then have

$$P_{rot} = \frac{dE_{rot}}{dt} = \frac{1}{2}(\dot{\omega}_B L_B + \omega_B \dot{L}_B)$$

Taking Eq. (15.1.11) into account we derive

$$\dot{\omega}_B L_B = \dot{\omega}_B(I_B \omega_B) = \left(\dot{\omega}_B^T I_B\right)\omega_B = (I_B \dot{\omega}_B)\omega_B = \dot{L}_B \omega_B = \omega_B \dot{L}_B = \omega_B T_B$$

Therefore

$$P_{rot} = \frac{1}{2}(\omega_B T_B + \omega_B T_B) = \omega T \quad \text{rotational power} \tag{15.1.16}$$

The latter generalization holds because the rotational power is a scalar and therefore is independent of any chosen reference frame.

15.2
Torque-free Motion

With the physical relations found so far, let's derive the essential characteristics of the rotational dynamics of a body. Under the assumption that the body is not subject to any external force, i.e. $T_x, T_y, T_z = 0$ (such freely rotating bodies are also called *free gyros*), we can apply the torque-free Eulerian equation of motion.

15.2.1
Basic Considerations

Let's have a look at an arbitrarily shaped body from the point of view of an external observer in an inertial reference frame. Equation $L = I\omega$ states that in general the rotation ω is not parallel to L and hence (see above) not constant – the body tumbles. What is the more profound reason for this? The reason is as follows. An arbitrary mass particle i of the body has instantaneous tangential velocity v_i. If the particle with position vector r_i from the center of mass (origin) by chance does not lie on the plane vertical to the rotation axis through the center of mass, then its angular momentum $L_i = m_i r_i \times v_i$ is not parallel to the rotation axis (see Fig. 15.1). Due to the body's rigid bonding, the particle is forced on a circular orbit around the rotation axis (Fig. 15.1, dashed circle); L_i rotates around the rotation axis. The total angular momentum of the rigid body is $L = \sum_i L_i$. If the mass particles are arranged symmetrically with regard to the rotation axis, i.e. if the rotation axis is a body symmetry axis (principal axis), all the L_i add up to a total momentum parallel to the rotation axis. However, this is not generally the case, so the total angular momentum L is generally not parallel to the rotation axis w. The implication on the movement of the body in these two cases is as follows:

1. One of the principal axes of inertia is initially aligned parallel to L.

   In this case $I \equiv I_P = \mathrm{diag}\,(I_x, I_y, I_z) = const$. This implies $L \parallel w$. The law of conservation angular momentum, $L = const$, directly results in $w = const$. So the body continues to rotate around this principal axis of inertia.

2. Not one of the principal axes of inertia is initially aligned parallel to L.

   As mentioned earlier we then have

   $$L_i = I_{ii}(t)w_i(t) + I_{ij}(t)w_j(t) + I_{ik}(t)w_k(t) = const, \quad i = 1, 2, 3$$

   This gives the body the freedom to change the rotational rate continuously. So $w \neq const$. This implies that it may tumble wildly.

Our objective in the next three sections is to find out with the equations of motion what $w = w(t)$ looks like for a given inertial characteristic of the body near a principal axis. In the last section we are going to qualitatively describe general rotational motions.

15.2.2
Stability and Nutation

We start out with the first case $L \parallel w$, where the body rotates around one of its principal axes. Theoretically we have $w = const$. However, this does not necessarily imply that this rotation is stable under the impact of a small perturbation. We therefore want to find stability criteria and modes of motion for rotations around one of the three principal axes of inertia.

Stability Criteria

Let's assume the body has an initial rotation of w_0 around an arbitrary principal axis of inertia, which we call z axis, i.e.: $w_z = w_0$ and $w_x = w_y = 0$. This choice is in accordance with Euler's equation of motion (Eq. (15.1.13)), without external torques. Note that by employing Euler's equation we automatically switch into the principal axes system. So we observe the body to be stationary while the world around turns, including L and w. Now we give the body a kick sideways and thereby perturb the rotation by the tiny amount ε, which causes small rotation rates around the x and y axes by the amounts $w_x, w_y \propto \varepsilon$, while $w_z \approx w_0 \approx const$ roughly remains constant. For this perturbed case, the

Euler's equations (15.1.13) take on the form

$$I_x \dot{\omega}_x = (I_y - I_z) \omega_y \omega_0$$
$$I_y \dot{\omega}_y = (I_z - I_x) \omega_x \omega_0$$
$$I_z \dot{\omega}_z = (I_x - I_y) \omega_x \omega_y \approx 0$$

Terms of the order ε^2 can be neglected with regard to ε. This is why $(I_x - I_y) \omega_x \omega_y \propto \varepsilon^2 \approx 0$. Therefore from the last equation follows $\dot{\omega}_z \approx 0$, which in retrospect confirms our initial assumption $\omega_z \approx \omega_0 \approx const$. We rewrite the first two equations as

$$\dot{\omega}_x = \frac{I_y - I_z}{I_x} \omega_0 \omega_y \tag{15.2.1a}$$

$$\dot{\omega}_y = \frac{I_z - I_x}{I_y} \omega_0 \omega_x \tag{15.2.1b}$$

These are coupled differential equations, which we decouple by differentiating Eq. (15.2.1a) and inserting the result into Eq. (15.2.1b), or the other way round. This gives us

$$\ddot{\omega}_x = -\lambda^2 \omega_x$$
$$\ddot{\omega}_y = -\lambda^2 \omega_y \tag{15.2.2a}$$

with

$$\lambda := \omega_0 \sqrt{\frac{I_z - I_y}{I_y} \frac{I_z - I_x}{I_x}} \tag{15.2.2b}$$

The general solutions of these differential equations are well known:

$$\omega_x(t) = \omega_{x0} e^{i(\lambda t + \varphi)}$$
$$\omega_y(t) = \omega_{y0} e^{i(\lambda t + \phi)}$$

We now come to the stability criteria of this perturbed rotation. We get no exponentially diverging solutions only if λ is real. This means that, to achieve a stable rotation around the z axis ($\omega_z = const$), the radicand of the root in Eq. (15.2.2b) has to be positive. This is the case if

$$I_z > I_x, I_y \quad \text{or} \quad I_z < I_x, I_y$$

For a symmetrical gyro, i.e. if the two principal axes $I_x = I_y \neq I_z$, we get $\lambda = \omega_0 (I_z - I_y) / I_y$. In this case, λ is always real and thus the three principal axes are always stable. We summarize these important results:

> A rotation is only stable if it takes place along the major or minor principal axis. A rotation around the principal axes with mean moment of inertia is not stable. If two principal moments of inertia are the same (symmetrical gyro), all the principal axes are stable.

This feature is often used to give a symmetrical satellite a fixed orientation in space by spinning it around a principal axis of inertia (spin stabilization).

Nutation

We now want to know how the spin axis of a body moves. If λ is real, the general solution for Eq. (12.2.2) is

$$\omega_x(t) = \omega_{x0} \sin(\lambda t + \varphi) \qquad (15.2.3a)$$

and by inserting it into Eq. (15.2.1b) and integrating we get for the other component

$$\omega_y(t) = -\omega_{y0} \cos(\lambda t + \varphi) \qquad (15.2.3b)$$

with

$$\omega_{y0} = \omega_{x0} \sqrt{\frac{I_z - I_x}{I_z - I_y} \frac{I_x}{I_y}}$$

In the body-fixed principal axes system, for arbitrary principal moments of inertia the tip of the rotation vector therefore describes an elliptical motion with semi-axes $\omega_{x0} \neq \omega_{y0}$. That is, rotation axis and rotation speed both change continuously. If $I_z \gg I_x, I_y$ (flat body), we get $\omega_{y0} = \omega_{x0}\sqrt{I_x/I_y}$; and for $I_z \ll I_x, I_y$ (elongated body), we get $\omega_{y0} = \omega_{x0} \cdot I_x/I_y$.

In the important case of a *symmetrical gyro* $I_x = I_y$, we get $\omega_{x0} = \omega_{y0}$, i.e. the tip rotates on a circle (see Fig. 15.6). In addition, the following is valid:

$$\omega_{xy}^2 := \omega_x^2 + \omega_y^2 = \omega_{x0}^2 = const$$

with ω_{xy} as the turning component in the x-y plane perpendicular to the x axis. In total we have

$$|\omega| = \sqrt{\omega_x^2 + \omega_y^2 + \omega_z^2} = \sqrt{\omega_{xy}^2 + \omega_0^2} = const \qquad (15.2.4)$$

So, a symmetrical gyro maintains a constant rate of rotation, only the rotation axis moves conically.

With this we get the following stable motion as viewed from a body-fixed reference system. If initially the rotation ω is along one of the stable principal

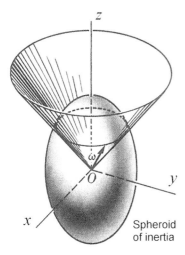

Figure 15.6 The motion of ω of a symmetrical gyro in the principal axes system with its spheroid of inertia.

axes, ω is maintained and the world around turns evenly around this axis. A small external perturbation (slight touch of the body = torque), provokes the vector ω to generally describe an elliptical (with changing length) or for symmetrical gyros a conical (with constant length) rotation around the principal axis. From an external point of view, the body first rotates evenly about the stable principal axis (= body axis) – it rotates stationary. The kick provokes the rotation axis to turn elliptically (with changing rate of rotation) or conically (with constant rotation rate) around the angular momentum vector now fixed in space. This motion is called *nutation*, so it pitches or, in colloquial English, it tumbles. But this nutation of the rotation axis is not what we see, because the rotation axis is imperceptible to us. What we see is the motion of the body axis u_z. To illustrate its motion we need to know what the relation of u_z with L, ω is, to which we come now.

15.2.3
Nutation of a Torque-Free Symmetrical Gyro

The torque-free symmetrical gyro, $I_x = I_y \neq I_z$, is a case in point to study the motion of the body axis. Due to the conspicuous symmetrical form, we call the u_z body axis the *figure axis*. For an external observer (inertial frame) the rotation axis ω of a symmetric gyro moves on a cone, the so-called nutation cone around the angular momentum vector (see Fig. 15.7) with frequency (see

Eq. (15.2.2a))

$$\nu_N = \frac{\lambda}{2\pi} = \frac{\omega_0}{2\pi}\frac{I_z - I_x}{I_x} \quad \text{nutation frequency} \quad (15.2.5)$$

The nutation of the figure axis, which is directly coupled to the nutation of the rotation axis, can be illustrated as follows. Angular momentum and rotation axis define two constant angles with the figure axis (see Fig. 15.7)

$$\tan\theta := \tan\angle(L, u_z) = \frac{L_{xy}}{L_z} = \frac{I_x\omega_{xy}}{I_z\omega_z} = const \quad \text{nutation angle} \quad (15.2.6)$$

$$\tan\gamma := \tan\angle(\omega, u_z) = \frac{\omega_{xy}}{\omega_y} = const$$

Hence

$$\tan\theta = \frac{I_x}{I_z}\tan\gamma$$

So, for these angles the following is valid:

$\theta > \gamma \quad @ \; I_x = I_y > I_z \quad$ (elongated "prolate" body)

$\theta < \gamma \quad @ \; I_x = I_y < I_z \quad$ (flat "oblate" body)

In addition, it can be shown that L, ω, u_z are coplanar. These properties make it possible to illustrate the nutation for prolate and oblate symmetrical gyros as follows. The figure axis of the spinning gyro turns with frequency ν_N and nutation angle θ around the angular momentum axis L, which is fixed in space. The instantaneous rotation axis, which lies in between (prolate) or on the opposite side (oblate), turns with the same rate. These conical rotations can be also interpreted as if two cones either side by side (prolate) or inside each other (oblate) mutually roll on each other (Fig. 15.7) without slip, whereby the line of contact is just the instantaneous rotation axis.

This example shows clearly why it is so difficult to understand the rotational dynamics of a body. When we look at a rotating body our eyes are fixed on the conspicuous figure axis. From a physical point of view, however, this figure axis does not play a role, because Euler's equations assign a physical meaning only to the rotation. But the location of the rotation axis in space is non-perceptible to us. Merely the angular momentum axis can be imagined fictitiously as a steady axis, because the figure axis co-rotates with the rotation axis around it. It is this co-rotation around this fictitious axis that makes the rotational dynamics become clear to us.

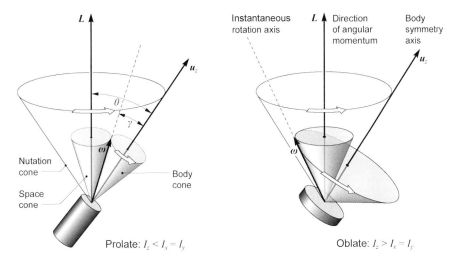

Figure 15.7 Nutation of a free symmetrical gyro for a prolate (left) and oblate body (right) in an inertial system.

Clarification of Terms

The terms "prolate" and "oblate" require some explanation. One has to distinguish between the prolate/oblate geometry of a symmetrical body (Fig. 15.8) and the prolate/oblate rotation of such a body. A prolate (oblate) geometry means that the body is elongated (flat) with regard to its axis of symmetry. A prolate (oblate) rotation, on the other hand, means that a body rotates around the minor (major) principal axis. If a cigarette rolls down a slope, its prolate shape also rotates in a prolate way. If you put the cigarette flat on a table and turn it, the prolate cigarette rotates in an oblate way. On the other hand, a coin rotating in a perpendicular position on a table is an oblate body rotating in a prolate way. The friction with the table causes the rotation to slowly turn over into a flat rotation along the outer edge of the coin, i.e. an oblate rotation, until the coin comes to rest.

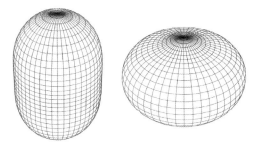

Figure 15.8 Prolate (left) and oblate (right) geometry.

Further confusion arises from the terms *nutation* and *precession*. The pitch around an axis is generally denoted as precession. But precession in physics has the special meaning of the pitch of the angular momentum around the axis of an applying external force (see Section 15.3). Though precession and nutation for a spinning top look quite the same, the physics is much different. Nevertheless, the term "precession" is often used for nutation in the literature, which adds to the confusion of these two effects. Therefore we refrain here from using the term "precession" or "precess" for the nutational motion.

15.2.4
Nutation and Energy Dissipation

If a satellite is exposed to residual atmosphere, this leads to a decrease of its rotational energy due to drag. The loss of energy is called *energy dissipation*. The rotational energy of a satellite can also be dissipated by internal friction, for instance by fuel sloshing or deliberately by nutation dampers (see Fig. 15.9). Energy dissipation at a constant angular momentum has quite different impacts on nutation. To see why, let's have a look at the kinetic energy of rotation around an arbitrary instantaneous rotation axis ω. According to Eq. (15.1.9) and $L_\omega = \omega \cdot I_\omega = L = const$ (see "Introductory considerations" in Section 15.1.1) it amounts to

$$E_{rot,\omega} = \frac{1}{2} I_\omega \omega^2 = \frac{1}{2} \frac{L^2}{I_\omega} \qquad (15.2.7)$$

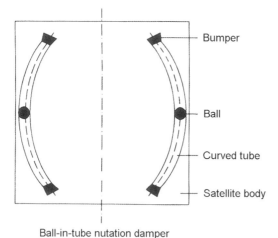

Ball-in-tube nutation damper

Figure 15.9 Rotations around the axis perpendicular to the image plane are damped by this nutation damper due to the friction of the accelerated ball with the viscous fluid in the tube.

As the moments of inertia depend on the rotation axis, so does the corresponding rotational energy. We also recall from "Introductory considerations" in Section 15.1.1 that $I_\omega = \sum_i m_i R_{i\omega}^2$, where $R_{i\omega}$ is the distance between the body particle with mass m_i and the rotation axis. This shows that the moment of inertia is smallest and, because of Eq. (15.2.7), its energy is largest when the body rotates in a prolate way. (Strictly speaking, when it rotates along the minor principal axis. But because most satellites are symmetrical gyros, we adopt this denomination from now on.) For a body rotating in an oblate way, the corresponding quantities are of course exactly the opposite. In summary

$$E_{max} = \frac{1}{2}\frac{L^2}{I_{min}} \quad @ \text{ prolate rotation}$$

$$E_{min} = \frac{1}{2}\frac{L^2}{I_{max}} \quad @ \text{ oblate rotation}$$

So, if a satellite loses rotational energy at $L = const$ it has to pass from a rather prolate to a more oblate nutational rotation. This implies nothing other than changing its nutation angle.

Nutation Angle Rate

To be able to determine the nutation angle rate quantitatively, for a symmetrical gyro we are looking for a relation between E_{rot} and θ. We use Eqs. (15.1.7) and (15.1.9) to form the expression

$$L^2 - 2E_{rot}I_x = \omega_z^2 I_z (I_z - I_x) = L^2 \cos^2\theta \frac{I_z - I_x}{I_z} \tag{15.2.8}$$

The latter results from Eq. (15.2.6) because of

$$\cos\theta = \frac{1}{\sqrt{1+\tan^2\theta}} = \frac{L_z}{\sqrt{L_z^2 + L_{xy}^2}} = \frac{L_z}{L} = \frac{I_z\omega_z}{L}$$

Differentiating Eq. (15.2.8) with regard to time delivers

$$\dot\theta = \frac{2}{L^2 \sin 2\theta} \frac{I_z I_x}{I_z - I_x} \dot{E}_{rot} \tag{15.2.9}$$

Modes of Motion

A rotating body dissipating energy at constant angular momentum can therefore behave in two different ways:

1. Initially, the body rotates in a *prolate way*.

   In this case $I_z < I_x$. Let's assume that initially there is no nutation along the minor principal axis, $\theta = 0$, so the body has maximal rotational energy. Then for even the slightest energy dissipation $\Delta E_{rot} < 0$, we get $\Delta\theta > 0$: the body starts to nutate and increases its nutation amplitude further and further.

2. Initially, the body rotates or nutates in a *nearly oblate way*.

   Now $I_z > I_x$. The rotational energy then is close to minimal at the beginning. According to Eq. (15.2.9) the body can achieve a further reduction of the rotational energy $\Delta E_{rot} < 0$ only by reducing the nutation, $\Delta\theta < 0$: nutation is dampened. When the nutation has at some time died out this way, the satellite has reached its state of minimal energy at constant angular momentum. By spinning evenly along the major principal axis (so-called *flat spin*) there is no way to dissipate energy internally any further. Only external residual atmosphere can further reduce rotational energy by decelerating the spin by simultaneously reducing the angular momentum. But there will never be nutation again.

We therefore derive the important satellite design rule

> When a satellite should be spin-stabilized, its spin should be around its major principal axis. In this case, the spin is not only stable, but any energy dissipation such as deceleration by the remaining atmosphere dampens out any possible initial nutation: the satellite spins evenly also in the long term.

NASA unfortunately had to learn this major-principal-axis rule, when on February 1, 1958, Explorer 1 was launched into space as the first US satellite. The highly elongated (prolate) satellite was spin-stabilized in a prolate way. Because of its four long wire aerials causing energy dissipation due to atmospheric drag, the satellite unexpectedly turned to a flat spin after just a few hours. This was only noted because it interfered with communications with the satellite.

The behavior patterns derived here are only valid for fully rigid bodies. Dual-spin satellites, where two parts of the body are rotatively decoupled, have a partially different behavior. They can be spin-stable also along their prolate axis. As the significance of dual-spin satellites is rather small today, we do not want to go into details here.

15.2.5
General Torque-Free Motion

Until now we have examined only rotational motion near the principal axes. We finally want to investigate the most general case: the torque-free motion of an arbitrarily shaped body. Because its mathematical description is far too complex, we want to restrict the discussion to only a qualitative geometrical representation developed by Poinsot, which instead is very clear.

For our representation we adopt the body-fixed reference system of the principal axis. What can we say about the motion of L and the rotation vector ω in this frame in the most general case? Well, their motion is restricted only by two constraints:

- The first is the *conservation of angular momentum*, which is written in the principal axes frame as

$$\frac{L_x^2}{L^2} + \frac{L_y^2}{L^2} + \frac{L_z^2}{L^2} = 1 \quad L \text{ sphere} \tag{15.2.10}$$

- The second is the *conservation of energy*, which reads in the principal axes frame according to Eq. (15.1.9) with Eq. (15.1.6)

$$\frac{L_x^2}{2E_{rot}I_x} + \frac{L_y^2}{2E_{rot}I_y} + \frac{L_z^2}{2E_{rot}I_z} = 1 \quad \begin{matrix}\textbf{Poinsot ellipsoid}\\(E \text{ ellipsoid})\end{matrix} \tag{15.2.11}$$

The first equation is the functional equation for the surface of a sphere with radius L, on which the tip of the angular momentum vector is free to move. The second functional equation describes a triaxial ellipsoid (Poinsot ellipsoid or E ellipsoid) with the semi-axes $\sqrt{2E_{rot}I_x}$, $\sqrt{2E_{rot}I_y}$, $\sqrt{2E_{rot}I_z}$. Because the tip of L has also to move on the surface of the ellipsoid, L can only move on the intersection line of the two surfaces, the so-called *polhode* (see Fig. 15.10a).

We are now going to describe the possible paths of L under energy dissipation. If the energy is maximum then the E ellipsoid encloses the L sphere and touches it tangentially in the direction of the minor principal axis – the polhode is just a point. In this case the body rotates nutation-free along the principal axis, which we will denote as the x axis, with the smallest moment of inertia. When the rotational energy lowers, the polhode point quickly becomes a circle. Now L turns on this circle around the x axis. (Or reciprocally, the x axis nutates around L with nutation angle θ. This latter view was adopted in Fig. 15.7 where, for an external observer, L was fixed and the principal body axis nutated around it.). This situation is depicted in Fig. 15.10a and in addition in Fig. 15.10b with the direction of motion. The more the energy decreases the larger becomes the polhode (nutation circle). In addition it slowly buckles until it transits into a diagonal circle limit when L travels through the mean

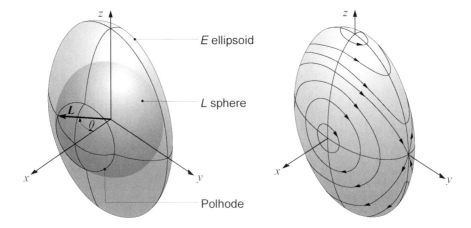

Figure 15.10 The fixed L sphere and E ellipsoid in the principal axis system (a, left). Their intersection line makes up the polhode, which is depicted for different rotational energies on the right (b).

principal axis. In this case the E ellipsoid touches the L sphere at the mean principal axis. But because L smoothly passes through this principal y axis, this rotation is not stable (cf. Section 15.2.2): the body tumbles maximally. When the energy is further reduced, L begins to nutate at large angles around the major principal axis. The nutation angle shrinks with further decline of the energy until at minimal energy the body spins nutation-free around this major principal axis. In this limiting case the L sphere encloses the E ellipsoid and touches it tangentially at the crossing with the z body axis.

15.3
Gyro under External Torque

Until now we have looked at the momentum of a rigid body unaffected by its environment. The general result was a rotation with nutation. Now, we assume that there are external forces acting on the rotating body. They can be not only gravitational forces, such as the Moon, other planets, or the Sun, affecting the Earth, but also the magnetic field of the Earth interacting with some internal magnetic fields of an S/C. We want to know the reaction of the body to these forces. Let's imagine an asymmetrical ($I_x \neq I_y \neq I_z$) or symmetrical ($I_x = I_y \neq I_z$) gyro rotating in a stable way along its major or minor principal axis. An external force F impacting with a lever arm at distance r from its center of mass causes the external torque $T = r \times F$ on the gyro. As, because of Eq. (15.1.11), $dL/dt = T$ is valid, we can generally observe the following:

1. If the external torque is parallel to the angular momentum $T \parallel L$, then due to $dL = T \cdot dt \parallel L$ also dL is parallel to L, and the angular momentum only changes its absolute value, but not its direction. Depending on the sign of dL, the body rotates faster or slower. If the body spins along its minor or major principal axis with $L \parallel \omega$, then according to Eq. (15.1.16) the rotational power is maximal with $P_{rot} = \omega T$.

2. If T and therefore dL are perpendicular to L, L changes only its direction, but not its absolute value. The induced rotation of L as depicted in Fig. 15.11 is called the *precession* of the body. For $L \parallel \omega$ the rotational energy then does not change, and the rotational power is $P_{rot} = 0$.

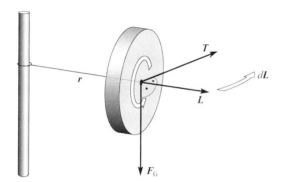

Figure 15.11 Precession of a gyro under the impact of an external force (here gravitational force F_G).

Let's have a closer look at the precession of a gyro. According to Eq. (15.1.10) the external force (in our example it is the gravitational force) F_G produces an external torque $T = r \times F_G$ that is perpendicular to F_G and r. As, because of $dL = T \cdot dt$, the change of the torque is always in the direction of the plane perpendicular to F_G, L will also always move in circles in this plane. This is the precession motion.

If a nutating gyro is subject to an external torque, nutation and precession superimpose and its figure axis produces a motion which may look like to the one depicted in Fig. 15.12. Let's take a concrete example. The Earth's polar axis nutates around its angular momentum axis with a nutation amplitude of 19 arcseconds and a period of

$$T_N = \frac{2\pi}{\omega_N} = 18.6 \text{ years}$$

Because of their gravitative impact on the equatorial bulge, the Sun and the Moon create a torque on the Earth that induces an additional precession with

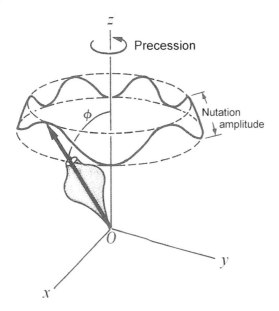

Figure 15.12 Superimposed nutation and precession of a symmetrical gyro.

the period

$$T_P = \frac{2\pi}{\omega_P} = 25\,800 \text{ years}$$

where the precession angle of course is the angle between the plane in which the torque applies and the polar axis, and hence is $\phi = 23.5°$. The polar axis under these circumstances describes a compound motion much close to that in Fig. 15.12.

15.4
Gravity-Gradient Stabilization

In low Earth orbit the gravitation gradient of the Earth may have a significant impact on the spatial orientation of the satellite and its rotational behavior, which may be either undesirable or harnessed skillfully as we will see in this section. If the satellite elongates a lot in the radial direction to the Earth, the satellite as a whole (i.e. with its center of mass) moves in a force-balanced way around the Earth. Those satellite parts which are by Δr further away from the center of mass, but nevertheless move with the same angular orbit velocity ω, are subject to the additional centrifugal force $\Delta F_Z = m\omega^2 \Delta r$, and, apart from that, a smaller gravitational force. Those parts of the satellite that

15.4 Gravity-Gradient Stabilization

are closer to the center of mass experience just the opposite effect: a smaller centrifugal force and a larger gravitational force. All in all, this results in a gravity-gradient (GG) torque, which tends to align the satellite along the radial vector (i.e. in the direction of the geocenter). The question here is this: What are the effects of this force gradient on the attitude and the rotational behavior of the satellite?

To answer this question, first of all we have to calculate the torque of the gravity gradient on the S/C. This torque will then be applied to Euler's equations (15.1.13) to study the dynamic behavior and the dynamic stability of the S/C.

15.4.1
Gravity-Gradient Torque

According to Fig. 15.13 the infinitesimally small torque acting with the lever arm r on each particle with mass dm of the body is given by

$$dT = r \times dF = r \times \left[-\frac{\mu \cdot dm}{|R+r|^3} (R+r) \right]$$

As $R \gg r$ holds very well, the denominator can be approximated with very high accuracy

$$\frac{1}{|R+r|^3} = \frac{1}{(R^2 + 2R \cdot r + r^2)^{3/2}} = \frac{1}{R^3 (1 + 2R \cdot r/R^2)^{3/2}}$$

$$\approx \frac{1}{R^2} \left(1 - 3\frac{R \cdot r}{R^2} \right)$$

Thus, the total gravity-gradient torque is

$$T_{GG} = \int r \times dF = -\frac{\mu}{R^3} \int r \times (R+r) \cdot \left(1 - 3\frac{R \cdot r}{R^2} \right) dm$$

Because of $r \times r = 0$ we get

$$T_{GG} = -\frac{3\mu}{R^5} R \times \int r \cdot \left(1 - 3\frac{R \cdot r}{R^2} \right) dm$$

But as r measures the distance with regard to the center of mass, $\int r \, dm = 0$ is valid, and thus

$$T_{GG} = -\frac{3\mu}{R^5} R \times \int (R \cdot r) r \, dm$$

Because of the vector relation $r \times (R \times r) = (r \cdot r) R - (R \cdot r) r$ we get

$$T_{GG} = \frac{3\mu}{R^5} R \times \int \left[r \times (R \times r) - (r \cdot r) R \right] dm$$

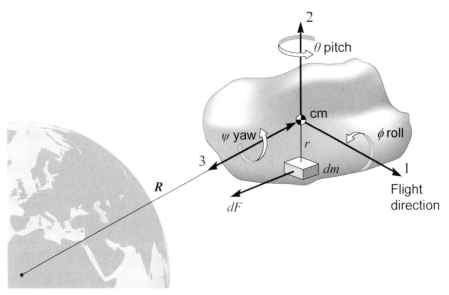

Figure 15.13 Definition of the inverted RSW reference frame (cf. Figs. 15.3 and 15.4; cm is the center of mass.

As $R \times R = 0$, the second term of the integral becomes zero, and we get

$$T_{GG} = \frac{3\mu}{R^5} R \times \int r \times (R \times r)\, dm$$

According to Eqs. (15.1.2) and (15.1.3) we have

$$\int r \times (R \times r)\, dm = IR$$

with I as the inertia tensor. So for the gravity-gradient torque, we finally get the simple expression

$$T_{GG} = \frac{3\mu}{R^5} R \times IR = 3\omega_\nabla^2 u_R \times Iu_R \tag{15.4.1}$$

with u_R as the unit vector from the geocenter to the center of mass of the S/C and the circular orbit frequency $\omega_\nabla = \sqrt{\mu/R^3}$ of the satellite.

15.4.2
Gravity-Gradient Induced Attitude Changes

Now that we know the torque on the S/C, we want to know if there are stable attitudes of the S/C under this influence. To figure this out we employ the principal axis body system x, y, z in which the inertia tensor is diagonal, $I_P =$

diag (I_x, I_y, I_z). In this reference system with the body at rest and oriented towards the Earth ($x, y, z = 1, 2, 3$), the vector from the geocenter to the body center of mass is given as $\boldsymbol{R} = (R_x, R_y, R_z)^T$. With this Eq. (15.4.1) results in

$$T_{GG} = \frac{3\omega_\nabla^2}{R^2} \begin{pmatrix} R_x \\ R_y \\ R_z \end{pmatrix} \times \begin{pmatrix} I_x R_x \\ I_y R_y \\ I_z R_z \end{pmatrix} = \frac{3\omega_\nabla^2}{R^2} \begin{bmatrix} R_y R_z (I_z - I_y) \\ R_x R_z (I_x - I_z) \\ R_x R_y (I_y - I_x) \end{bmatrix} \quad (15.4.2)$$

From this we recognize two conditions:

1. The S/C is not subject to a torque around the principal i axis if the S/C is rotationally symmetrical around this axis, i.e. if $I_j = I_k$. This is reasonable because, as the axis is rotationally symmetrical, there is no imbalance on which the force can act to turn the body.

2. The S/C is not subject to a torque around *any* axis, if
   (i) the S/C is fully symmetrical, i.e. $I_x = I_y = I_z$, or
   (ii) one of the body axes points exactly to nadir, i.e. $\boldsymbol{R} = R(\pm 1, 0, 0)^T$ or $\boldsymbol{R} = R(0, \pm 1, 0)^T$ or $\boldsymbol{R} = R(0, 0, \pm 1)^T$.

Usually both these conditions are not fulfilled, and therefore a GG torque T_{GG} occurs, leading inevitably to a turn and hence to rotational dynamics of the S/C. There are a number of questions to be asked: What do these rotations look like? Are there any stable oscillations around certain axes? And, if this is the case, under what conditions? Or are the rotations all unstable? The "stability of an oscillation" expressed in simplified terms means that, with a small perturbation (external torque), the body carries out stable harmonic oscillations around this axis. If the oscillation is unstable, the state of motion exponentially degrades and passes to a stable oscillation around another axis, or even turns into a chaotic state of rotation between all the axes.

Let's investigate whether stable oscillations exist, and which ones are stable.

15.4.3
Stability of Gravity-Gradient Oscillations

We now permit small roll (ϕ), pitch (θ), and yaw (ψ) rotations as perturbations. With these the unit vector in the R direction in the principal axes frame is, according to Fig. 15.5,

$$u_{R,P} = -\sin\theta \cdot u_1 + \sin\phi \cos\theta \cdot u_2 + \cos\phi \cos\theta \cdot u_3$$
$$= (-\sin\theta, \sin\phi \cos\theta, \cos\phi \cos\theta)_{LVLH}$$

Inserting this into Eq. (15.4.1) results in

$$T_{GG,P} = \frac{3}{2}\omega_\nabla^2 \begin{bmatrix} (I_z - I_y)\cos^2\theta \sin 2\phi \\ -(I_x - I_z)\cos\phi \sin 2\theta \\ -(I_y - I_x)\sin\phi \sin 2\theta \end{bmatrix}_{LVLH}$$
(15.4.3)

$$\approx 3\omega_\nabla^2 \begin{bmatrix} (I_z - I_y)\phi \\ -(I_x - I_z)\theta \\ 0 \end{bmatrix}_{LVLH}$$

the latter for $\phi, \theta, \psi \to 0$. If we now apply Eq. (15.4.3) to Euler's equations (15.1.13), and consider from Eqs. (15.1.14) and (15.1.15) the Euler rate equations

$$\omega_P = (\dot\phi + \omega_\nabla \psi, \dot\theta + \omega_\nabla, \dot\psi - \omega_\nabla \phi)_{LVLH}$$

$$\dot\omega_P = (\ddot\phi + \omega_\nabla \dot\psi, \ddot\theta, \ddot\psi - \omega_\nabla \dot\phi)_{LVLH}$$

one gets the following equations of motion for rotations in the LVLH frame:

$$\ddot\phi + \omega_\nabla(1 - k_Y)\dot\psi + 4\omega_\nabla^2 k_Y\phi = 0 \tag{15.4.4a}$$

$$\ddot\psi + \omega_\nabla(k_R - 1)\dot\phi + \omega_\nabla^2 k_R\psi = 0 \tag{15.4.4b}$$

$$\ddot\theta + 3\omega_\nabla^2 k_P\theta = 0 \tag{15.4.4c}$$

with

$$k_P := \frac{I_x - I_z}{I_y}, \quad k_Y := \frac{I_y - I_z}{I_x}, \quad k_R := \frac{I_y - I_x}{I_z} \tag{15.4.5}$$

The first two equations describe a coupled yaw-roll motion. The last equation is decoupled and describes a pitch oscillation.

15.4.4
Pitch Oscillation

According to Eq. (15.4.4c) the S/C oscillates like a harmonic oscillator with frequency

$$\omega_2 = \omega_\nabla \sqrt{3(I_x - I_z)/I_y} \tag{15.4.6}$$

around the 2 axis, i.e. a pitch oscillation in the plane that is set up by the flight direction and nadir. This frequency is real, and thus the oscillation is stable, if the root is real, i.e. if

$$\boxed{I_z < I_x}$$ stability condition for pitch oscillation (15.4.7)

So we have a satellite, which circles the planet, with its z axis always swinging around the nadir direction (see Fig. 15.14). This remarkable effect is called *gravity-gradient stability*. We get the ideal state of an S/C circling the Earth by constantly pointing towards the geocenter as depicted in Fig. 15.15, if we dampen the oscillation so that it disappears (see Fig. 15.16). This stable state arouses a lot of practical interest, as it is possible to achieve a constant nadir orientation very easily without any control systems. Such a constant orientation can be used to align, for instance, a communication antenna or an Earth sensor always to the Earth's surface.

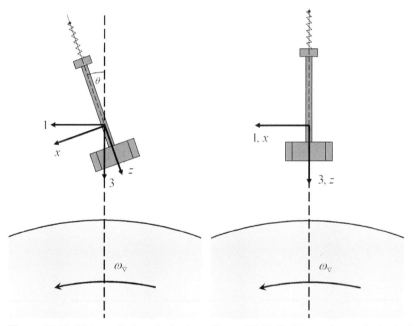

Figure 15.14 Pitch oscillation of a body in gravity-gradient mode.

Figure 15.15 Stable gravity-gradient orientation of a rotationally symmetrical prolate body.

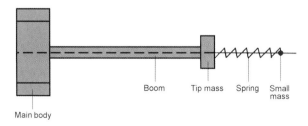

Figure 15.16 Spring libration damper for a gravity-gradient stabilized body.

This situation in which a circling body rotates about its own axis synchronously with its orbit period – a so-called a 1:1 orbital resonance – is by the way absolutely identical to the one of the Moon circling the Earth. Due to the gradient force of the Earth, the Moon is a bit elongated along the Earth–Moon connecting line. This prolate shape is aligned to Earth by the gradient force, leading to a coupled 1:1 motion with its revolution.

15.4.5
Coupled Roll-Yaw Oscillation

We are now looking for the stability criteria for rotations around the other two axes. If the S/C is rotationally symmetric, $I_x = I_y \neq I_z$, then according to Eq. (15.4.2) the body experiences no torque around the z axis, and the GG torque does not contribute to the stability along this axis: the satellite will freely rotate around its z axis. For the case of a nonsymmetrical S/C, $I_x \neq I_y \neq I_z$, we are looking for solutions for the first two coupled equations (15.4.3)a,b. We make the general ansatz

$$\psi = \psi_0 e^{s\omega_\nabla t}$$

$$\phi = \phi_0 e^{s\omega_\nabla t}$$

which we insert into Eqs. (15.4.4a) and (15.4.4b). Thus, we get the characteristic quartic equation

$$s^4 + s^2(1 + 3k_R + k_Y k_R) + 4k_Y k_R = 0$$

The solutions of this are

$$s_{1,2}^2 = -b \pm \sqrt{b^2 - 4k_Y k_R}$$

with

$$b = (1 + 3k_R + k_Y k_R)/2$$

To get periodic oscillations, all the solutions have to be imaginary. This is fulfilled if the root above is real, and $s_{1,2}^2 < 0$. Both conditions lead to

$$1 + 3k_R + k_Y k_R > 4\sqrt{k_Y k_R}$$

$$k_Y k_R > 0$$

(15.4.8)

In addition, the above pitch stability condition $I_z < I_x$ is equivalent to

$$k_Y < k_R$$

(15.4.9)

15.4 Gravity-Gradient Stabilization

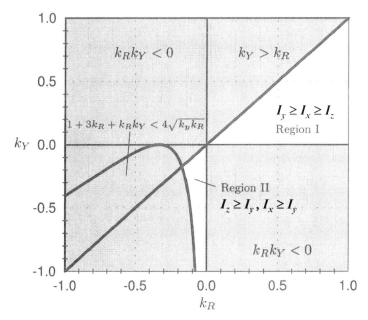

Figure 15.17 Stability zones of a roll–yaw oscillation in gravity-gradient mode.

These three stability conditions (15.4.8) and (15.4.9) are only fulfilled at the same time in the two white regions I and II of Figure 15.17.

In reality, region II is also not stable. One can see that from the stability conditions of region II: $I_z \geq I_y$, $I_x \geq I_y$. Because of Eq. (15.2.7)

$$E_{rot} - \frac{1}{2} I_\omega \omega^2 - \frac{1}{2} \frac{L^2}{I_\omega}$$

this pitch oscillation around the axis with minimal moment of inertia is a state of maximum oscillation energy. Even the smallest dissipation of energy (see Section 15.2.4), e.g. by residual atmosphere, leads to a transfer to a state of lowest energy, i.e. to the axis with the largest moment of inertia, and thus to region I. Therefore, for coupled roll-yaw oscillations of real dampened systems only region I is stable. This area is characterized by the condition

$\boxed{I_z \leq I_x \leq I_y}$ stability condition for roll-yaw oscillations   (15.4.10)

If the coupled roll-yaw oscillation is stable, it oscillates according to our ansatz $\psi = \psi_0 e^{s\omega_\nabla t}$, $\phi = \phi_0 e^{s\omega_\nabla t}$ with the angular frequency $|s|\omega_\nabla \approx \omega_\nabla$, that is, very slowly.

Appendix A
Astrodynamic Parameters

Table A.1 provides the most important parameters of the solar system planets for astrodynamics.

Table A.1 Characteristic parameters of the planets in the solar system: R = planet radius; $\langle r \rangle_t$ = time-averaged orbit radius; $\langle v \rangle_t$ = time-averaged orbital velocity. The values hold for the year 2010. An underscore indicates that this digit changes within a decade.

	R [km]	μ [km^{3}s^{-2}]	a [10^6 km]	e	$\langle r \rangle_t$ [10^6 km]	$\langle v \rangle_t$ [km s^{-1}]
Sun	696 000	1.327 1244×10^{11}	–	–	–	–
Mercury	2 440	2.2032×10^4	57.909 1$\underline{5}$	0.205 63	59.133	47.362
Venus	6 052	3.248 59×10^5	108.2089	0.006 77	108.21	35.020
Earth	6 378.14	3.986 006×10^5	149.5979	0.016 71	149.62	29.783
(Moon)	1 737.4	4.902 799×10^3	0.384 400	0.054 90	0.384 98	1.0175
Mars	3 396	4.2828×10^4	227.93$\underline{5}$	0.093 4$\underline{2}$	228.93	24.077
Jupiter	71 490	1.267 128×10^8	778.4$\underline{2}$	0.048 3$\underline{8}$	779.33	13.050
Saturn	60 270	3.794 06×10^7	1 426.$\underline{7}$	0.054 1$\underline{1}$	1 428.8	9.638
Uranus	25 560	5.794 56×10^6	2 870.9$\underline{9}$	0.047 1$\underline{5}$	2 874.2	6.795
Neptune	24 760	6.836 53×10^6	4 498.2$\underline{3}$	0.008 59	4 498	5.432
Pluto	1 196	830	5 906.3$\underline{6}$	0.248 81	6 089	4.666

A.1
Mean Orbit Radius

A.1.1
Titius–Bode Law

The orbit radii of the solar planets can be expressed empirically and approximately by the famous Titius–Bode law

$$\langle r \rangle [\mathrm{AU}] = 0.4 + 0.3 \times 2^n \quad \textbf{Titius–Bode law}$$

Astronautics. Ulrich Walter
Copyright © 2008 WILEY-VCH Verlag GmbH & Co. KGaA, Weinheim
ISBN: 978-3-527-40685-2

A Astrodynamic Parameters

with

Mercury	$n = -\infty$
Venus	$n = 0$
Earth	$n = 1$
Mars	$n = 2$
Asteroid belt	$n = 3$
Jupiter	$n = 4$
Saturn	$n = 5$
Uranus	$n = 6$

$1 \text{ AU} = 149.597\,870\,691 \times 10^6 \text{ km}$

A.1.2
Average over True Anomaly

$$\langle r \rangle_\theta = a\left(1 - e^2\right) \frac{1}{\pi} \int_0^\pi \frac{d\theta}{1 + e\cos\theta}$$

Since

$$\frac{1}{\pi} \int_0^\pi \frac{d\theta}{1 + e\cos\theta} = \frac{2}{\pi\sqrt{1-e^2}} \arctan\left(\sqrt{\frac{1-e}{1+e}} \cdot \tan\frac{\theta}{2}\right)\bigg|_0^\pi$$

$$= \frac{2}{\pi\sqrt{1-e^2}} \left(\frac{\pi}{2} - 0\right) = \frac{1}{\sqrt{1-e^2}}$$

then

$$\langle r \rangle_\theta = a\sqrt{1-e^2} = b \quad (b = \text{semi-minor axis}) \tag{A.1.1}$$

A.1.3
Time Average

$$\langle r \rangle_t = a\left(1 - e^2\right) \frac{2}{T} \int_0^{T/2} \frac{dt}{1 + e\cos\theta} = a\left(1 - e^2\right) \frac{2}{T} \int_0^\pi \frac{dt/d\theta}{1 + e\cos\theta} d\theta$$

Since

$$\frac{dt}{d\theta} = \frac{r^2}{h} = \frac{a^2\left(1 - e^2\right)^2}{(1 + e\cos\theta)^2 \sqrt{\mu a (1 - e^2)}} \quad \text{and} \quad T = 2\pi\sqrt{\frac{a^3}{\mu}}$$

it follows that

$$\langle r \rangle_t = a\left(1 - e^2\right)^{5/2} \frac{1}{\pi} \int_0^\pi \frac{d\theta}{(1 + e\cos\theta)^3}$$

$$\int \frac{d\theta}{(1+e\cos\theta)^3} = \frac{2+e^2}{(1-e^2)^{5/2}} \arctan\left(\sqrt{\frac{1-e}{1+e}} \cdot \tan\frac{\theta}{2}\right)$$
$$+ \frac{e\left(e^2 - 3e\cos\theta - 4\right)\sin\theta}{2\left(1-e^2\right)^2 (1+e\cos\theta)^2}$$

From this we get

$$\langle r \rangle_t = a\left(1-e^2\right)^{5/2} \frac{1}{\pi} \frac{\pi}{2} \frac{2+e^2}{(1-e^2)^{5/2}}.$$

and therefore

$$\langle r \rangle_t = a\left(1 + \frac{e^2}{2}\right) \tag{A.1.2}$$

A.2
Mean Orbital Velocity

Denote by

$$K(x) = \int_0^{\pi/2} \frac{dt}{\sqrt{1 - x^2 \sin^2 t}}$$
$$= \frac{\pi}{2}\left[1 + \left(\frac{1}{2}\right)^2 x^2 + \left(\frac{1\cdot 3}{2\cdot 4}\right)^2 x^4 + \left(\frac{1\cdot 3\cdot 5}{2\cdot 4\cdot 6}\right)^2 x^6 + \ldots\right] \quad @\ x \to 0$$

the *complete elliptic integral of the first kind*, and by

$$E(x) = \int_0^{\pi/2} \sqrt{1 - x^2 \sin^2 t} \cdot dt$$
$$= \frac{\pi}{2}\left[1 - \left(\frac{1}{2}\right)^2 x^2 - \left(\frac{1\cdot 3}{2\cdot 4}\right)^2 \frac{x^4}{3} + \left(\frac{1\cdot 3\cdot 5}{2\cdot 4\cdot 6}\right)^2 \frac{x^6}{5} + \ldots\right] \quad @\ x \to 0$$

the *complete elliptic integral of the second kind*.

A.2.1
Average over True Anomaly

$$\langle v \rangle_\theta = \frac{\mu}{h} \frac{1}{\pi} \int_0^\pi \sqrt{1 + 2e\cos\theta + e^2} \cdot d\theta = \sqrt{\frac{\mu}{a}} \sqrt{\frac{1+e}{1-e}} \frac{2}{\pi} E\left(\frac{2\sqrt{e}}{1+e}\right)$$

From this it follows that

$$\langle v \rangle_\theta = \sqrt{\frac{\mu}{a}} \left[1 + \frac{3}{4} e^2 + \frac{33}{64} e^4 + O(e^6) \right] \quad @\ e \to 0 \qquad (A.2.1)$$

A.2.2
Time Average

$$\langle v \rangle_t = \frac{\mu}{h} \frac{2}{T} \int_0^{T/2} \sqrt{1 + 2e \cos\theta + e^2} \cdot dt = \frac{\mu}{h} \frac{2}{T} \int_0^\pi \sqrt{1 + 2e \cos\theta + e^2} \frac{dt}{d\theta} d\theta$$

From this it follows that

$$\langle v \rangle_t = \sqrt{\frac{\mu}{a}} \left(1 - e^2\right) \cdot \frac{1}{\pi} \int_0^\pi \frac{\sqrt{1 + 2e \cos\theta + e^2}}{(1 + e \cos\theta)^2} d\theta$$

$$\int_0^\pi \frac{\sqrt{1 + 2e \cos\theta + e^2}}{(1 + e \cos\theta)^2} d\theta = \frac{1}{(1 - e^2)} \left[(1 - e) K\left(\frac{2\sqrt{e}}{1 + e}\right) + (1 + e) E\left(\frac{2\sqrt{e}}{1 + e}\right) \right]$$

$$\Rightarrow \langle v \rangle_t = \sqrt{\frac{\mu}{a}} \cdot \frac{1}{\pi} \left[(1 - e) K\left(\frac{2\sqrt{e}}{1 + e}\right) + (1 + e) E\left(\frac{2\sqrt{e}}{1 + e}\right) \right]$$

$$\Rightarrow \langle v \rangle_t = \sqrt{\frac{\mu}{a}} \left[1 - \frac{1}{4} e^2 - \frac{3}{64} e^4 - O(e^6) \right] \quad @\ e \to 0 \qquad (A.2.2)$$

Appendix B
Approximate Analytical Solution for Uneven Staging

Let $v_{*,i}$ and ε_i both be arbitrary. We define

$$v_{*,i} = v_* + \Delta v_{*,i} = v_* (1 + \delta_i)$$

and

$$\varepsilon_i = \varepsilon + \Delta \varepsilon_i = \varepsilon (1 + \Delta_i)$$

with arithmetic means

$$v_* := \frac{1}{n} \sum_i v_{*,i}$$

$$\varepsilon := \frac{1}{n} \sum_i \varepsilon_i$$

$$\Rightarrow \quad \sum_i \delta_i = 0 \qquad \sum_i \Delta_i = 0$$

Assuming $\delta_i \ll 1$ and $\Delta_i \ll 1$ we derive from Eq. (3.2.4)

$$\frac{\Delta v}{v_*} = \sum_{i=1}^{n} (1 + \delta_i) \ln \left[\frac{1}{\varepsilon_i} \left(1 - \frac{\alpha}{v_{*,i}} \right) \right]$$

$$= \sum_{i=1}^{n} (1 + \delta_i) \ln \varepsilon_i + \sum_{i=1}^{n} (1 + \delta_i) \ln \left(1 - \frac{\alpha}{v_{*,i}} \right)$$

Expanding $\ln (1 - \alpha/v_{*,i})$ for $v_{*,i} = v_* (1 + \delta_i)$ we get

$$\ln \left(1 - \frac{\alpha}{v_{*,i}} \right) \approx \ln \left(1 - \frac{\alpha}{v_*} \right) + \frac{\alpha/v_*}{1 - \alpha/v_*} \frac{\delta_i}{1 + \delta_i}$$

$$- \frac{1}{2} \left(\frac{\alpha/v_*}{1 - \alpha/v_*} \frac{\delta_i}{1 + \delta_i} \right)^2 + O\left(\delta_i^3\right)$$

With

$$\sum_i \delta_i = 0$$

Astronautics. Ulrich Walter
Copyright © 2008 WILEY-VCH Verlag GmbH & Co. KGaA, Weinheim
ISBN: 978-3-527-40685-2

B Approximate Analytical Solution for Uneven Staging

we get

$$\frac{\Delta v}{v_*} \approx \sum_i (1+\delta_i) \ln\left[\frac{1}{\varepsilon_i}\left(1-\frac{\alpha}{v_*}\right)\right] - \frac{1}{2}\left(\frac{\alpha/v_*}{1-\alpha/v_*}\right)^2 \sum_{i=1}^n \delta_i^2 + O\left(\sum_{i=1}^n \delta_i^3\right)$$

Due to

$$\sum_i (1+\delta_i) \ln\left[\frac{1}{\varepsilon_i}\left(1-\frac{\alpha}{v_*}\right)\right] = -\sum_i \delta_i \ln \varepsilon_i + \sum_i \ln\left[\frac{1}{\varepsilon_i}\left(1-\frac{\alpha}{v_*}\right)\right]$$

and

$$\sum_i \delta_i \ln \varepsilon_i = \sum_i \delta_i (\ln \varepsilon + \ln(1+\Delta_i)) = \sum_i \delta_i \Delta_i - \sum_i \delta_i \Delta_i^2$$

and also

$$\sum_i (1+\delta_i) \ln\left[\frac{1}{\varepsilon_i}\left(1-\frac{\alpha}{v_*}\right)\right] = n \ln \frac{1}{\varepsilon}\left(1-\frac{\alpha}{v_*}\right) - \sum_i \delta_i \Delta_i + \sum_i \delta_i \Delta_i^2$$

it follows that

$$\frac{\Delta v}{v_*} \approx n \ln \frac{1}{\varepsilon}\left(1-\frac{\alpha}{v_*}\right) - \frac{C}{v_*}$$

where

$$\frac{C}{v_*} = \sum_i \delta_i \left[\Delta_i + \frac{1}{2}\left(\frac{\alpha/v_*}{1-\alpha/v_*}\right)^2 \delta_i + O\left(\delta_i^2\right) + O\left(\Delta_i^2\right)\right]$$

From this it follows that

$$\frac{\alpha}{v_*} \approx 1 - \bar{\varepsilon} e^{\frac{\Delta v}{n v_*} + C} \tag{B.0.1}$$

From this we determine the quantity C as

$$C = \frac{1}{n}\sum_i \frac{\Delta v_{*,i}}{v_*}\left[\frac{\Delta \varepsilon_i}{\varepsilon_i} + \frac{1}{2}\left(1-\frac{1}{\bar{\varepsilon}}e^{-\frac{\Delta v}{n v_*}}\right)^2 \frac{\Delta v_{*,i}}{v_*}\right]$$

$$+ O\left[(\Delta\varepsilon_i)^2\right] + O\left[(\Delta v_{*,i})^2\right] \tag{B.0.2}$$

We therefore find a solution equivalent to Eq. (3.3.7) with the substitution

$$\frac{\Delta v}{n v_*} \rightarrow \frac{\Delta v}{n v_*} + C$$

Applying this substitution rule, we find the solutions equivalent to Eqs. (3.3.8) and (3.3.9)

$$\lambda_*^{1/n} = \frac{e_*^{1/nv_*-C} - \bar{\varepsilon}}{1-\varepsilon} \tag{B.0.3}$$

and

$$\Delta v = -nv_* \left\{ \ln\left[\lambda_*^{1/n}\left(\overline{1-\varepsilon}\right) + \bar{\varepsilon}\right] + C' \right\} \tag{B.0.4}$$

From Eq. (B.0.3) it follows that

$$e^{-\Delta v/nv_*} \approx \lambda_*^{1/n}\left(\overline{1-\varepsilon}\right) + \bar{\varepsilon}$$

and therefore

$$C' = \frac{1}{n}\sum_i \frac{\Delta v_{*,i}}{v_*}\left[\frac{\Delta\varepsilon_i}{\varepsilon_i} + \frac{\lambda_*^{2/n}}{2}\left(\frac{\overline{1-\varepsilon}}{\bar{\varepsilon}}\right)^2 \frac{\Delta v_{*,i}}{v_*}\right.$$

$$\left. + O\left[(\Delta\varepsilon_i)^2\right] + O\left[(\Delta v_{*,i})^2\right]\right] \tag{B.0.5}$$

With Eq. (B.0.1) inserted into Eq. (3.2.3) and with $e^{-C/nv_*} \approx 1 - C/nv_*$ we find after some elementary calculations

$$\lambda_{i,opt} = \frac{e^{-\Delta v/nv_*-C'} - \bar{\varepsilon}}{\frac{\bar{\varepsilon}}{\varepsilon_i} - e^{-\Delta v/nv_*-C'}} = \frac{\lambda_*^{1/n} + K}{\frac{\bar{\varepsilon}}{\varepsilon_i}\frac{1-\varepsilon_i}{1-\varepsilon} - \lambda_*^{1/n} - K} \tag{B.0.6}$$

with

$$K = \left(\lambda_*^{1/n} + \frac{\bar{\varepsilon}}{1-\varepsilon}\right)C'$$

Color Plates

Astronautics. Ulrich Walter
Copyright © 2008 WILEY-VCH Verlag GmbH & Co. KGaA, Weinheim
ISBN: 978-3-527-40685-2

Chapter 2

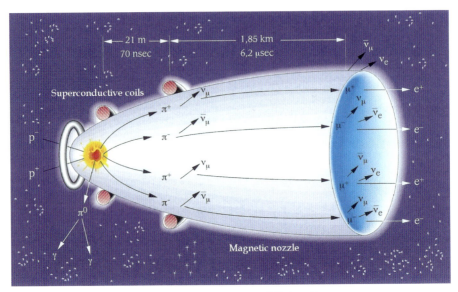

Figure 2.5 Working schema of a matter-antimatter annihilation drive. (See p. 37.)

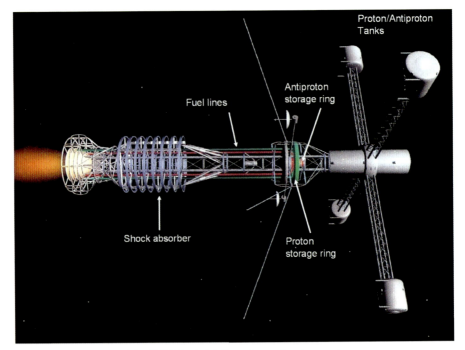

Figure 2.6 Artist view of the ICAN-II relativistic proton-antiproton annihilation drive rocket. (See p. 37.)

Chapter 11

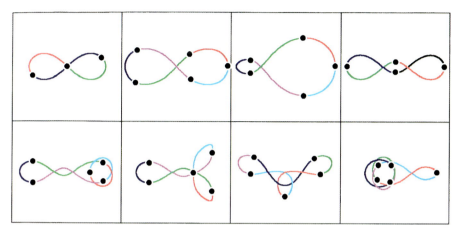

Figure 11.2 Some unstable co-orbital motions of three to five masses, showing the paths of the masses from their current position to their next position. (See p. 310.)

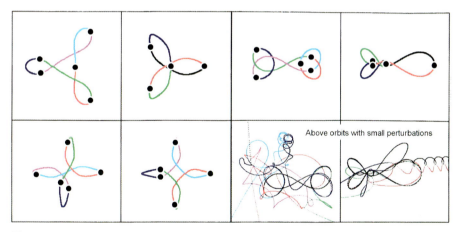

Figure 11.3 Further unstable co-orbital motions of four and five masses, showing the paths of the masses from their current position to their next position. (See p. 311.)

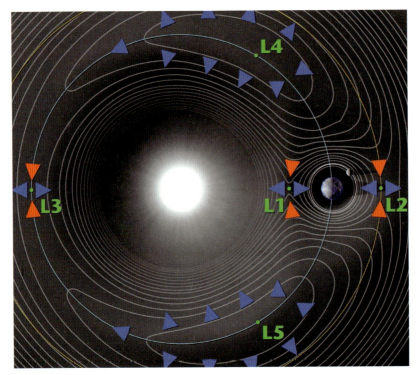

Figure 11.13 Equipotential lines of the effective potential U and the stabilizing (red) and destabilizing forces (blue) at the libration points resulting from it. (See p. 331.)

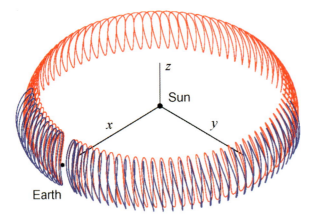

Figure 11.19 Three-dimensional representation of the horseshoe orbit of asteroid 2002 AA29 in the synodic Sun–Earth system. The looping is caused by a finite value of the z component of the initial velocity. The blue trajectory is the horseshoe orbit after reversal at the end points of the horseshoe. (See p. 338.)

Chapter 12

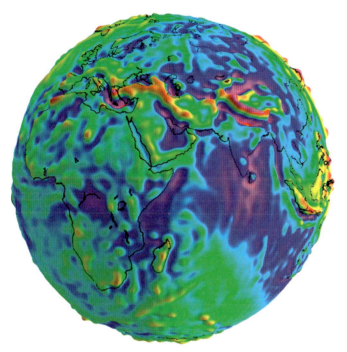

Figure 12.2 Geoid of the Earth. (See p. 345.)

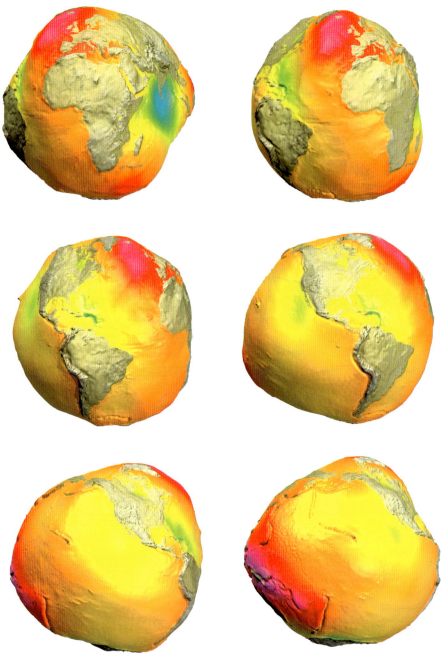

The potato residual of the geoid at different view angles in the equatorial plane. (See p. 350).

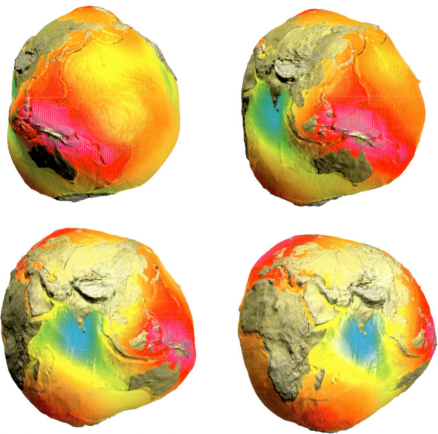

The potato residual of the geoid at different view angles in the equatorial plane (continued). (See p. 350).

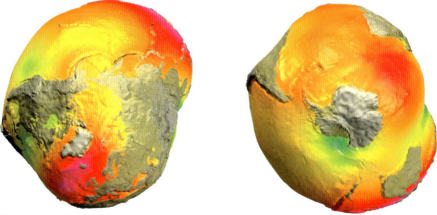

The potato residual of the geoid as vied from the North Pole (left) and the South Pole (right). (See p. 350.)

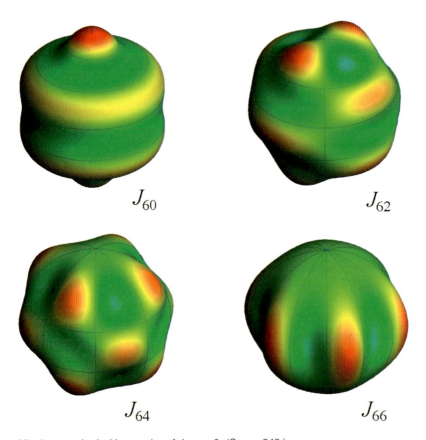

All relevant spherical harmonics of degree 6. (See p. 349.)

Color Plates | 471

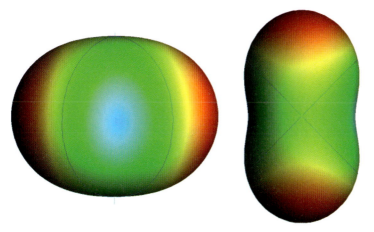

Figure 12.8 The sectorial perturbation term J_{22}: (left) observed from the terrestrial equatorial plane; (right) observed from the terrestrial pole. (See p. 359.)

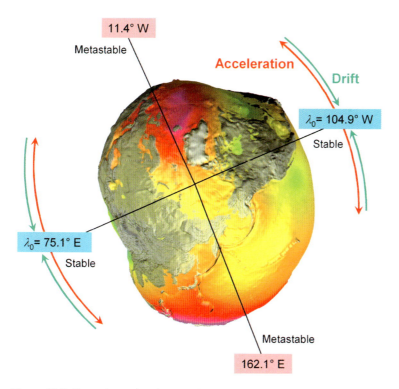

Figure 12.9 Dynamics at the abeam points of the triaxial "potato". (See p. 360.)

References

Barth, H. (2005). *Conrad Haas – Raketenpionier und Humanist*, Johannis Reeg Verlag. ISBN 3-937320-55-5

Bate, R. R., Mueller, D., and White, J. E. (1971). *Fundamentals of Astrodynamics*, Dover Publications. ISBN 0-486-60061-0

Battin, R. H. (1987). *An Introduction to the Mathematics and Methods of Astrodynamics*, AIAA Education Series. ISBN 0-930403-25-8

Berlin, P. (2005). *Satellite Platform Design*, Dept. Space Science of the Universities Luleå and Umeå. ISBN 91-631-4917-6

Brandenberger, R. H. and Vafa, C. (1989). *Superstrings in the Early Universe*, Nucl. Phys. B, **316**, 391

Brankin R., et al. (1989). Algorithm 670: a Runge–Kutta–Nyström code, *ACM Trans. Math. Software*, **15** (1), 31–40

Brown, Ch. D. (2002). *Elements of Spacecraft Design*, AIAA Education Series. ISBN 1-56347-524-3

Campan, G., Alby, F., and Gautier, H. (1995). *Station-keeping Techniques for Geostationary Satellites*, In *Spaceflight Dynamics*, Part II, Ed. J.-P. Carrou. ISBN 2-85428-37 5

Chobotov, V. A. (Ed.) (2002). *Orbital Mechanics*, 3rd Edition, AIAA Education Series. ISBN 1-56347-537-5

Cornelisse, J. W., Schöyer, H. F. R., and Wakker, K. F. (1979). *Rocket Propulsion and Spaceflight Dynamics*, Pitman Publishing. ISBN 0-273-01141-3

Curtis, H. D. (2005). *Orbital Mechanics for Engineering Students*, Elsevier. ISBN 0-7506-6169-0

Deser S., Jackiw, R., and T'Hooft, G. (1984). Three-dimensional Einstein gravity: Dynamics of flat space, *Ann. Phys.*, **152**, 220-235

Fortescue, P., Stark, J., and Swinerd, G. (2003). *Spacecraft Systems Engineering*, 3rd Edition, Wiley. ISBN 0-470-85102-3

Griffin, M. D., and French, J. R. (2004). *Space Vehicle Design*, 2nd Edition, AIAA Education Series. ISBN 1-56347-539-1-90-8

Groves, G. V. (1960). Motion of a satellite in the Earth's gravitational field, *Proc. R. Soc. London*, **254**, No. 1276, 48-65

Gurfil, P. (2007). Nonlinear feedback control of low-thrust orbital transfer in a central gravitational field, *Acta Astronautica*, **60**, 631-648

Gurzadyan, G. A. (1996). *Theory of Interplanetary Flights*, Gordon and Breach. ISBN 2-919875-15-9

Guthmann, A. (2000). *Einführung in die Himmelsmechanik und Ephemeridenrechnung*, 2nd Edition, Spektrum Akademischer Verlag. ISBN 3-8274-0574-2

Hale, F. J. (1994). *Introduction Into Space Flight*, Prentice-Hall. ISBN 0-13-481912-8

Hankey, W. L. (1988). *Re-Entry Aerodynamics*, AIAA Education Series. ISBN 0-930403-33-9

Hastings, D., and Garrett, H. (1996). *Spacecraft–Environment Interactions*, Cambridge University Press. ISBN 0-521-47128-1

Hill, P.G. and Peterson, R. (1992). M*echanics and Thermodynamics of Propulsion*, 2nd Edition, Addison-Wesley. ISBN 0-201-14659-2

Kaplan, M. H. (1976). *Modern Spacecraft Dynamics and Control*, Wiley. ISBN 0-471-45703-5

King-Hele, D. (1987). *Theory of Satellite Orbits in an Atmosphere*, Springer, ISBN 0216922526

Larson, W. J., and Wertz, J. R. (1999). *Space Mission Analysis and Design*, 3rd Edition, Microcosm Press. ISBN 1-881883-10-8

Loh, W. H. T. (1968). *Re-Entry and Planetary Entry: Physics and Technology*, vol. I: *Dynamics, Physics, Radiation, Heat, Transfer and Ablation*, and vol. II: *Advanced Concepts, Experiments, Guidance-Control and Technology*, Springer-Verlag

Messerschmid, E., and Fasoulas, S. (2000). *Raumfahrtsysteme*, Springer-Verlag. ISBN 3-540-66803-9

Montenbruck, O., and Gill, E. (2000). *Satellite Orbits – Models, Methods, Applications*, Springer-Verlag. ISBN 3-540-67280-X

Murray, C. D., and Dermott, S. F. (1999). *Solar Systems Dynamics*, Cambridge University Press. ISBN 0-521-57295-9

Noton, M. (1998). *Spacecraft Navigation and Guidance*, Springer-Verlag. ISBN 3-540-76248-5

Pisacane, V. L., and Moore, R. C. (1994). *Fundamentals of Space Systems*, Oxford University Press. ISBN 0-19-507497-1

Prussing, J. E., and Conway, B. A. (1993). *Orbital Mechanics*, Oxford University Press. ISBN 0-19-507834-9

Regan, F. J. (1984). *Re-Entry Vehicle Dynamics*, AIAA Education Series. ISBN 0-915928-78-7

Reif, F. (1965). *Fundamentals of Statistical and Thermal Physics*, McGraw-Hill. ISBN 0070518009, [Appendix A10]

Roy, A. E. (2005). *Orbital Motion*, 4th Edition, Institute of Physics Publishing. ISBN 0-7503-1015-4

Ruppe, H. O. (1966). *Introduction to Astronautics*, vol. 1, Academic Press

Schaub, H., and Junkins, J. L. (2003). *Analytical Mechanics of Space Systems*, AIAA Education Series. ISBN 1-56347-563-4

Sellers, J. J. (2005). *Understanding Space*, 3rd Edition, McGraw-Hill. ISBN 0-07-340775-5

Sidi, M. J. (1997). *Spacecraft Dynamics and Control*, Cambridge University Press. ISBN 0-521-55072-6

Steiner, W., and Schagerl, M. (2004). *Raumflugmechanik*, Springer-Verlag. ISBN 3-540-20761-9

Sutton, G. P., and Biblarz, O. (2001). *Rocket Propulsion Elements*, 7th Edition, John Wiley & Sons. ISBN 0-471-32642-9

Tapley, B. C., Schulz, B. E., and Born, G. H. (2004). *Statistical Orbit Determination*, Academic Press. ISBN 0126836302

Tegmark, M. (1997). On the Dimensionality of Spacetime, *Class. Quantum Grav.*, **14**, L69–75

Thomson, W. T. (1986). *Introduction to Space Dynamics*, Dover Books. ISBN 0-486-65113-4

Tribble, A. C. (2003). *The Space Environment – Implications for Spacecraft Design*, Princeton University Press. ISBN 0-691-10299-6

Turner, M. J. L. (2006). *Rocket and Spacecraft Propulsion*, 2nd Edition, Springer-Verlag. ISBN 3540221905

Vallado, D. A. (2001). *Fundamentals of Astrodynamics and Applications*, 2nd Edition, McGraw-Hill. ISBN 0-07-066829-9

Walter, U. (2006). Relativistic Rocket and Space Flight, *Acta Astronautica*, **59**, 453-516

Wiesel, W. E. (1997). *Spaceflight Dynamics*, McGraw-Hill. ISBN 0-07-070110-5

Index

a

acceleration   XIX, 32, 66, 80, 95, 96
– centrifugal   117
– constant   **40**
– ion   **97**
– proper   XXII
– voltage   98, 100, 102
– zone   97
adiabatic index   XXII, **67**, 69
adiabatic process   67, 73
adiabatic state change   66
aerobraking   228
aiming radius   XXII, **167**, 228, 233, 234, **234**, 235–238
altitude   **108**
altitude variable
– dimensionless   XXII
ambient pressure   5, 95
angle of attack   XXII, **118**, 254, 269, 271, 296, 297, 301, 303, 304
angular frequency   335
angular momentum   XX, 23, 142, **142**, 143, 144, 151–154, 168, 185, 202, 243, 422, **422**, 428
– specific   XX
angular velocity   XXIII, 117, **143**, 162, 219, 313, 318, 323, 361, 422
– mean motion   243
annihilation rocket   **36**, 41
anomaly   162
– eccentric   161, **161**
– – elliptic   XX
– – hyperbolic   XX
– mean   **162**
– true   162
AOA, *see* angle of attack
apoapsis   XVII, **159**, 160, 166, 187, 193, 194, 201, 202, 205, **206**, 366, 379
apocenter   159
Apollo missions   289, 290
Apollo 11   258, 290, 291
approach hyperbola   227
argument of periapsis   XXIII, **154**

arrival orbit   **226**
ascending node   154
ascent   18, 22, 27, 59, **107**
– final conditions   121
– initial   121
– optimum   **121**
– phases   **119**
– trajectory   121, 122, 129
astrodynamic parameters   455
atmosphere   XVII, 116
atmospheric
– density   **108**, 112, 126, 255
– maneuvers   390
– model   **111**
– – piecewise-exponential   **113**
– pressure   **108**
– temperature profile   **109**

b

ballistic coefficient   XIX, **375**, 385
ballistic entries   283
ballistic radiation coefficient   366
ballistic reentry   253, **270**, 299
bank angle   296, 298, 301
bank reversals   301, 302
Barker's equation   158, 182
barometric formula   110, **110**, 255, 272
before flyby   235
behind flyby   235
bi-elliptic transfer   199, 202, 207, 247
blackout phase   303, 306
blackout time   254
body system   XVII, 426, 429
Burdet transformation   174, 182

c

capture orbit   228
center of mass   XVII, XIX, 14, 119, 139–141, 230, 307, 308, 312, 313, 317, 323, 334, 340, 422, 446, 447
center of pressure   13
centrifugal force   123, 125, 145, 211, 212, 267, 295, 296, 326, 328, 331, 332

Astronautics. Ulrich Walter
Copyright © 2008 WILEY-VCH Verlag GmbH & Co. KGaA, Weinheim
ISBN: 978-3-527-40685-2

– perturbation 276
chamber pressure 81–83, **83**, 85
– ratio 87
characteristic energy XIX, **167**, 215
characteristic velocity XIX, **85**, 103
chemical propellant 65
Child–Langmuir law 99
cicular restricted three-body system (CR3BP) 328
CIRA-72 **114**
CIRA-86 111
circle 147
circular orbit 152, **156**, 162, 185, 189, 191, 192, 201, 203, 207, **208**, 228, *see also* circle
– kick-burn 208
– transfer 187
circularization time **381**, 385
co-orbital objects 338
co-rotating coordinate system 143
coasting phase **120**
combustion chamber 5, 6, 8, 9, 65, 66, 69, 78, 79, 85
– design **85**
combustion efficiency 80
combustion enthalpy 80
conic section 148
conical nozzles 91
connecting vector 139
conservation
– laws 135, 141, 172
– of angular momentum 135, **136**, 172, 443
– of charge 98
– of energy 70, 72, 135, **136**, 172, 230, 443
– of mass 10
– of momentum 1, **2**, 3, 4, **136**, 308
– of total angular momentum 308
conservative force 144
constant flight path angle rate
– steering law 127
constant pitch rate 128
– steering law 128
constant-pitch-rate maneuver 129
constellation angle 219
continuity equation 8, 10, **10**, 71
– charge 98
continuous thrust transfer 191, 192, **195**
contraction mapping 183
conversion of orbital elements **412**
coordinate system 115
– equatorial 154
– inertial 115
– rotating 116
– space 393
Coordinated Universal Time 400
coordinates, time 399

Coriolis force 326, 332, 334–336
critical acceleration 263, 288, 295
critical altitude 282, 298
critical deceleration 278, **282**, 283, **298**, 299
critical velocity 284, 298, 299
cross section 10, 66, 71, 85

d

deceleration 268, 304, 306
declination 398
deflection angle 231, 234, 236, 237
degree of freedom **67**, 69, 77, 119
– oscillatory 68
– rotational 68, 97
– translatory 68
– vibrational 97
delta-v 18, 19, 20, 28, 124, 192, 195, 200, 227, 229, 237, 238, 390
– budget XIX, **23**, 24, 187, 188, 193, 198, 199, 202, 210
– demand 228
– losses 129
deorbit **255**, 256, 300, 305
– burn 255
departure hyperbolas **216**
departure orbit 213, **214**
departure velocity 214
Descartes' rule of signs 158, 182
Dirac's delta function XXII
divergence 79, 91
– loss factor 92
double dip reentry 290
drag XIX, 13, **13**, 61, 111, 114, 116, 185, 228, 238, 255, **260**, 267, 269, **271**, 301, 302, 304, **374**
– coefficient XIX, **375**
– – dimensionless XXII, 123, 261
– force XVII, XX
– losses 27, 123, 125, 126, 129
drift velocity 69, *see also* flow velocity
dual-propellant propulsion 62
dynamic stability 334

e

E ellipsoid 443
Earth → Mars 221, 223, 226
Earth → Mars transit 216
Earth → Moon transfer 248
Earth's
– atmosphere **107**
– flattening 355
– rotation 116
– rotational gain 129
Earth–Mars 219, 221
– constellation 250

Earth–Moon
- orbit   **244**
- system   245, 322, 328–330, 335, 336
east–west drift   362
east–west station keeping   363
eccentric anomaly   222, 224
- elliptical   167
- hyperbola   167
eccentricity   XX, 149, 152, 153, **154**, 159, 160, 165, 170, 187, 193, 205, 215, 226, 228, 367, 370, 379
- vector   **147**, 181
ecliptic latitude   398
ecliptic longitude   398
effective exhaust velocity   **8**, 12, 13, 29, 34, 51, 60, 61
effective potential   325, 331, 335, 466
efficiency   38
eigentime   XXII, **31**
Einstein field equations   132
electric generator   95
electrical potential   XXI
electrical propulsion   195
- optimize
- - calculation scheme   104
electrothermal propulsions   12
ellipse   147, 159
- entry   256
elliptical orbit   **159**, 170, 185, 187, 228, see also ellipse
emissivity   253, 274, 300
energy conservation   73–75, 77, **144**, 243, 308, 323
energy dissipation   440, 443, 453
energy supply system   95
engine
- design   11, 82, **85**, 86
- - optimum   86
- efficiency   95, 101, 102
- ion   **96**
- performance   **80**
- staging   62
enthalpy   XX, 67, 70
entry angle   262, 274, 285, 291, 299, 302
entry corridor   258, 290
entry energy   288
entry interface   255, **255**, 256, 259, 260, 269, 300, 302
entry velocity   260
epoch   XXI, **155**, 357, 398, 401
- standard   155
equation of motion   **114**, 116, 117, 119, 122, **131**, 135, 138, 139, **140**, 141, 146, 155, 169, **260**, 270, **296**, 307, 311, 316, 326, 353, 358, 361, 428
- normalized   261, 305

- reduced   263, 305
equations of rotational motion   **428**, 429
equilibrium glide   300, 301, see also lifting reentry
escape velocity   24, 159, **159**, 229
Euler configuration   309, **311**, 328, 339, 340
Euler's equations   **7**, 10, **429**, 435
Euler's rate equations   **432**, 450
Eulerian points   319, **319**, **320**, 328, 332, 336, 339
excess velocity   214, 215, 217, 218, 225
excitation energy   101
excited degrees of freedom   68
exhaust velocity   4, **8**, 15, 25, 47, 54, 73, 77, 80, 98, 102, 104
- effective   2
- - relativistic   36, 38
- mean effective   60
- uniform   **53**
exit
- cross section   97
- velocity   288
exosphere   110
expansion ratio   XXII, 86, **86**, 87, 90
external efficiency   **27**, **38**
- relativistic   39
external energy   81
external forces   2, 13, **13**, 18, 19, 21, 114, 118
external torque   444

f

Farquhar transfer orbit   245
figure of merit   12, 13, 22, 28, 80–82, 85
final conditions   122
first cosmic velocity   157
First Point of Aries   395
flat spin   442
flattening of the Earth   356, 357
flight mechanics   107, 119
flight path angle   XXII, 22, 123–125, 127, 150, **150**, 253, 255, 259, 260, 275, **276**, 287, 291, 302, 376
- equation   **278**
- rate   124
flow velocity   70–75, 77, 78
flyby   **228**, 230, 237, 238, 240
- before   231
- behind   231
- delta-v 236
- from inside   231
- from outside   231
- maneuver   24, 229, 323
- plane   230, 231
- potential   243
focal point   149, 159, 162, 163, 166

forces, external   137
free gyros   433
free return
– orbits   214
– trajectory   **246**
fuel demand   122, *see also* propellant demand

g
Gaussian variational equations   203, **343**, 344, 368, 376, **387**
general two-body system   140
GEO   59, 195, 197, *see also* geostationary orbit
geocentric equatorial reference   396
geocentric system   218
geoid   **345**
geostationary orbit   155, 185, 341, 358, 360, 362, 363
gliding reentry, *see* lifting reentry
gravitation   123
– perturbation   276
gravitational acceleration   108
gravitational constant   XX, 132, **141**
gravitational field   18, **20**, 21, **134**, 135, 138, 209, 210
gravitational force   **13**, 18, 21, 22, **134**, 135, 144, 145, 176, 210, 211, 295, 296
gravitational loss   **21**, 22, 27, 44, 122–125, 129
gravitational parameter
– standard   XXII
gravitational potential   XXI, 131–134, 152, 173, 179, **346**
– anisotropic   349
gravity-assist maneuver   209, 229
gravity gradient   XVII, 446
– stabilization   **446**
– torque   447, **448**
gravity turn   **124**
– maneuver   129
Greenwich Mean Time (GMT)   400
Gregorian calendar   401
guiding center   371
– system   373
gyro   444

h
halo orbit   **332**, 333
harmonics
– sectorial   349
– spherical   349
– tesseral   349
– zonal   349
Harris–Priester model   111
heat flow rate   XXI

heat flux   XXI, 253, 254
– maximum   **273**
heat load   254
heating, maximum   **273**, *see also* maximum heat flux
heliocentric ecliptical reference system   395
heliocentric reference frame   394
heliocentric reference system   230, 236
heliocentric system   229, 240, 243
heliocentric system of coordinates   217
heterosphere   **110, 111**, 255
Hill curve   328, 329, 336, 337
Hill periodic system   308
Hohmann   XVII
Hohmann ellipse   213
Hohmann orbit   119
Hohmann transfer   **185**, 186, **187**, 188–190, 192, 193, 195, 196, 199, 201, 207, 217, 218, 251
– Earth–Mars   188
– orbit   216, 223
– orbital elements   120
homosphere   **110**, 255
homotopy continuation method   408
horseshoe orbits   **337**
hyperbola   147, 166, 230, 247
– approach   **226**
hyperbolic excess velocity   167, **167**, 230
hyperbolic orbit   **166**, *see also* hyperbola

i
ideal gas law   **67**, 108
ideally adapted nozzle   9, 10, 79, 81–84
impulsive maneuver   **20**, 21, 27
inclination   XX, **154**, 201
inertia tensor   **423**, 448
inertial coordinate system   117
inertial force   137
inertial (physical) systems   427
inertial reference frame   **393**, 425, 428, 433
– sideral   394
inertial reference system   XVII
inertial tensor   XX
infinite-expansion coefficient   XIX, **82**, 88
initial conditions   121, 122
injection   201, 215, 217, 223, 225–228, 250
– transfer   190, 218
injection burn   195
inner orbit   187, 190, 193
inner planet   217, 219, 223, 225, 231
integral efficiency, *see* external efficiency
integral of motion   153, 308, 323
– Jacobi   325
intensive thermodynamic variables   67
intercontinental ballistic missiles   120

internal degrees of freedom   65
internal efficiency   **26**, 39, **81**
– ion   **101**
internal energy   67–69
– of a gas   XXI
International Atomic Time   400
international celestial reference frame (ICFR)   394
International Space Station   157, 259, 375, 383, 384
interplanetary flight   59, 95, 153, **209**, 216
interplanetary reentry   258
ion engine   95
isentropic   67
isentropic process   67
isomass configurations   309, **310**
ISS, *see* International Space Station

j
Jacchia 1977   111
Jacobi constant   322, 325, 329, 336, 337, 340
Jacobi integral   **325**, 327
jet power   **11**
jet straying   91
Julian centuries   402
Julian Date   XX, 401
– modified   XXI

k
Kepler's
– first law   **159**
– second law   **143**, 144, 160
– third law   160, **160**, 313
Keplerian elements   153, **153**, 156, 343, 355, 403, 412, 415
Keplerian equation   163, **163**, 164, 171
Keplerian orbits   119
Keplerian problem   **155**, 158, 161, 163
kick-burn phase   **120**
kinetic energy   23, 26, 67, 68, 70, 75, 101, 137, 152, 157, 167, 186, 267, 272, 288, 412
– specific   145
Kronecker symbol   XXII, 424

l
L sphere   443
Lagrange configuration   309, **316**, 334, 340
Lagrange multiplier method   **48**, 50
Lagrange's planetary equations   **354**
Lagrange's quintic equation   315
Lagrangian points   319, **319**, 322, 328, 334–336
Laguerre method   164
Lambert's method   **409**
Lambert's problem   420
Laplace–Runge–Lenz vector   146

Laval nozzle   73, 78, 79
LEO   195, 253, 260
– parking orbit   214
– reentry   258
libration damper   451
libration points   246, 249, 319, **319**, 322, 327, 331, 336, 340, 466
lift   XVIII, 13, **13**, 114, 116, 123, 255, **260**, 267, 271, **274**, 281, 301, 304
– coefficient   XIX
– – dimensionless   123, 261
– force   XX
– negative   275, 281
– vector   390
lifting reentry   294, 295, **295**
line of apsides   187, 188, 216, 370
line of nodes   154
Lissajous orbits   334
local horizon   22, **150**
local vertical local horizontal reference system   XVIII, *see also* LVLH frame
low Earth orbit   186, 210, 341, 446
lunisolar perturbation   364
LVLH frame   430, 431, 450

m
Mach number   XX, **74**
major or minor principal   444
major principal axis   **424**, 436, 439, 442, 444
maneuver   25
– constant flight path angle rate   127
– constant-pitch-rate   126
– gravity-turn   **124**
– one-impulse   **200**
many-body system   152
mass   **17**
– power plant   103
mass conservation   72
mass flow density   **71**, 72, 95, 97
mass flow rate   XX, 3, 4, 8, 10, 18, 60, 65, 71, 78, 80, 82, 84, 85, 99, 100, 196
mass flux density   XXII
mass ratio   XXII, **29**, **46**
mass-specific power output   XXII
max-q   126
maximum aerodynamic pressure   126
maximum heat flux   300
maximum heat load   299
mean anomaly   168, 357
mean motion   XXI, **162**, 219
mean number of excited degrees   XXI
mean orbit radius   455
mean orbital velocity   457
mean solar day   400

mechanical efficiency, *see* external efficiency
mechanical energy 144
Mercury capsule 254
mesosphere 110
Michielsen diagram 214
minor principal axis **424**, 436, 439, 442
mission HITEN 249
Modified Julian Date 402
molar enthalpy XX, 65, 80
molar mass 67
moment of inertia 423, **424**
momentum thrust **8, 9**, 10, 15, 73, 77, 93
Moon mission 219, 258
MSIS-86 111
MSISE-90 111
multi-body system 229, 248
multi-stepping, *see* staging, serial
multistaging, *see* staging, serial

n

n-body problem 309
n-body system 181
negative lift 290
Newton's
– equation of motion **174**
– first law 137
– law of gravitation 135
– laws 135
– method 50, 55, **163**, 164, 168, 171, 183, 207
– second law 4, 14, 137
– third law 118, 137
nodal point 201
Noether's theorem 136, 172
non-Hohmann transfers **220**
north–south drift 364
north–south station keeping 364
NOZOMI spacecraft 250
nozzle XVIII, 8, 18, 65, 66, 69, 73, **73**, 75, 78, 84, 87, 88
– aperture angle 92
– bell-shaped 91
– coefficient XIX, *see also* nozzle efficiency
– cross section 61, 72–74, 76–78
– design **85**, 90
– efficiency XIX, 77, 87, 88, **88**, 90
– exit 66, 71, 79, 85, 86
– – cross section 83, 85, 87
– ideally adapted **77**, 81, 90
– shape **90**
number of stages 48, 53
nutation **436**, 437, 440, 442, 445
– angle 438, 441
– angle rate 441
– damper 440
– frequency 438

o

Oberth maneuver 167
oblate body 439
oblate geometry 439
oblate rotation 439, 441
oblate way 442
optimized ascent trajectory 129
optimum engine design **87**
optimum exhaust velocity **79**
optimum trajectory 121
optimum-ascent problem 122
optimum-ascent trajectory 123
orbit 214
– angle XXIII, 149, 155, 162, 188
– determination **403**, 415
– equation **147**, 155, 161, 163, 164, 168, 169, **171**, 175
– estimation **408**
– inclination 124
– inner 185, 221
– lifetime **383**, 384, 385, 390
– number XX, **412**
– outer 185, 221
– phasing 208, **218**
– transfers 20
– transitions **185**
– velocity 23, **149**, 157, 201, 203, 217
orbital elements 120, 121, 153, **153**, 158, 185, 200, 202–205, 208, 242, 256, 376, 388
orbital energy 151–153, 157, 159, 160, 166, 186, 201, 202
– specific XXII
orbital period XXI, **156**, 160, **160**, 205, 357
orbital plane 154, 199, 201, 202
orbital radius XXI
orbital resonance 358
osculating elements 343
osculating orbit 353
outer orbit 187, 190, 193
outer planet 217, 219, 223, 225, 231
over-expansion 79, 88

p

parabola 147, 157
parabolic orbit **157**, *see also* parabola
parallel staging 59, 61
parking orbit 214, 216, 225, 226
partial rocket **45**, 47
patched conics **210**, 212
– method 308
path flight angle 117
payload XVIII, 45
– mass **17**, 28, 46, 95, 103

– ratio   XXII, **29**, 43, **46**, 48–51, 54, 60, 102–104
– – optimized   58
– – total   **46**
– true   **45**
perigee/apogee   159
periapsis   XVIII, 149, 154, **159**, 160, 162, 163, 168, 187, 193, 205, 206, **206**, 357, 366, 377, 379, 381
pericenter   159
perifocal reference frame   397
perturbations   341, 368, 376, 449
– Cowell's method   350
– Encke's method   353
– gravitational   **345**
– lunisolar   **362**
– progression of epoch   356
– progression of line of apsides   356
– regression of nodes   355
– secular   355
– triaxiality   **357**
perturbed rotation   435
photon propulsion   34
photon rocket   36
phygoid modes   291
phygoid period   294
pitch   431, 449, 451
– angle   XXIII, 125
– maneuver   125
– oscillation   450
– rate   129
planet
– inner   218
– outer   218
planetary capture   **227**
planetary constellation   219
planetocentric reference system   230, 234
pogo   80
Poinsot ellipsoid (E ellipsoid)   443
Poisson equation   98, 133, 173, 179, 180
polhode   443
potential energy   23, 134, 152, 157, 167, 186, 412
– specific   145
power plant mass   103, 104
power supply system   103
precession   440, 445, **445**
– angle   446
pressure thrust   **8**, 9, **9**, 10, 15, 18, 20, 73, 77, 78
principal axes   **424**, 435, 436, 443, 448
– frame   449
– system   XVIII, 427, 429, 431, 434, 437
principal moments of inertia   **424**
progression of the line of apsides   357
prolate body   439

prolate geometry   439
prolate rotation   439, 441
prolate way   442
propagator, universal   419
propellant demand   **23**, 24, 25, 102
propellant force   **4**, 5, 6
propellant gas   5, 10, 65, **66**, 85
propellant gas density   100
propellant mass   10, 17, 20, 103, 104
proper acceleration   **32**
proper reference frame   34
proper speed   XXII, **32**, 33, 34, 39
proper time   **31**
propulsion
– demand   24, **24**, 29, 43, 47, 48, 50, 60, 104, 185, 190, 192, 197, 207, 228, 247, 257
– electric   **95**
– – optimization   **102**
– ion   96
– thermal   **65**

r
radial vector   131, 138, 139, 149, 150
range rate   404, 405
rapidity   39
rapprochement   214
– orbits   **244**, 245
rectilinear orbit   152, **168**, 169
reduced mass   XXII
reentry   107, **253**
– high-lift   305
– low-lift   279, 305
– phase   255, 269, 270
– skip   **285**
– trajectory   262, 281, 291
reference frame   31, 39
– inertial   32
– proper   32
reference system   12, 27
– co-rotational   150
– heliocentric   238
– inertial   2
– physical   425
– planetocentric   232
reflection   263, **285**
reflection altitude   285, 291
reflection point   281
regression of nodes   357
relativistic rocket   25, **30**, 35
relativistic rocket equation   34
rest frame   31, 32
restricted three-body problem   309, **318**
– circular   318, **323**
restricted three-body system   243
Reynolds number   XXI, 119, 260
right ascension   398

– of ascending node   XXIII, **154**, **206**
rigid body dynamics   **421**
rocket   **17**
– annihilation   26
– efficiency   **26**
– equation   **19**, 27, 28, 35, 196
– – parallel staging   61
– – relativistic   **34**, 35
– equation of motion   **13**, 14
– in free space   **19**
– performance   **11**, 80
– principle   1, **2**
– stage number   XXI
– staging   63
– thrust   **4**
roll   431, 449
roll angle   296, see also bank angle
roll reversals   301
roll-yaw   452, 453
root-mean-square velocity   68
rotation   **421**
rotation axis   421, 423
rotational energy   146, 324, 328, 440–443
rotational kinetic energy   **427**
rotational potential   **145**, 324
rotational power   **432**
RSW reference frame   430, 448
Runge–Kutta method   262
Runge–Lenz vector   146

s
Saint Venant equation   71
Saturn V   4, 55, 62
scale height   XX, **110**, 111, 272, 381, 382, 384
second cosmic velocity   159, 258
sectorial harmonics   349
sectorial terms   358
secular changes   382
secular component   363, 364
secular equations of variation   379
secular perturbations   350
secular variation   379
semi-latus rectum   XXI, **147**, 151, 158
semi-major axis   XIX, 145, 147, 151, **154**, 157, **159**, 160, 169, 186, 188, 190, 194, 201, 205, **205**, 218, 224, 228, 243, 379
semi-minor axis   XIX, **159**, 189
sensitivity analysis   193, **215**
serial staging   47, 61
serial-stage rocket equation   **47**, 51
serial-staged rocket   52
series expansions   **165**, 183, 340
shock attenuation   **79**
shock waves   66, 79
sidereal day   399

single-stage rocket   45, 47, 52
site-track algorithm   408
skip   290
skip reentry, see skipping
skipping   271, 291, 301, 302
skips   263
slant range   404, 405
SOI   XVIII, XXI, **210**, 214, 216, 230, 235
– Moon   213
solar day   399
solar radiation pressure   341, **365**
solar system planets
– parameters   455
sound velocity   65, **71**, 72, 74
Soyuz capsule   260
Space Shuttle   15, 41, 82, 119, 121–124, 126, 129, 197, 254, 259, 291, 294, 297, 299, 301, 376, 390
– 120° roll maneuver   124
– propulsion   63
– reentry   **300**
– SSME   82, 90
space-time transformations   39, **39**, 40
specific gas constant of air   XXI
specific impulse   XX, **12**, 13, 22, 80, 83, 95
– mass   12
specific impulse variation   84
specific orbital energy   145, 156
specific power   103
specific thermal heat capacity   67
speed of sound   XIX
sphere of influence, see SOI
spherical harmonics   349
spheroid of inertia   437
spin stabilization   436, 442
stability   309, 310, **328**, **334**, 359, 434, 435, 449, 450, 452, 453
– dynamical   361
– gravity-gradient   451
stabilization
– dual-spin   442
stage number   50, **56**, 57, 58, 60
stage optimization   **47**
staged rocket   30
staging   **43**
– parallel   43, **60**
– serial   **44**
– uneven   **54**
– uniform   **51**
stagnation point   255, 274, 300
standard atmosphere
– specific gas constant   109
standard epoch   398, 402
standard gravitational parameter   **134**, 160, 170
Stanton number   XXI

state vector   149, 153, 243, 269, 403, 407, 409, 412, 415
– propagation   **415**
stationary points   315
steering   123
steering angle   4
steering losses   27, 123, 124, 129
Stefan–Boltzmann constant   253
Stefan–Boltzmann law   253
stratosphere   110
structural   XVIII
structural factor   XXII, 51
structural mass   **17**, 28, 29, 46, 95, 103
structural ratio   29, **29**, **46**, 47, 52, 54, 59, 60
subsonic   75, **75**, 76
Sundman's inequality   181
Sundman's theorem   182
supersonic   75, **75**, 76–78
surface reflectivity   XXII
swing-by maneuver   229
symmetrical gyro   435–437, 441
synodic   XVIII
synodic period   **219**
synodic system   243, 315, 323–326, 330, 332, 334, 338
system *LVLH*   431

t
tadpole orbits   **337**
tandem staging, *see* staging, serial
tangential thrust transfer   18
tangential velocity   116
tank staging   62
terrestrial reference frames   396
tesseral harmonics   349
tesseral terms   358
thermal efficiency   **69**
thermal engines   72, 80
thermodynamic state variables   69
thermodynamic variables   73, 78
thermosphere   110
three-body problem   **307**
three-body system   249
– Earth–Moon–S/C   247
three-impulse transfer   199
three-stage rocket   52
throat   XVIII, 6, 7, 66, 72, **72**, 73, 85, 91
– cross section   65, 83, 85, 86
– pressure   89
thrust   **4**, 77, 78, 80, **82**, 83, 85, 87, 89, 97, 114, 116, 122
– angle   XXII, **118**, 119, 124, 128
– brief   **21**
– coefficient   XIX, **86**, 87, 88, 90
– effective   XIX
– ion engine   **99**

– loss factor   92, 93
– maneuver   19–21
– performance   **87**
– phase   **120**, 121, 123, 129
– power   **11**
– relativistic   **35**
– variation
– – relative   84
– vector   118
Tisserand Relation   244
Titius–Bode law   455
topocentric reference frame   396, 397, 422
topocentric system   426
torque-free motion   **433**, 443
total impulse   12, **12**
total launch mass   **17**, 46, 47
total payload ratio   46, 48, 52, 53, 59
total rocket efficiency   26
total thrust   73
trajectory   118, 120
– optimum-ascent   122, **128**
transfer   186
– bi-elliptic   **197**
– Earth–Mars   192
– injection   194
– *n*-impulse   199
– orbit   119, 122, 129, 217, 223, 244, 247–249
– – optimum   119
– three-impulse   **197**, 198
– time   188, 196, 197, 249
– – Hohmann   218
transit time   223, 226
transition time   225
transmitted spacecraft power   **12**
triaxiality   389
Trojans   335, 338
troposphere   110
true anomaly   XXIII, **149**, **154**, 222, 223, 225, 456, 457
turn angle   XXII
two-body problem   138, 139
two-body system   229, 230, 307
two-impulse transfer   119
two-stage rocket   52, 62

u
under-expansion   79, 88
uneven staging   **459**
uniform payload ratio   56
uniform staging   56
universal gas constant   XXI, 67
Universal Time (UT)   400

v
Vandenkerckhove function   83

velocity   117, **143**, 150, 160, 227
vernal equinox   154, 394, 396, 398, 401
vertical ascent   22
virial theorem   **152**, 181
vis-viva equation   **145**, 148, 151, 152, 167, 171, **173**, 244

w
waiting time   219, 251
weak stability boundary
– maneuvers   209
– transfers   **247**

weight-specific impulse   12
wetted area, effective   119

y
yaw   431, 449

z
zero-velocity curve   328
Zeuner–Wantzel equation   70
Ziolkowski equation   19
zonal harmonics   349
zonal terms   358